PREFACE

1. Scope

The Joint Publication 1-02, *Department of Defense Dictionary of Military and Associated Terms* sets forth standard US military and associated terminology to encompass the joint activity of the Armed Forces of the United States. These military and associated terms, together with their definitions, constitute approved Department of Defense (DOD) terminology for general use by all DOD components.

2. Purpose

This publication supplements standard English-language dictionaries and standardizes military and associated terminology to improve communication and mutual understanding within DOD, with other federal agencies, and among the United States and its allies.

3. Application

This publication applies to the Office of the Secretary of Defense, the Services, the Joint Staff, combatant commands, DOD agencies, and all other DOD components. It is the primary terminology source when preparing correspondence, to include policy, strategy, doctrine, and planning documents. Criteria for inclusion of terminology in JP 1-02 is enumerated in Department of Defense Instruction (DODI) 5025.12, *Standardization of Military and Associated Terminology,* and Chairman of the Joint Chiefs of Staff Instruction (CJCSI) 5705.01, *Standardization of Military and Associated Terminology.*

4. Publication Format

This edition of JP 1-02 has been published in two basic parts:

a. Terms and definitions. These are annotated with the source publication.

b. Abbreviations and acronyms. The source publication establishes the authoritative context for proper understanding and management of the associated term.

5. JP 1-02 Online Availability and Update Schedule

JP 1-02 is accessible online as a searchable database and in PDF format at the following Internet address: http://www.dtic.mil/doctrine/dod_dictionary and at the following NIPRNET address: https://jdeis.js.mil. The contents of JP 1-02 are updated on a monthly basis to include any terminology additions, modifications, or deletions made within the previous calendar month in accordance with CJCSI 5705.01.

For the Chairman of the Joint Chiefs of Staff:

WILLIAM E. GORTNEY
Vice Admiral, USN
Director, Joint Staff

TABLE OF CONTENTS

Intentionally Blank

A

acceptability — The joint operation plan review criterion for assessing whether the contemplated course of action is proportional, worth the cost, consistent with the law of war; and is militarily and politically supportable. See also **adequacy; feasibility.** (JP 5-0)

access — In counterintelligence and intelligence use, a. a way or means of approach to identify a target; or b. exploitable proximity to or ability to approach an individual, facility, or information that enables target to carry out the intended mission. (JP 2-01.2)

access to classified information — The ability and opportunity to obtain knowledge of classified information by persons with the proper security clearance and a need to know of specified classified information. (JP 2-01)

accompanying supplies — Unit supplies that deploy with forces. (JP 4-01.5)

accountability — The obligation imposed by law or lawful order or regulation on an officer or other person for keeping accurate record of property, documents, or funds. (JP 1)

acoustic intelligence — Intelligence derived from the collection and processing of acoustic phenomena. Also called **ACINT.** (JP 2-0)

acquisition and cross-servicing agreement — Agreement, negotiated on a bilateral basis with United States allies or coalition partners, that allow United States forces to exchange most common types of support, including food, fuel, transportation, ammunition, and equipment. Also called **ACSA.** See also **cross-servicing.** (JP 4-08)

actionable intelligence — Intelligence information that is directly useful to customers for immediate exploitation without having to go through the full intelligence production process. (JP 2-01.2)

action phase — In amphibious operations, the period of time between the arrival of the landing forces of the amphibious force in the operational area and the accomplishment of their mission. See also **amphibious force; amphibious operation; landing force; mission.** (JP 3-02)

activation — Order to active duty (other than for training) in the federal service. See also **active duty; federal service.** (JP 4-05)

active air defense — Direct defensive action taken to destroy, nullify, or reduce the effectiveness of hostile air and missile threats against friendly forces and assets. See also **air defense.** (JP 3-01)

active communications satellite — See **communications satellite.**

active defense — The employment of limited offensive action and counterattacks to deny a contested area or position to the enemy. See also **passive defense.** (JP 3-60)

active duty — Full-time duty in the active military service of the United States. This includes members of the Reserve Component serving on active duty or full-time training duty, but does not include full-time National Guard duty. Also called **AD.** See also **active duty for training; inactive duty training**. (JP 4-05)

active duty for special work — A tour of active duty for reserve personnel authorized from military and reserve personnel appropriations for work on active or reserve component programs. Also called **ADSW.** (JP 1-0)

active duty for training — A tour of active duty which is used for training members of the Reserve Components to provide trained units and qualified persons to fill the needs of the Armed Forces in time of war or national emergency and such other times as the national security requires. The member is under orders that provide for return to non-active status when the period of active duty for training is completed. This includes annual training, special tours of active duty for training, school tours, and the initial duty for training performed by nonprior service enlistees. Also called **ADT.**

Active Guard and Reserve — National Guard and Reserve members who are on voluntary active duty providing full-time support to National Guard, Reserve, and Active Component organizations for the purpose of organizing, administering, recruiting, instructing, or training the Reserve Components. Also called **AGR.** (CJCSM 3150.13)

active status — Status of all Reserves except those on an inactive status list or in the Retired Reserve. Reservists in an active status may train for points and/or pay and may be considered for promotion.

activity — 1. A unit, organization, or installation performing a function or mission. 2. A function, mission, action, or collection of actions. Also called **ACT.** (JP 3-0)

act of mercy — In personnel recovery, assistance rendered to evaders by an individual or elements of the local population who sympathize or empathize with the evaders' cause or plight. See also **evader; evasion; recovery; recovery operations.** (JP 3-50)

acute radiation dose — Total ionizing radiation dose received at one time and over a period so short that biological recovery cannot occur. (JP 3-11)

acute radiation syndrome — An acute illness caused by irradiation of the body by a high dose of penetrating radiation in a very short period of time. Also called **ARS.** (JP 3-11)

Adaptive Planning and Execution system — A Department of Defense system of joint policies, processes, procedures, and reporting structures, supported by communications and information technology, that is used by the joint planning and execution community to monitor, plan, and execute mobilization, deployment, employment, sustainment, redeployment, and demobilization activities associated with joint operations. Also called **APEX system.** (JP 5-0)

adequacy — The joint operation plan review criterion for assessing whether the scope and concept of planned operations can accomplish the assigned mission and comply with the planning guidance provided. See also **acceptability; feasibility.** (JP 5-0)

adjust — An order to the observer or spotter to initiate an adjustment on a designated target.

administrative contracting officer — Contracting officer whose primary duties involve contract administration. Also called **ACO.** See also **contracting officer; procuring contracting officer.** (JP 4-10)

administrative control — Direction or exercise of authority over subordinate or other organizations in respect to administration and support. Also called **ADCON.** (JP 1)

administrative loading — A loading method that gives primary consideration to achieving maximum utilization of troop and cargo space without regard to tactical considerations. Equipment and supplies must be unloaded and sorted before they can be used. Also called **commercial loading.** (JP 3-02.1)

advanced base — A base located in or near an operational area whose primary mission is to support military operations. (JP 3-34)

advanced operations base — In special operations, a small temporary base established near or within a joint special operations area to command, control, and/or support training or tactical operations. Facilities are normally austere. The base may be ashore or afloat. If ashore, it may include an airfield or unimproved airstrip, a pier, or an anchorage. An advanced operations base is normally controlled and/or supported by a main operations base or a forward operations base. Also called **AOB.** See also **forward operations base; main operations base.** (JP 3-05.1)

advance force — A temporary organization within the amphibious task force which precedes the main body to the objective area, for preparing the objective for the main assault by conducting such operations as reconnaissance, seizure of supporting positions, mine countermeasures, preliminary bombardment, underwater demolitions, and air support. (JP 3-02)

advance guard — Detachment sent ahead of the main force to ensure its uninterrupted advance; to protect the main body against surprise; to facilitate the advance by removing obstacles and repairing roads and bridges; and to cover the deployment of the main body if it is committed to action. (JP 3-07.2)

adversary — A party acknowledged as potentially hostile to a friendly party and against which the use of force may be envisaged. (JP 3-0)

adversary template — A model based on an adversary's known or postulated preferred methods of operation illustrating the disposition and activity of adversary forces and assets conducting a particular operation unconstrained by the impact of the operational environment. (JP 2-01.3)

Aegis — A ship-based combat system that can detect, track, target, and engage air, surface, and subsurface threats, including ballistic missiles on some modified ships. (JP 3-01)

aerial port — An airfield that has been designated for the sustained air movement of personnel and materiel as well as an authorized port for entrance into or departure from the country where located. Also called **APORT**. See also **port of debarkation; port of embarkation**. (JP 3-17)

aerial port squadron — An Air Force organization that operates and provides the functions assigned to aerial ports. (JP 4-01.5)

aeromedical evacuation — The movement of patients under medical supervision to and between medical treatment facilities by air transportation. Also called **AE**. (JP 4-02)

aeromedical evacuation control team — A core team assigned to a component-numbered air force air operations center air mobility division that provides operational planning, scheduling, and execution of theater aeromedical evacuation missions and positioning of aeromedical evacuation ground forces. Also called **AECT**. See also **aeromedical evacuation; air mobility division**. (JP 3-17)

aeromedical evacuation unit — An operational medical organization concerned primarily with the management and control of patients being transported via an aeromedical evacuation system or system echelon. (JP 4-02)

aeronautical chart — A specialized representation of mapped features of the Earth, or some part of it, produced to show selected terrain, cultural and hydrographic features, and supplemental information required for air navigation, pilotage, or for planning air operations. (JP 2-03)

aerospace defense — 1. All defensive measures designed to destroy or nullify attacking enemy aircraft and missiles and also negate hostile space systems. 2. An inclusive term encompassing air defense, ballistic missile defense, and space defense. See also **air defense; space defense**. (JP 3-27)

afloat pre-positioning force — Shipping maintained in full operational status to afloat pre-position military equipment and supplies in support of combatant commanders' operation plans, consisting of the three maritime pre-positioning ships squadrons, the Army's afloat pre-positioning stocks-3 ships, and the Defense Logistics Agency, and the Air Force ships. Also called **APF**. See also **maritime pre-positioning ships**. (JP 4-01.2)

afloat pre-positioning operations — Pre-positioning of ships, preloaded with equipment and supplies that provides for an alternative to land-based programs. Also called **APO**. See also **operation.** (JP 4-01.6)

agency — In intelligence usage, an organization or individual engaged in collecting and/or processing information. Also called **collection agency.** See also **agent; intelligence process; source.** (JP 2-01)

agent — In intelligence usage, one who is authorized or instructed to obtain or to assist in obtaining information for intelligence or counterintelligence purposes. (JP 2-01.2)

aimpoint — 1. A point associated with a target and assigned for a specific weapon impact. 2. A prominent radar-significant feature used to assist an aircrew in navigating and delivering their weapons. See also **desired point of impact.** (JP 3-60)

air and missile defense — Direct [active and passive] defensive actions taken to destroy, nullify, or reduce the effectiveness of hostile air and ballistic missile threats against friendly forces and assets. Also called **AMD.** (JP 3-01)

air and space expeditionary task force — A deployed numbered air force or command echelon immediately subordinate to a numbered air force provided as the United States Air Force component command committed to a joint operation. Also called **AETF.** (JP 3-30)

air and space operations center — The senior agency of the Air Force component commander that provides command and control of Air Force air and space operations and coordinates with other components and Services. Also called **AOC.** (JP 3-30)

air apportionment — The determination and assignment of the total expected effort by percentage and/or by priority that should be devoted to the various air operations for a given period of time. (JP 3-0)

air assault — The movement of friendly assault forces by rotary-wing aircraft to engage and destroy enemy forces or to seize and hold key terrain. See also **assault.** (JP 3-18)

air assault force — A force composed primarily of ground and rotary-wing air units organized, equipped, and trained for air assault operations. (JP 3-18)

air assault operation — An operation in which assault forces, using the mobility of rotary-wing assets and the total integration of available firepower, maneuver under the control of a ground or air maneuver commander to engage enemy forces or to seize and hold key terrain. (JP 3-18)

airborne — 1. In relation to personnel, troops especially trained to effect, following transport by air, an assault debarkation, either by parachuting or touchdown. 2. In relation to equipment, pieces of equipment that have been especially designed for use by airborne troops during or after an assault debarkation as well as some aeronautical equipment used to accomplish a particular mission. 3. When applied to materiel, items that form an integral part of the aircraft. 4. The state of an aircraft, from the instant it becomes entirely sustained by air until it ceases to be so sustained. Also called **ABN.** (JP 3-17)

airborne alert — A state of aircraft readiness wherein combat-equipped aircraft are airborne and ready for immediate action to reduce reaction time and to increase survivability. See also **combat air patrol; ground alert.** (JP 3-01)

airborne assault — The use of airborne forces to parachute into an objective area to attack and eliminate armed resistance and secure designated objectives. (JP 3-18)

airborne early warning — The detection of enemy air or surface units by radar or other equipment carried in an airborne vehicle, and the transmitting of a warning to friendly units. Also called **AEW.** (JP 3-52)

airborne mission coordinator — The designated individual that serves as an airborne extension of the component commander or supported commander responsible for the personnel recovery mission. Also called **AMC.** See also **combat search and rescue; combat search and rescue task force; personnel recovery coordination cell.** (JP 3-50)

airborne operation — An operation involving the air movement into an objective area of combat forces and their logistic support for execution of a tactical, operational, or strategic mission. See also **assault; assault phase.** (JP 3-18)

air-breathing missile — A missile with an engine requiring the intake of air for combustion of its fuel, as in a ramjet or turbojet. (JP 3-01)

air-capable ship — A ship other than an aircraft carrier, nuclear; amphibious assault ship (general purpose); or amphibious assault ship (multipurpose) from which aircraft can take off, be recovered, or routinely receive and transfer logistic support. Also called **ACS.** (JP 3-04)

air component coordination element — An Air Force component element that interfaces and provides liaison with the joint force land component commander, or commander Army forces. The air component coordination element is the senior Air Force element assisting the joint force land component commander, or commander Army forces in planning air component supporting and supported requirements. Also called **ACCE.** (JP 3-30)

air corridor — A restricted air route of travel specified for use by friendly aircraft and established for the purpose of preventing friendly aircraft from being fired on by friendly forces. (JP 3-52)

aircraft carrier — A warship designed to support and operate aircraft, engage in attacks on targets afloat or ashore, and engage in sustained operations in support of other forces. Also called **CV or CVN.** (JP 3-32)

air defense — Defensive measures designed to destroy attacking enemy aircraft or missiles in the atmosphere, or to nullify or reduce the effectiveness of such attack. Also called **AD.** See also **active air defense; aerospace defense; passive air defense.** (JP 3-01)

air defense area — 1. **overseas** — A specifically defined airspace for which air defense must be planned and provided. 2. **United States** — Airspace of defined dimensions designated by the appropriate agency within which the ready control of airborne vehicles is required in the interest of national security during an air defense emergency. (JP 3-01)

air defense artillery — Weapons and equipment for actively combating air targets from the ground. Also called **ADA.** (JP 3-01)

air defense identification zone — Airspace of defined dimensions within which the ready identification, location, and control of airborne vehicles are required. Also called **ADIZ.** (JP 3-52)

air defense region — A geographical subdivision of an air defense area. (JP 3-01)

air defense sector — A geographical subdivision of an air defense region. (JP 3-01)

air defense warning condition — An air defense warning given in the form of a color code corresponding to the degree of air raid probability with yellow standing for when an attack by hostile aircraft or missiles is probable; red for when an attack by hostile aircraft or missiles is imminent or is in progress; and white for when an attack by hostile aircraft or missiles is improbable. Also called **ADWC.** (JP 3-01)

air domain — The atmosphere, beginning at the Earth's surface, extending to the altitude where its effects upon operations become negligible. (JP 3-30)

airdrop — The unloading of personnel or materiel from aircraft in flight. See also **air movement; free drop; free fall; high velocity drop; low velocity drop.** (JP 3-17)

airfield — An area prepared for the accommodation (including any buildings, installations, and equipment), landing, and takeoff of aircraft. See also **departure airfield; landing area; landing site.** (JP 3-17)

Air Force special operations component — The Air Force component of a joint force special operations component. Also called AFSOC. See also **Army special operations component; Navy special operations component.** (JP 3-05.1)

Air Force special operations detachment — A squadron-size headquarters that could be a composite organization composed of different Air Force special operations assets, normally subordinate to an Air Force special operations component. Also called **AFSOD.** (JP 3-05)

Air Force special operations forces — Those Active and Reserve Component Air Force forces designated by the Secretary of Defense that are specifically organized, trained, and equipped to conduct and support special operations. Also called **AFSOF.** (JP 3-05)

airhead — 1. A designated area in a hostile or potentially hostile operational area that, when seized and held, ensures the continuous air landing of troops and materiel and provides the maneuver space necessary for projected operations. Also called **a lodgment area**. (JP 3-18) 2. A designated location in an operational area used as a base for supply and evacuation by air. See also **beachhead.** (JP 3-17)

airhead line — A line denoting the limits of the objective area for an airborne assault. See also **airhead; assault phase; objective area.** (JP 3-18)

air interdiction — Air operations conducted to divert, disrupt, delay, or destroy the enemy's military surface capabilities before it can be brought to bear effectively against friendly forces, or to otherwise achieve objectives that are conducted at such distances from friendly forces that detailed integration of each air mission with the fire and movement of friendly forces is not required. (JP 3-03)

airland — Move by air and disembark, or unload, after the aircraft has landed or while an aircraft is hovering. See also **air movement.** (JP 3-17)

air land operation — An operation involving movement by air with a designated destination for further ground deployment of units and personnel and/or further ground distribution of supplies. See also **airland.** (JP 3-17)

air liaison officer — The senior tactical air control party member attached to a ground unit who functions as the primary advisor to the ground commander on air power. An air liaison officer is usually an aeronautically rated officer. Also called **ALO.** See also **liaison.** (JP 3-09.3)

airlift capability — The total capacity expressed in terms of number of passengers and/or weight/cubic displacement of cargo that can be carried at any one time to a given destination by available airlift. See also **airlift requirement.** (JP 3-17)

airlift control team — A core team within the joint air operations center with intratheater airlift functional expertise to plan, coordinate, manage, and execute intratheater airlift operations in support of the joint force air component commander. Also called **ALCT.** See also **air and space operations center; air mobility division; intratheater airlift.** (JP 3-17)

airlift mission commander — A commander designated when airlift aircraft are participating in airlift operations specified in the implementing directive. See also **joint force air component commander.** (JP 3-17)

airlift requirement — The total number of passengers and/or weight/cubic displacement of cargo required to be carried by air for a specific task. See also **airlift capability.** (JP 3-17)

airmiss — See **near miss.**

air mobility — The rapid movement of personnel, materiel and forces to and from or within a theater by air. See also **air refueling.** (JP 3-17)

Air Mobility Command — The Air Force component command of the United States Transportation Command. Also called **AMC.** (JP 3-17)

air mobility control team — A core team within the joint air operations center that directs or redirects air mobility forces in response to requirements changes, higher priorities, or immediate execution requirements. Also called **AMCT.** See also **air and space operations center; air mobility; air mobility division.** (JP 3-17)

air mobility division — Located in the joint air operations center to plan, coordinate, task, and execute the air mobility mission consisting of the air mobility control team, airlift control team, air refueling control team, and aeromedical evacuation control team. Also called **AMD.** See also **air mobility; joint air operations center.** (JP 3-17)

air mobility liaison officer — A rated United States Air Force mobility air forces officer selected, trained, and equipped to assess, train, advise, and assist with mobility air forces and ground force integration for air movement and sustainment. Also called **AMLO.** (JP 3-17)

air movement — Air transport of units, personnel, supplies, and equipment including airdrops and air landings. See also **airdrop; airland.** (JP 3-17)

air refueling — The refueling of an aircraft in flight by another aircraft. Also called **AR.** (JP 3-17)

air refueling control team — A core team within the joint air operations center that coordinates aerial refueling to support combat air operations or to support a strategic airbridge. Also called **ARCT.** See also **air and space operations center; air mobility division; air refueling.** (JP 3-17)

air route — The navigable airspace between two points, identified to the extent necessary for the application of flight rules. (JP 3-52)

air route traffic control center — The principal facility exercising en route control of aircraft operating under instrument flight rules within its area of jurisdiction. Approximately 26 such centers cover the United States and its possessions. Each has a communication capability to adjacent centers.

air sovereignty — A nation's inherent right to exercise absolute control and authority over the airspace above its territory. (JP 3-27)

airspace control — A process used to increase operational effectiveness by promoting the safe, efficient, and flexible use of airspace. (JP 3-52)

airspace control area — Airspace that is laterally defined by the boundaries of the operational area, and may be subdivided into airspace control sectors. (JP 3-01)

airspace control authority — The commander designated to assume overall responsibility for the operation of the airspace control system in the airspace control area. Also called **ACA.** See also **airspace control; airspace control area; airspace control system; control; operation.** (JP 3-52)

airspace control order — An order implementing the airspace control plan that provides the details of the approved requests for airspace coordinating measures. It is published either as part of the air tasking order or as a separate document. Also called **ACO.** (JP 3-52)

airspace control plan — The document approved by the joint force commander that provides specific planning guidance and procedures for the airspace control system for the joint force operational area. Also called **ACP.** See also **airspace control system; joint force commander.** (JP 3-52)

airspace control procedures — Rules, mechanisms, and directions that facilitate the control and use of airspace of specified dimensions. See also **airspace control authority; airspace control order; airspace control plan.** (JP 3-52)

airspace control sector — A sub-element of the airspace control area, established to facilitate the control of the overall area. Airspace control sector boundaries normally coincide with air defense organization subdivision boundaries. Airspace control sectors are designated in accordance with procedures and guidance contained in the airspace control plan in consideration of Service component, host nation, and multinational airspace control capabilities and requirements. See also **airspace control area.** (JP 3-52)

airspace control system — An arrangement of those organizations, personnel, policies, procedures, and facilities required to perform airspace control functions. Also called **ACS.** (JP 3-52)

airspace coordinating measures — Measures employed to facilitate the efficient use of airspace to accomplish missions and simultaneously provide safeguards for friendly forces. Also called **ACMs.** See also **airspace control area; airspace control sector; airspace coordination area; high-density airspace control zone; weapons engagement zone.** (JP 3-52)

airspace coordination area — A three-dimensional block of airspace in a target area, established by the appropriate ground commander, in which friendly aircraft are reasonably safe from friendly surface fires. The airspace coordination area may be formal or informal. Also called **ACA.** (JP 3-09.3)

airspace management — The coordination, integration, and regulation of the use of airspace of defined dimensions. (JP 3-52)

airspace reservation — The airspace located above an area on the surface of the land or water, designated and set apart by Executive Order of the President or by a state, commonwealth, or territory, over which the flight of aircraft is prohibited or restricted for the purpose of national defense or for other governmental purposes.

air superiority — That degree of dominance in the air battle by one force that permits the conduct of its operations at a given time and place without prohibitive interference from air and missile threats. (JP 3-01)

air support coordination section — In amphibious operations, the section of the Navy tactical air control center designated to coordinate, control, and integrate all direct support aircraft (i.e., close air support) and assault support operations. Also called **ASCS.** (JP 3-02)

air support operations center — The principal air control agency of the theater air control system responsible for the direction and control of air operations directly supporting the ground combat element. It coordinates air missions requiring integration with other supporting arms and ground forces. It normally collocates with the Army tactical headquarters senior fire support coordination center within the ground combat element. Also called **ASOC.** See also **close air support; operation; tactical air control center.** (JP 3-09.3)

air support request — A means to request preplanned and immediate close air support, air interdiction, air reconnaissance, surveillance, escort, helicopter airlift, and other aircraft missions. Also called **AIRSUPREQ.** (JP 3-30)

air supremacy — That degree of air superiority wherein the opposing force is incapable of effective interference within the operational area using air and missile threats. (JP 3-01)

air tasking order — A method used to task and disseminate to components, subordinate units, and command and control agencies projected sorties, capabilities and/or forces to targets and specific missions. Normally provides specific instructions to include call signs, targets, controlling agencies, etc., as well as general instructions. Also called **ATO.** (JP 3-30)

air terminal — A facility on an airfield that functions as an air transportation hub and accommodates the loading and unloading of airlift aircraft and the intransit processing of traffic. (JP 3-17)

air traffic controller — An air controller specially trained and certified for civilian air traffic control. (JP 3-52)

air traffic control section — In amphibious operations, the section of the Navy tactical air control center designed to provide initial safe passage, radar control, and surveillance for close air support aircraft in the operational area. Also called **ATCS.** (JP 3-02)

airways station — A ground communication installation established, manned, and equipped to communicate with aircraft in flight, as well as with other designated airways installations, for the purpose of expeditious and safe movements of aircraft. These stations may or may not be located on designated airways.

alert force — Specified forces maintained in a special degree of readiness. (JP 3-10)

alerting service — A service provided to notify appropriate organizations regarding aircraft in need of search and rescue aid, and assist such organizations as required. (JP 3-52)

alert order — 1. A crisis action planning directive from the Secretary of Defense, issued by the Chairman of the Joint Chiefs of Staff, that provides essential guidance for planning and directs the initiation of execution planning for the selected course of action authorized by the Secretary of Defense. 2. A planning directive that provides essential planning guidance, directs the initiation of execution planning after the directing authority approves a military course of action, but does not authorize execution. Also called **ALERTORD.** See also **course of action; execution planning.** (JP 5-0)

all appropriate action — Action taken in self-defense that is reasonable in intensity, duration, and magnitude, based on all the facts known to the commander at the time. (JP 3-68)

alliance — The relationship that results from a formal agreement between two or more nations for broad, long-term objectives that further the common interests of the members. See also **coalition; multinational.** (JP 3-0)

allocation —Distribution of limited forces and resources for employment among competing requirements. See also **apportionment.** (JP 5-0)

allocation request — A message used to provide an estimate of the total air effort, to identify any excess and joint force general support aircraft sorties, and to identify unfilled air requirements. This message is used only for preplanned missions and is transmitted on a daily basis, normally 24 hours prior to the start of the next air tasking day. Also called **ALLOREQ.** (JP 3-30)

allotment — The temporary change of assignment of tactical air forces between subordinate commands. The authority to allot is vested in the commander having combatant command (command authority). See also **combatant command (command authority).**

allowable cabin load — The maximum payload that can be carried on an individual sortie. Also called **ACL.** (JP 3-17)

all-source intelligence — 1. Intelligence products and/or organizations and activities that incorporate all sources of information in the production of finished intelligence. 2. In intelligence collection, a phrase that indicates that in the satisfaction of intelligence requirements, all collection, processing, exploitation, and reporting systems and resources are identified for possible use and those most capable are tasked. See also **intelligence.** (JP 2-0)

American Forces Radio and Television Service — A worldwide radio and television broadcasting organization that provides United States military commanders overseas and at sea with sufficient electronic media resources to effectively communicate theater, local, Department of Defense, and Service-unique command information to their personnel and family members. Also called **AFRTS.** (JP 3-61)

ammunition controlled supply rate — In Army usage, the amount of ammunition estimated to be available to sustain operations of a designated force for a specified time if expenditures are controlled at that rate. It is expressed in terms of rounds per weapon per day for ammunition items fired by weapons, and in terms of units of measure per organization per day for bulk allotment ammunition items. Tactical commanders use this rate to control expenditures of ammunition during tactical operations at planned intervals. It is issued through command channels at each level. It is determined based on consideration of the required supply rates submitted by subordinate commanders and ammunition assets available.

ammunition lot — A quantity of homogeneous ammunition, identified by a unique lot number, which is manufactured, assembled, or renovated by one producer under uniform conditions and which is expected to function in a uniform manner. (JP 3-04)

amphibian — A small craft, propelled by propellers and wheels or by air cushions for the purpose of moving on both land and water. (JP 4-01.6)

amphibious assault — The principal type of amphibious operation that involves establishing a force on a hostile or potentially hostile shore. See also **assault; assault phase.** (JP 3-02)

amphibious assault ship (multipurpose) — A naval ship designed to embark, deploy, and land elements of a landing force in an assault by helicopters, landing craft, amphibious vehicles, and by combinations of these methods. Also called **LHD.** (JP 3-02)

amphibious breaching — The conduct of a deliberate breaching operation specifically designed to overcome antilanding defenses in order to conduct an amphibious assault. (JP 3-02)

amphibious bulk liquid transfer system — Hosereel system providing capability to deliver fuel and/or water from ship to shore. Also called **ABLTS.** (JP 4-01.6)

amphibious construction battalion — A permanently commissioned naval unit, subordinate to the commander, naval beach group, designed to provide an administrative unit from which personnel and equipment are formed in tactical elements and made available to appropriate commanders to operate pontoon causeways, transfer barges, warping tugs, and assault bulk fuel systems, and to meet salvage requirements of the naval beach party. Also called **PHIBCB.** (JP 3-02)

amphibious defense zone — The area encompassing the amphibious objective area and the adjoining airspace required by accompanying naval forces for the purpose of air defense. Also called an **ADZ.** (JP 3-02)

amphibious demonstration — A type of amphibious operation conducted for the purpose of deceiving the enemy by a show of force with the expectation of deluding the enemy into a course of action unfavorable to him. (JP 3-02)

amphibious force — An amphibious task force and a landing force together with other forces that are trained, organized, and equipped for amphibious operations. Also called **AF.** See also **amphibious operation; amphibious task force; landing force.** (JP 3-02)

amphibious lift — The total capacity of assault shipping utilized in an amphibious operation, expressed in terms of personnel, vehicles, and measurement or weight tons of supplies. (JP 3-02)

amphibious objective area — A geographical area (delineated for command and control purposes in the initiating directive) within which is located the objective(s) to be secured by the amphibious force. This area must be of sufficient size to ensure accomplishment of the amphibious force's mission and must provide sufficient area for conducting necessary sea, air, and land operations. Also called **AOA.** See also **amphibious force; mission.** (JP 3-02)

amphibious operation — A military operation launched from the sea by an amphibious force, embarked in ships or craft with the primary purpose of introducing a landing force ashore to accomplish the assigned mission. See also **amphibious force; landing force; mission; operation.** (JP 3-02)

amphibious planning — The process of planning for an amphibious operation, distinguished by the necessity for concurrent, parallel, and detailed planning by all participating forces. The planning pattern is cyclical in nature, composed of a series of analyses and judgments of operational situations, each stemming from those that have preceded. (JP 3-02)

amphibious raid — A type of amphibious operation involving swift incursion into or temporary occupation of an objective followed by a planned withdrawal. See also **amphibious operation.** (JP 3-02)

amphibious shipping — Organic Navy ships specifically designed to transport, land, and support landing forces in amphibious assault operations and capable of being loaded or unloaded by naval personnel without external assistance in the amphibious objective area. (JP 3-02)

amphibious squadron — A tactical and administrative organization composed of amphibious assault shipping to transport troops and their equipment for an amphibious assault operation. Also called **PHIBRON.** (JP 3-02)

amphibious task force — A Navy task organization formed to conduct amphibious operations. The amphibious task force, together with the landing force and other forces, constitutes the amphibious force. Also called **ATF.** See also **amphibious force; amphibious operation; landing force.** (JP 3-02)

amphibious tractor — See **amphibious vehicle.**

amphibious transport dock — A ship designed to transport and land troops, equipment, and supplies by means of embarked landing craft, amphibious vehicles, and helicopters. Designated as **LPD.** (JP 3-02)

amphibious transport group — A subdivision of an amphibious task force composed primarily of transport ships. The size of the transport group will depend upon the scope of the operation. Ships of the transport group will be combat-loaded to support the landing force scheme of maneuver ashore. A transport unit will usually be formed to embark troops and equipment to be landed over a designated beach or to embark all helicopter-borne troops and equipment. (JP 3-02)

amphibious vehicle — A wheeled or tracked vehicle capable of operating on both land and water. See also **landing craft.** (JP 3-02)

amphibious vehicle availability table — A tabulation of the type and number of amphibious vehicles available primarily for assault landings and for support of other elements of the operation. (JP 3-02)

JP 1-02

15

amphibious vehicle employment plan — A plan showing in tabular form the planned employment of amphibious vehicles in landing operations, including their employment after the initial movement to the beach. (JP 3-02)

amphibious vehicle launching area — An area, in the vicinity of and to seaward of the line of departure, to which landing ships proceed and launch amphibious vehicles. (JP 3-02)

amphibious withdrawal — A type of amphibious operation involving the extraction of forces by sea in ships or craft from a hostile or potentially hostile shore. See also **amphibious operation.** (JP 3-02)

analysis and production — In intelligence usage, the conversion of processed information into intelligence through the integration, evaluation, analysis, and interpretation of all source data and the preparation of intelligence products in support of known or anticipated user requirements. See also **intelligence process.** (JP 2-01)

annual screening — One day of active duty for training required each year for Individual Ready Reserve members so the Services can keep current on each member's physical condition, dependency status, military qualifications, civilian occupational skills, availability for service, and other information.

annual training — The minimal period of training reserve members must perform each year to satisfy the training requirements associated with their Reserve Component assignment. Also called **AT.**

antemortem data — Medical records, samples, and photographs taken prior to death. These include (but are not limited to) fingerprints, dental x-rays, body tissue samples, photographs of tattoos, or other identifying marks. These "pre-death" records would be compared against records completed after death to help establish a positive identification of human remains. See also **mortuary affairs.** (JP 4-06)

antideficiency violations — The incurring of obligations or the making of expenditure (outlays) in violation of appropriation law as to purpose, time, and amounts as specified in the defense appropriation or appropriations of funds. (JP 1-06)

antiradiation missile — A missile which homes passively on a radiation source. Also called **ARM.** See also **guided missile.** (JP 3-01)

antisubmarine warfare — Operations conducted with the intention of denying the enemy the effective use of submarines. Also called **ASW.** (JP 3-32)

antiterrorism — Defensive measures used to reduce the vulnerability of individuals and property to terrorist acts, to include rapid containment by local military and civilian forces. Also called **AT.** See also **counterterrorism; terrorism.** (JP 3-07.2)

anti-vehicle land mine — A mine designed to immobilize or destroy a vehicle. Also called **AVL**. (JP 3-15)

application — 1. The system or problem to which a computer is applied. 2. In the intelligence context, the direct extraction and tailoring of information from an existing foundation of intelligence and near real time reporting. (JP 2-0)

apportionment — In the general sense, distribution of forces and capabilities as the starting point for planning, etc. See also **allocation.** (JP 5-0)

approach schedule — In amphibious operations, this schedule indicates, for each scheduled wave, the time of departure from the rendezvous area, from the line of departure, and from other control points and the time of arrival at the beach. (JP 3-02)

approach time — The time at which an aircraft is expected to commence approach procedure.

apron — A defined area on an airfield intended to accommodate aircraft for purposes of loading or unloading passengers or cargo, refueling, parking, or maintenance. (JP 3-34)

area air defense commander — The component commander with the preponderance of air defense capability and the required command, control, and communications capabilities who is assigned by the joint force commander to plan and execute integrated air defense operations. Also called **AADC.** (JP 3-01)

area command — A command which is composed of those organized elements of one or more of the Armed Services, designated to operate in a specific geographical area, which are placed under a single commander. See also **command.** (JP 3-10)

area damage control — Measures taken before, during, or after hostile action or natural or manmade disasters, to reduce the probability of damage and minimize its effects. Also called **ADC.** (JP 3-10)

area of influence — A geographical area wherein a commander is directly capable of influencing operations by maneuver or fire support systems normally under the commander's command or control. (JP 3-0)

area of interest — That area of concern to the commander, including the area of influence, areas adjacent thereto, and extending into enemy territory. This area also includes areas occupied by enemy forces who could jeopardize the accomplishment of the mission. Also called **AOI.** See also **area of influence.** (JP 3-0)

area of operations — An operational area defined by the joint force commander for land and maritime forces that should be large enough to accomplish their missions and protect their forces. Also called **AO.** See also **area of responsibility; joint operations area; joint special operations area.** (JP 3-0)

area of responsibility — The geographical area associated with a combatant command within which a geographic combatant commander has authority to plan and conduct operations. Also called **AOR.** See also **combatant command.** (JP 1)

area search — Visual reconnaissance of limited or defined areas. (JP 3-50)

Armed Forces of the United States — A term used to denote collectively all components of the Army, Marine Corps, Navy, Air Force, and Coast Guard (when mobilized under Title 10, United States Code, to augment the Navy). See also **United States Armed Forces.** (JP 1)

arming — As applied to explosives, weapons, and ammunition, the changing from a safe condition to a state of readiness for initiation. (JP 3-15)

arms control agreement — The written or unwritten embodiment of the acceptance of one or more arms control measures by two or more nations. (JP 2-01)

Army air-ground system — The Army system which provides for interface between Army and tactical air support agencies of other Services in the planning, evaluating, processing, and coordinating of air support requirements and operations. It is composed of appropriate staff members, including G-2 air and G-3 air personnel, and necessary communication equipment. Also called **AAGS.** (JP 3-09)

Army corps — An intermediate headquarters between divisions and the theater army. A corps may consist of two or more divisions together with supporting brigades. Corps headquarters are capable of serving as either a joint task force or joint force land component command headquarters. (JP 3-31)

Army Service component command — Command responsible for recommendations to the joint force commander on the allocation and employment of Army forces within a combatant command. Also called **ASCC.** (JP 3-31)

Army special operations component — The Army component of a joint force special operations component. Also called **ARSOC.** See also **Air Force special operations component; Navy special operations component.** (JP 3-05.1)

Army special operations forces — Those Active and Reserve Component Army forces designated by the Secretary of Defense that are specifically organized, trained, and equipped to conduct and support special operations. Also called **ARSOF.** (JP 3-05)

Army support area — The specific support area for a theater Army that is outside of a division or corps's operational area established primarily for the positioning, employment, and protection of theater support units; and where the majority of the sustaining operations occur. (JP 3-31)

arrival zone — In counterdrug operations, the area in or adjacent to the United States where smuggling concludes and domestic distribution begins (by air, an airstrip; by sea, an offload point on land, or transfer to small boats). See also **transit zone.** (JP 3-07.4)

ascent phase — That portion of the flight of a ballistic missile or space vehicle that begins after powered flight and ends just prior to apogee. (JP 3-01)

assault — 1. In an amphibious operation, the period of time between the arrival of the major assault forces of the amphibious task force in the objective area and the accomplishment of the amphibious task force mission. (JP 3-02) 2. To make a short, violent, but well-ordered attack against a local objective, such as a gun emplacement, a fort, or a machine gun nest. (JP 3-18) 3. A phase of an airborne operation beginning with delivery by air of the assault echelon of the force into the objective area and extending through attack of assault objectives and consolidation of the initial airhead. See also **assault phase.** (JP 3-18)

assault breaching — A part of amphibious breaching in support of an amphibious assault involving a fire support mission using precision guided munitions to neutralize mines and obstacles in the surf zone and on the beach. (JP 3-02)

assault craft — A landing craft or amphibious vehicle primarily employed for landing troops and equipment in the assault waves of an amphibious operation. (JP 3-02)

assault craft unit — A permanently commissioned naval organization, subordinate to the commander, naval beach group, that contains landing craft and crews necessary to provide lighterage required in an amphibious operation. Also called **ACU.** (JP 3-02)

assault echelon — In amphibious operations, the element of a force comprised of tailored units and aircraft assigned to conduct the initial assault on the operational area. Also called **AE.** See also **amphibious operation.** (JP 3-02)

assault fire — 1. That fire delivered by attacking troops as they close with the enemy. 2. In artillery, extremely accurate, short-range destruction fire at point targets.

assault follow-on echelon — In amphibious operations, that echelon of the assault troops, vehicles, aircraft, equipment, and supplies that, though not needed to initiate the assault, is required to support and sustain the assault. In order to accomplish its purpose, it is normally required in the objective area no later than five days after commencement of the assault landing. Also called **AFOE.** (JP 3-02)

assault phase — In an airborne operation, a phase beginning with delivery by air of the assault echelon of the force into the objective area and extending through attack of assault objectives and consolidation of the initial airhead. See also **assault.** (JP 3-18)

assault schedule — In amphibious operations, this schedule provides the formation, composition, and timing of waves landing over the beach. (JP 3-02)

assault shipping — Shipping assigned to the amphibious task force and utilized for transporting assault troops, vehicles, equipment, and supplies to the objective area. (JP 3-02)

assault wave — See **wave.**

assembly area — 1. An area in which a command is assembled preparatory to further action. 2. In a supply installation, the gross area used for collecting and combining components into complete units, kits, or assemblies. (JP 4-09)

assessment — 1. A continuous process that measures the overall effectiveness of employing joint force capabilities during military operations. 2. Determination of the progress toward accomplishing a task, creating a condition, or achieving an objective. 3. Analysis of the security, effectiveness, and potential of an existing or planned intelligence activity. 4. Judgment of the motives, qualifications, and characteristics of present or prospective employees or "agents." (JP 3-0)

assessment agent — The organization responsible for conducting an assessment of an approved joint publication. Also called **AA.** (CJCSM 5120.01)

asset validation — In intelligence use, the process used to determine the asset authenticity, reliability, utility, suitability, and degree of control the case officer or others have. (JP 2-01.2)

asset visibility — Provides users with information on the location, movement, status, and identity of units, personnel, equipment, and supplies, which facilitates the capability to act upon that information to improve overall performance of the Department of Defense's logistics practices. Also called **AV.** (JP 3-35)

assign — 1. To place units or personnel in an organization where such placement is relatively permanent, and/or where such organization controls and administers the units or personnel for the primary function, or greater portion of the functions, of the unit or personnel. 2. To detail individuals to specific duties or functions where such duties or functions are primary and/or relatively permanent. See also **attach.** (JP 3-0)

assistance in kind — The provision of material and services for a logistic exchange of materials and services of equal value between the governments of eligible countries. Also called **AIK.** (JP 1-06)

assumption — A supposition on the current situation or a presupposition on the future course of events, either or both assumed to be true in the absence of positive proof, necessary to enable the commander in the process of planning to complete an estimate of the situation and make a decision on the course of action. (JP 5-0)

asymmetric — In military operations the application of dissimilar strategies, tactics, capabilities, and methods to circumvent or negate an opponent's strengths while exploiting his weaknesses. (JP 3-15.1)

atmospheric environment — The envelope of air surrounding the Earth, including its interfaces and interactions with the Earth's solid or liquid surface. (JP 3-59)

attach — 1. The placement of units or personnel in an organization where such placement is relatively temporary. 2. The detailing of individuals to specific functions where such functions are secondary or relatively temporary. See also **assign.** (JP 3-0)

attack assessment — An evaluation of information to determine the potential or actual nature and objectives of an attack for the purpose of providing information for timely decisions. See also **damage estimation.** (JP 3-14)

attack group — A subordinate task organization of the Navy forces of an amphibious task force. It is composed of assault shipping and supporting naval units designated to transport, protect, land, and initially support a landing group. (JP 3-02)

attack heading — 1. The interceptor heading during the attack phase that will achieve the desired track-crossing angle. 2. The assigned magnetic compass heading to be flown by aircraft during the delivery phase of an air strike. (JP 3-09.3)

attack position — The last position occupied by the assault echelon before crossing the line of departure.

attack the network operations — Lethal and nonlethal actions and operations against networks conducted continuously and simultaneously at multiple levels (tactical, operational, and strategic) that capitalize on or create key vulnerabilities and disrupt activities to eliminate the enemy's ability to function in order to enable success of the operation or campaign. Also called **AtN operations.** (JP 3-15.1)

augmentation forces — Forces to be transferred from a supporting combatant commander to the combatant command (command authority) or operational control of a supported combatant commander during the execution of an operation order approved by the President and Secretary of Defense. (JP 5-0)

authenticate — A challenge given by voice or electrical means to attest to the authenticity of a person, message, or transmission. (JP 3-50)

authentication — 1. A security measure designed to protect a communications system against acceptance of a fraudulent transmission or simulation by establishing the validity of a transmission, message, or originator. 2. A means of identifying individuals and verifying their eligibility to receive specific categories of information. 3. Evidence by proper signature or seal that a document is genuine and official. 4. In personnel recovery missions, the process whereby the identity of an isolated person is confirmed. See also **evader; evasion; recovery operations; security.** (JP 3-50)

authenticator — A symbol or group of symbols, or a series of bits, selected or derived in a prearranged manner and usually inserted at a predetermined point within a message or transmission for the purpose of attesting to the validity of the message or transmission. (JP 3-13.3)

authorized departure — A procedure, short of ordered departure, by which mission employees or dependents or both, are permitted to leave post in advance of normal rotation when the national interests or imminent threat to life require it. (JP 3-68)

automated identification technology — A suite of tools for facilitating total asset visibility source data capture and transfer. Automated identification technology includes a variety of devices, such as bar codes, magnetic strips, optical memory cards, and radio frequency tags for marking or "tagging" individual items, multi-packs, equipment, air pallets, or containers, along with the hardware and software required to create the devices, read the information on them, and integrate that information with other logistic information. Also called **AIT.** (JP 3-35)

Automated Repatriation Reporting System — The Defense Manpower Data Center uses this system to track the status of noncombatant evacuees after they have arrived in an initial safe haven in the United States. (JP 3-68)

automatic approach and landing — A control mode in which the aircraft's speed and flight path are automatically controlled for approach, flare-out, and landing. (JP 3-52)

autonomous operation — In air defense, the mode of operation assumed by a unit after it has lost all communications with higher echelons forcing the unit commander to assume full responsibility for control of weapons and engagement of hostile targets. (JP 3-01)

available-to-load date — A date specified for each unit in a time-phased force and deployment data indicating when that unit will be ready to load at the point of embarkation. Also called **ALD.** (JP 5-0)

avenue of approach — An air or ground route of an attacking force of a given size leading to its objective or to key terrain in its path. Also called **AA.** (JP 2-01.3)

aviation medicine — The special field of medicine which is related to the biological and psychological problems of flight. (JP 4-02)

axis of advance — A line of advance assigned for purposes of control; often a road or a group of roads, or a designated series of locations, extending in the direction of the enemy. (JP 3-03)

B

backfill — Reserve Component units and individuals recalled to replace deploying active units and/or individuals in the continental United States and outside the continental United States. See also **Reserve Component.** (JP 4-05)

bale cubic capacity — The space available for cargo measured in cubic feet to the inside of the cargo battens, on the frames, and to the underside of the beams. (JP 4-01.2)

ballistic missile — Any missile which does not rely upon aerodynamic surfaces to produce lift and consequently follows a ballistic trajectory when thrust is terminated. See also **guided missile.** (JP 3-01)

bare base — A base having minimum essential facilities to house, sustain, and support operations to include, if required, a stabilized runway, taxiways, and aircraft parking areas. A bare base must have a source of water that can be made potable. Other requirements to operate under bare base conditions form a necessary part of the force package deployed to the bare base. See also **base.** (JP 3-05.1)

barrage — 1. A prearranged barrier of fires, except that delivered by small arms, designed to protect friendly troops and installations by impeding enemy movements across defensive lines or areas. 2. A type of electronic attack intended for simultaneous jamming over a wide area of frequency spectrum. See also **electronic warfare; fires.**

barrier — A coordinated series of natural or man-made obstacles designed or employed to channel, direct, restrict, delay, or stop the movement of an opposing force and to impose additional losses in personnel, time, and equipment on the opposing force. (JP 3-15)

barrier combat air patrol — One or more divisions or elements of fighter aircraft employed between a force and an objective area as a barrier across the probable direction of enemy attack. See also **combat air patrol.** (JP 3-01)

barrier, obstacle, and mine warfare plan — A comprehensive, coordinated plan that includes responsibilities; general location of unspecified and specific barriers, obstacles, and minefields; special instructions; limitations; coordination; and completion times; and may designate locations of obstacle zones or belts. (JP 3-15)

base — 1. A locality from which operations are projected or supported. 2. An area or locality containing installations which provide logistic or other support. 3. Home airfield or home carrier. See also **facility.** (JP 4-0)

base boundary — A line that delineates the surface area of a base for the purpose of facilitating coordination and deconfliction of operations between adjacent units, formations, or areas. (JP 3-10)

base cluster — In base defense operations, a collection of bases, geographically grouped for mutual protection and ease of command and control. (JP 3-10)

base cluster commander — In base defense operations, a senior base commander designated by the joint force commander responsible for coordinating the defense of bases within the base cluster and for integrating defense plans of bases into a base cluster defense plan. (JP 3-10)

base cluster operations center — A command and control facility that serves as the base cluster commander's focal point for defense and security of the base cluster. Also called **BCOC**. (JP 3-10)

base commander — In base defense operations, the officer assigned to command a base. (JP 3-10)

base defense — The local military measures, both normal and emergency, required to nullify or reduce the effectiveness of enemy attacks on, or sabotage of, a base, to ensure that the maximum capacity of its facilities is available to US forces. (JP 3-10)

base defense forces — Troops assigned or attached to a base for the primary purpose of base defense and security as well as augmentees and selectively armed personnel available to the base commander for base defense from units performing primary missions other than base defense. (JP 3-10)

base defense operations center — A command and control facility, with responsibilities similar to a base cluster operations center, established by the base commander to serve as the focal point for base security and defense. It plans, directs, integrates, coordinates, and controls all base defense efforts. Also called **BDOC**. (JP 3-10)

base defense zone — An air defense zone established around an air base and limited to the engagement envelope of short-range air defense weapons systems defending that base. Base defense zones have specific entry, exit, and identification, friend or foe procedures established. Also called **BDZ**. (JP 3-52)

base development (less force beddown) — The acquisition, development, expansion, improvement, construction and/or replacement of the facilities and resources of a location to support forces. (JP 3-34)

baseline costs — The continuing annual costs of military operations funded by the operations and maintenance and military personnel appropriations. (JP 1-06)

base operating support — Directly assisting, maintaining, supplying, and distributing support of forces at the operating location. Also called **BOS**. (JP 4-0)

base operating support-integrator — The designated Service component or joint task force commander assigned to synchronize all sustainment functions for a contingency base. Also called **BOS-I**. (JP 4-0)

base plan — A type of operation plan that describes the concept of operations, major forces, sustainment concept, and anticipated timelines for completing the mission without annexes or time-phased force and deployment data. Also called **BPLAN**. (JP 5-0)

base support installation — A Department of Defense Service or agency installation within the United States and its territories tasked to serve as a base for military forces engaged in either homeland defense or defense support of civil authorities. Also called **BSI**. (JP 3-28)

basic encyclopedia — A compilation of identified installations and physical areas of potential significance as objectives for attack. Also called **BE**. (JP 2-01)

basic load — The quantity of supplies required to be on hand within, and which can be moved by, a unit or formation. It is expressed according to the wartime organization of the unit or formation and maintained at the prescribed levels. (JP 4-09)

basic tactical organization — The conventional organization of landing force units for combat, involving combinations of infantry, supporting ground arms, and aviation for accomplishment of missions ashore. This organizational form is employed as soon as possible following the landing of the various assault components of the landing force.

battalion landing team — In an amphibious operation, an infantry battalion normally reinforced by necessary combat and service elements; the basic unit for planning an assault landing. Also called **BLT**. (JP 3-02)

battle damage assessment — The estimate of damage composed of physical and functional damage assessment, as well as target system assessment, resulting from the application of lethal or nonlethal military force. Also called **BDA**. See also **combat assessment**. (JP 3-0)

battle damage repair — Essential repair, which may be improvised, carried out rapidly in a battle environment in order to return damaged or disabled equipment to temporary service. Also called **BDR**. (JP 4-09)

battlefield coordination detachment — An Army liaison located in the air operations center that provides selected operational functions between the Army forces and the air component commander. Also called **BCD**. See also **air and space operations center; liaison**. (JP 3-03)

battle injury — Damage or harm sustained by personnel during or as a result of battle conditions. Also called **BI**. (JP 4-02)

battle management — The management of activities within the operational environment based on the commands, direction, and guidance given by appropriate authority. Also called **BM.** (JP 3-01)

beach — 1. The area extending from the shoreline inland to a marked change in physiographic form or material, or to the line of permanent vegetation (coastline). 2. In amphibious operations, that portion of the shoreline designated for landing of a tactical organization. (JP 3-02)

beach group — See **naval beach group; shore party.**

beachhead — A designated area on a hostile or potentially hostile shore that, when seized and held, ensures the continuous landing of troops and materiel, and provides maneuver space requisite for subsequent projected operations ashore. (JP 3-02)

beachmaster unit — A commissioned naval unit of the naval beach group designed to provide to the shore party a Navy component known as a beach party, which is capable of supporting the amphibious landing of one division (reinforced). Also called **BMU.** See also **beach party; naval beach group; shore party.** (JP 4-01.6)

beach party — The Navy component of the landing force support party under the tactical control of the landing force support party commander. See also **beachmaster unit; shore party.** (JP 3-02)

beach party commander — The naval officer in command of the naval component of the shore party.

beach photography — Vertical, oblique, ground, and periscope coverage at varying scales to provide information of offshore, shore, and inland areas. It covers terrain that provides observation of the beaches and is primarily concerned with the geological and tactical aspects of the beach.

beach support area — In amphibious operations, the area to the rear of a landing force or elements thereof, established and operated by shore party units, which contains the facilities for the unloading of troops and materiel and the support of the forces ashore; it includes facilities for the evacuation of wounded, enemy prisoners of war, and captured materiel. Also called **BSA.** (JP 3-02)

beach survey — The collection of data describing the physical characteristics of a beach; that is, an area whose boundaries are a shoreline, a coastline, and two natural or arbitrary assigned flanks.

beach width — The horizontal dimensions of the beach measured at right angles to the shoreline from the line of extreme low water inland to the landward limit of the beach (the coastline).

begin morning civil twilight — The period of time at which the sun is halfway between beginning morning and nautical twilight and sunrise, when there is enough light to see objects clearly with the unaided eye. At this time, light intensification devices are no longer effective, and the sun is six degrees below the eastern horizon. Also called **BMCT.**

begin morning nautical twilight — The start of that period where, in good conditions and in the absence of other illumination, enough light is available to identify the general outlines of ground objects and conduct limited military operations. Light intensification devices are still effective and may have enhanced capabilities. At this time, the sun is 12 degrees below the eastern horizon. Also called **BMNT.**

believed-to-be — In mortuary affairs, the status of any human remains until a positive identification has been determined. Used interchangeably with tentative identification. Also called **BTB.** (JP 4-06)

berm — The nearly horizontal portion of a beach or backshore having an abrupt fall and either formed by deposition of material by wave action at the limit of ordinary high tide or constructed to protect materials handling equipment during air cushion vehicle operations. (JP 4-01.6)

bill — A ship's publication listing operational or administrative procedures. (JP 3-04)

biological agent — A microorganism (or a toxin derived from it) that causes disease in personnel, plants, or animals or causes the deterioration of materiel. See also **chemical agent.** (JP 3-11)

biological hazard — An organism, or substance derived from an organism, that poses a threat to human or animal health. (JP 3-11)

biometrics — The process of recognizing an individual based on measurable anatomical, physiological, and behavioral characteristics. (JP 2-0)

biometrics-enabled intelligence — The intelligence derived from the processing of biologic identity data and other all-source for information concerning persons of interest. Also called **BEI.** (JP 2-0)

blister agent — A chemical agent that injures the eyes and lungs, and burns or blisters the skin. Also called **vesicant agent.** (JP 3-11)

blood agent — A chemical compound, including the cyanide group, that affects bodily functions by preventing the normal utilization of oxygen by body tissues. (JP 3-11)

blood chit — A small sheet of material depicting an American flag and a statement in several languages to the effect that anyone assisting the bearer to safety will be rewarded. See also **evasion aid.** (JP 3-50)

Blue Bark — US military personnel, US citizen civilian employees of the Department of Defense, and the dependents of both categories who travel in connection with the death of an immediate family member. It also applies to designated escorts for dependents of deceased military members. Furthermore, the term is used to designate the personal property shipment of a deceased member. (JP 4-06)

boat group — The basic organization of landing craft. One boat group is organized for each battalion landing team (or equivalent) to be landed in the first trip of landing craft or amphibious vehicles. (JP 3-02)

boat lane — A lane for amphibious assault landing craft, which extends from the line of departure to the beach. (JP 3-02)

boat space — The space and weight factor used to determine the capacity of boats, landing craft, and amphibious vehicles. With respect to landing craft and amphibious vehicles, it is based on the requirements of one person with individual equipment. The person is assumed to weigh 224 pounds and to occupy 13.5 cubic feet of space. (JP 3-02)

boat wave — See **wave.**

bona fides — 1. In personnel recovery, the use of verbal or visual communication by individuals who are unknown to one another, to establish their authenticity, sincerity, honesty, and truthfulness. See also **evasion; recovery; recovery operations.** (JP 3-50) 2. The lack of fraud or deceit: a determination that a person is who he/she says he/she is. (JP 2-01.2)

boost phase — That portion of the flight of a ballistic missile or space vehicle during which the booster and sustainer engines operate. See also midcourse phase; terminal phase. (JP 3-01)

bottom mine — A mine with negative buoyancy which remains on the seabed. See also **mine.** (JP 3-15)

boundary — A line that delineates surface areas for the purpose of facilitating coordination and deconfliction of operations between adjacent units, formations, or areas. (JP 3-0)

branch — 1. A subdivision of any organization. 2. A geographically separate unit of an activity, which performs all or part of the primary functions of the parent activity on a smaller scale. 3. An arm or service of the Army. 4. The contingency options built into the base plan used for changing the mission, orientation, or direction of movement of a force to aid success of the operation based on anticipated events, opportunities, or disruptions caused by enemy actions and reactions. See also **sequel.** (JP 5-0)

breakbulk cargo — Any commodity that, because of its weight, dimensions, or incompatibility with other cargo, must be shipped by mode other than military van or military container moved via the sea. See also **breakbulk ship.** (JP 4-09)

breakbulk ship — A ship with conventional holds for stowage of breakbulk cargo, below or above deck, and equipped with cargo-handling gear. Ships also may be capable of carrying a limited number of containers, above or below deck. See also **breakbulk cargo.** (JP 4-09)

brevity code — A code which provides no security but which has as its sole purpose the shortening of messages rather than the concealment of their content. (JP 3-04)

brigade combat team — As combined arms teams, brigade combat teams form the basic building block of the Army's tactical formations. They are the principal means of executing engagements. Three standardized brigade combat teams designs exist; heavy, infantry, and Stryker. Battalion-sized maneuver, fires, reconnaissance, and sustainment units are organic to a brigade combat team. Also called **BCT.** (JP 3-31)

broken stowage — The space lost in the holds of a vessel because of the contour of the ship and the shape of the cargo. Dunnage, ladders, and stanchions are included in broken stowage. (JP 3-02.1)

broken stowage factor — A factor applied to the available space for embarkation due to the loss between boxes, between vehicles, around stanchions, and over cargo. The factor will vary, depending on the type and size of vehicles, type and size of general cargo, training and experience of loading personnel, type of loading, method of stowage, and configuration of compartments. (JP 3-02.1)

buddy-aid — Acute medical care (first aid) provided by a non-medical Service member to another person. (JP 4-02)

buffer zone — 1. A defined area controlled by a peace operations force from which disputing or belligerent forces have been excluded. Also called **area of separation** in some United Nations operations. Also called **BZ.** See also **line of demarcation; peace operations.** (JP 3-07.3) 2. A designated area used for safety in military operations. (JP 3-01)

building systems — Structures assembled from manufactured components designed to provide specific building configurations. (JP 3-34)

bulk cargo — That which is generally shipped in volume where the transportation conveyance is the only external container; such as liquids, ore, or grain. (JP 4-01.5)

bulk petroleum product — A liquid petroleum product transported by various means and stored in tanks or containers having an individual fill capacity greater than 250 liters. (JP 4-03)

bulk storage — 1. Storage in a warehouse of supplies and equipment in large quantities, usually in original containers, as distinguished from bin storage. 2. Storage of liquids, such as petroleum products in tanks, as distinguished from drum or packaged storage. (JP 4-03)

C

cache — A source of subsistence and supplies, typically containing items such as food, water, medical items, and/or communications equipment, packaged to prevent damage from exposure and hidden in isolated locations by such methods as burial, concealment, and/or submersion, to support isolated personnel. See also **evader; evasion; recovery; recovery operations.** (JP 3-50)

call sign — Any combination of characters or pronounceable words, which identifies a communication facility, a command, an authority, an activity, or a unit; used primarily for establishing and maintaining communications. Also called **CS.** (JP 3-50)

campaign — A series of related major operations aimed at achieving strategic and operational objectives within a given time and space. See also **campaign plan.** (JP 5-0)

campaign plan — A joint operation plan for a series of related major operations aimed at achieving strategic or operational objectives within a given time and space. See also **campaign; campaign planning.** (JP 5-0)

campaign planning — The process whereby combatant commanders and subordinate joint force commanders translate national or theater strategy into operational concepts through the development of an operation plan for a campaign. See also **campaign; campaign plan.** (JP 5-0)

canalize — To restrict operations to a narrow zone by use of existing or reinforcing obstacles or by fire or bombing. (JP 3-15)

candidate target list — A list of objects or entities submitted by component commanders, appropriate agencies, or the joint force commander's staff for further development and inclusion on the joint target list and/or restricted target list, or moved to the no-strike list. Also called **CTL.** See also joint integrated prioritized target list; target, target nomination list. (JP 3-60)

capstone publication — The top joint doctrine publication in the hierarchy of joint publications that links joint doctrine to national strategy and the contributions of other government departments and agencies, multinational partners, and reinforces policy for command and control. See also **joint publication; keystone publications.** (CJCSM 5120.01)

capstone requirements document — A document that contains performance-based requirements to facilitate development of individual operational requirements documents by providing a common framework and operational concept to guide their development. Also called **CRD.**

---

cargo increment number — A seven-character alphanumeric field that uniquely describes a non-unit-cargo entry (line) in the Joint Operation Planning and Execution System time-phased force and deployment data. (JP 3-35)

carrier air wing — Two or more aircraft squadrons formed under one commander for administrative and tactical control of operations from a carrier. Also called **CVW**. (JP 3-32)

carrier control zone — The airspace within a circular limit defined by 5 miles horizontal radius from the carrier, extending upward from the surface to and including 2,500 feet unless otherwise designated for special operations, and is under the cognizance of the air officer during visual meteorological conditions. (JP 3-52)

carrier strike group — A standing naval task group consisting of a carrier, embarked air wing, surface combatants, and submarines as assigned in direct support, operating in mutual support with the task of destroying hostile submarine, surface, and air forces within the group's assigned operational area and striking at targets along hostile shore lines or projecting power inland. Also called **CSG**. (JP 3-32)

cartridge-actuated device — Small explosive devices used to eject stores from launched devices, actuate other explosive systems, or provide initiation for aircrew escape devices. Also called **CAD**. (JP 3-04)

CARVER — A special operations forces acronym used throughout the targeting and mission planning cycle to assess mission validity and requirements. The acronym stands for criticality, accessibility, recuperability, vulnerability, effect, and recognizability. (JP 3-05.1)

case officer — A professional employee of an intelligence or counterintelligence organization, who is responsible for providing directions for an agent operation and/or handling intelligence assets. (JP 2-01.2)

casualty — Any person who is lost to the organization by having been declared dead, duty status – whereabouts unknown, missing, ill, or injured. See also **hostile casualty**. (JP 4-02)

casualty evacuation — The unregulated movement of casualties that can include movement both to and between medical treatment facilities. Also called **CASEVAC**. See also **casualty; evacuation; medical treatment facility**. (JP 4-02)

casualty receiving and treatment ship — In amphibious operations, a ship designated to receive, provide treatment for, and transfer casualties. (JP 3-02)

catastrophic event — Any natural or man-made incident, including terrorism, which results in extraordinary levels of mass casualties, damage, or disruption severely affecting the population, infrastructure, environment, economy, national morale, and/or government functions. (JP 3-28)

causeway — A craft similar in design to a barge, but longer and narrower, designed to assist in the discharge and transport of cargo from vessels. (JP 4-01.6)

causeway launching area — An area located near the line of departure but clear of the approach lanes to an area located in the inner transport area. (JP 3-02)

C-day — The unnamed day on which a deployment operation commences or is to commence. (JP 5-0)

cell — A subordinate organization formed around a specific process, capability, or activity within a designated larger organization of a joint force commander's headquarters. (JP 3-33)

center — An enduring functional organization, with a supporting staff, designed to perform a joint function within a joint force commander's headquarters. (JP 3-33)

center of gravity — The source of power that provides moral or physical strength, freedom of action, or will to act. Also called **COG.** See also **decisive point.** (JP 5-0)

centigray — A unit of absorbed dose of radiation (one centigray equals one rad). (JP 3-11)

central control officer — The officer designated by the amphibious task force commander for the overall coordination of the waterborne ship-to-shore movement. The central control officer is embarked in the central control ship. Also called **CCO.** (JP 3-02)

centralized control — 1. In air defense, the control mode whereby a higher echelon makes direct target assignments to fire units. (JP 3-01) 2. In joint air operations, placing within one commander the responsibility and authority for planning, directing, and coordinating a military operation or group/category of operations. See also **decentralized control.** (JP 3-30)

chaff — Radar confusion reflectors, consisting of thin, narrow metallic strips of various lengths and frequency responses, which are used to reflect echoes for confusion purposes. (JP 3-13.1)

chain of command — The succession of commanding officers from a superior to a subordinate through which command is exercised. Also called **command channel.** (JP 1)

Chairman of the Joint Chiefs of Staff instruction — A replacement document for all types of correspondence containing Chairman of the Joint Chiefs of Staff policy and guidance that does not involve the employment of forces, which is of indefinite duration and is applicable to external agencies, or both the Joint Staff and external agencies. Also called **CJCSI.** See also **Chairman of the Joint Chiefs of Staff manual.** (CJCSM 5120.01)

Chairman of the Joint Chiefs of Staff manual — A document containing detailed procedures for performing specific tasks that do not involve the employment of forces, which is of indefinite duration and is applicable to external agencies or both the Joint Staff and external agencies. Also called **CJCSM.** See also **Chairman of the Joint Chiefs of Staff instruction.** (CJCSM 5120.01)

chalk number — The number given to a complete load and to the transporting carrier. (JP 3-17)

change detection — An image enhancement technique that compares two images of the same area from different time periods and eliminates identical picture elements in order to leave the signatures that have undergone change. (JP 2-03)

channel airlift — Airlift provided for movement of sustainment cargo, scheduled either regularly or depending upon volume of workload, between designated ports of embarkation and ports of debarkation over validated contingency or distribution routes. (JP 3-17)

chemical agent — A chemical substance that is intended for use in military operations to kill, seriously injure, or incapacitate mainly through its physiological effects. See also **chemical warfare; riot control agent.** (JP 3-11)

chemical, biological, radiological, and nuclear consequence management — Actions taken to plan, prepare, respond to, and recover from chemical, biological, radiological, and nuclear incidents. Also called **CBRN CM.** (JP 3-41)

chemical, biological, radiological, and nuclear defense — Measures taken to minimize or negate the vulnerabilities to, and/or effects of, a chemical, biological, radiological, or nuclear hazard or incident. Also called **CBRN defense.** (JP 3-11)

chemical, biological, radiological, and nuclear environment — An operational environment that includes chemical, biological, radiological, and nuclear threats and hazards and their potential resulting effects. Also called **CBRN environment.** (JP 3-11)

chemical, biological, radiological, and nuclear hazard — Chemical, biological, radiological, and nuclear elements that could create adverse effects due to an accidental or deliberate release and dissemination. Also called **CBRN hazard.** (JP 3-11)

chemical, biological, radiological, and nuclear passive defense — Passive measures taken to minimize or negate the vulnerability to, and effects of, chemical, biological, radiological, or nuclear attacks. This mission area focuses on maintaining the joint force's ability to continue military operations in a chemical, biological, radiological, or nuclear environment. Also called **CBRN passive defense**. (JP 3-40)

chemical, biological, radiological, or nuclear incident — Any occurrence, resulting from the use of chemical, biological, radiological and nuclear weapons and devices; the emergence of secondary hazards arising from counterforce targeting; or the release of toxic industrial materials into the environment, involving the emergence of chemical, biological, radiological and nuclear hazards. (JP 3-11)

chemical, biological, radiological, or nuclear weapon — A fully engineered assembly designed for employment to cause the release of a chemical or biological agent or radiological material onto a chosen target or to generate a nuclear detonation. Also called **CBRN weapon.** (JP 3-11)

chemical hazard — Any chemical manufactured, used, transported, or stored that can cause death or other harm through toxic properties of those materials, including chemical agents and chemical weapons prohibited under the Chemical Weapons Convention as well as toxic industrial chemicals. (JP 3-11)

chemical warfare — All aspects of military operations involving the employment of lethal and incapacitating munitions/agents and the warning and protective measures associated with such offensive operations. Also called **CW.** See also **chemical agent; chemical weapon; riot control agent.** (JP 3-11)

chemical weapon — Together or separately, (a) a toxic chemical and its precursors, except when intended for a purpose not prohibited under the Chemical Weapons Convention; (b) a munition or device, specifically designed to cause death or other harm through toxic properties of those chemicals specified in (a), above, which would be released as a result of the employment of such munition or device; (c) any equipment specifically designed for use directly in connection with the employment of munitions or devices specified in (b), above. See also **chemical agent; chemical warfare; riot control agent.** (JP 3-11)

chief of mission — The principal officer (the ambassador) in charge of a diplomatic facility of the United States, including any individual assigned to be temporarily in charge of such a facility. The chief of mission is the personal representative of the President to the country of accreditation. The chief of mission is responsible for the direction, coordination, and supervision of all US Government executive branch employees in that country (except those under the command of a US area military commander). The security of the diplomatic post is the chief of mission's direct responsibility. Also called **COM.** (JP 3-08)

chief of staff — The senior or principal member or head of a staff who acts as the controlling member of a staff for purposes of the coordination of its work or to exercise command in another's name. Also called **COS.** (JP 3-33)

chief of station — The senior United States intelligence officer in a foreign country and the direct representative of the Director National Intelligence, to whom the officer reports through the Director Central Intelligence Agency. Usually the senior representative of the Central Intelligence Agency assigned to a US mission. Also called **COS**. (JP 2-01.2)

civil administration — An administration established by a foreign government in (1) friendly territory, under an agreement with the government of the area concerned, to exercise certain authority normally the function of the local government; or (2) hostile territory, occupied by United States forces, where a foreign government exercises executive, legislative, and judicial authority until an indigenous civil government can be established. Also called **CA**. (JP 3-05)

civil affairs — Designated Active and Reserve Component forces and units organized, trained, and equipped specifically to conduct civil affairs operations and to support civil-military operations. Also called **CA**. See also **civil-military operations.** (JP 3-57)

civil affairs operations — Actions planned, executed, and assessed by civil affairs forces that enhance awareness of and manage the interaction with the civil component of the operational environment; identify and mitigate underlying causes of instability within civil society; or involve the application of functional specialty skills normally the responsibility of civil government. Also called **CAO**. (JP 3-57)

civil augmentation program — Standing, long-term external support contacts designed to augment Service logistic capabilities with contracted support in both preplanned and short notice contingencies. Examples include US Army Logistics Civil Augmentation Program, Air Force Contract Augmentation Program, and US Navy Global Contingency Capabilities Contracts. Also called **CAP**. See also **contingency; contingency contract; external support contract.** (JP 4-10)

civil authorities — Those elected and appointed officers and employees who constitute the government of the United States, the governments of the 50 states, the District of Columbia, the Commonwealth of Puerto Rico, United States territories, and political subdivisions thereof. (JP 3-28)

civil authority information support — Department of Defense information activities conducted under a designated lead federal agency or other United States civil authority to support dissemination of public or other critical information during domestic emergencies. Also called **CAIS**. (JP 3-13.2)

civil emergency — Any occasion or instance for which, in the determination of the President, federal assistance is needed to supplement state and local efforts and capabilities to save lives and to protect property and public health and safety, or to lessen or avert the threat of a catastrophe in any part of the United States. (JP 3-28)

civilian internee — A civilian who is interned during armed conflict, occupation, or other military operation for security reasons, for protection, or because he or she committed an offense against the detaining power. Also called **CI.** (DODD 2310.01E)

civil information — Relevant data relating to the civil areas, structures, capabilities, organizations, people, and events of the civil component of the operational environment used to support the situational awareness of the supported commander. (JP 3-57)

civil information management — Process whereby data relating to the civil component of the operational environment is gathered, collated, processed, analyzed, produced into information products, and disseminated. Also called **CIM.** (JP 3-57)

civil-military medicine — A discipline within operational medicine comprising public health and medical issues that involve a civil-military interface (foreign or domestic), including military medical support to civil authorities (domestic), medical elements of cooperation activities, and medical civil-military operations. (JP 4-02)

civil-military operations — Activities of a commander performed by designated civil affairs or other military forces that establish, maintain, influence, or exploit relations between military forces, indigenous populations, and institutions, by directly supporting the attainment of objectives relating to the reestablishment or maintenance of stability within a region or host nation. Also called **CMO.** See also **civil affairs; operation.** (JP 3-57)

civil-military operations center — An organization, normally comprised of civil affairs, established to plan and facilitate coordination of activities of the Armed Forces of the United States within indigenous populations and institutions, the private sector, intergovernmental organizations, nongovernmental organizations, multinational forces, and other governmental agencies in support of the joint force commander. Also called **CMOC.** See also **civil-military operations; operation.** (JP 3-57)

civil-military team — A temporary organization of civilian and military personnel task-organized to provide an optimal mix of capabilities and expertise to accomplish specific operational and planning tasks. (JP 3-57)

civil reconnaissance — A targeted, planned, and coordinated observation and evaluation of specific civil aspects of the environment such as areas, structures, capabilities, organizations, people, or events. Also called **CR.** (JP 3-57)

Civil Reserve Air Fleet — A program in which the Department of Defense contracts for the services of specific aircraft, owned by a United States entity or citizen, during national emergencies and defense-oriented situations when expanded civil augmentation of military airlift activity is required. Also called **CRAF.** See also **reserve.** (JP 3-17)

civil search and rescue — Search and/or rescue operations and associated civilian services provided to assist persons in potential or actual distress and protect property in a non-hostile environment. Also called **civil SAR.** (JP 3-50)

clandestine — Any activity or operation sponsored or conducted by governmental departments or agencies with the intent to assure secrecy and concealment. (JP 2-01.2)

clandestine intelligence collection — The acquisition of protected intelligence information in a way designed to conceal the nature of the operation and protect the source. (JP 2-01.2)

clandestine operation — An operation sponsored or conducted by governmental departments or agencies in such a way as to assure secrecy or concealment. A clandestine operation differs from a covert operation in that emphasis is placed on concealment of the operation rather than on concealment of the identity of the sponsor. In special operations, an activity may be both covert and clandestine and may focus equally on operational considerations and intelligence-related activities. See also **covert operation; overt operation.** (JP 3-05.1)

classes of supply — The ten categories into which supplies are grouped in order to facilitate supply management and planning. I. Rations and gratuitous issue of health, morale, and welfare items. II. Clothing, individual equipment, tentage, tool sets, and administrative and housekeeping supplies and equipment. III. Petroleum, oils, and lubricants. IV. Construction materials. V. Ammunition. VI. Personal demand items. VII. Major end items, including tanks, helicopters, and radios. VIII. Medical. IX. Repair parts and components for equipment maintenance. X. Nonstandard items to support nonmilitary programs such as agriculture and economic development. See also **petroleum, oils, and lubricants.** (JP 4-09)

classification — The determination that official information requires, in the interests of national security, a specific degree of protection against unauthorized disclosure, coupled with a designation signifying that such a determination has been made. See also **security classification.** (JP 2-01.2)

classified information — Official information that has been determined to require, in the interests of national security, protection against unauthorized disclosure and which has been so designated. (JP 2-01.2)

clearance capacity — An estimate expressed in agreed upon units of cargo measurement per day of the cargo or people that may be transported inland from a beach or port over the available means of inland communication, including roads, railroads, airlift, and inland waterways. See also **throughput capacity.** (JP 4-01.5)

clearance decontamination — The final level of decontamination that provides the decontamination of equipment and personnel to a level that allows unrestricted transportation, maintenance, employment, and disposal. (JP 3-11)

clearing operation — An operation designed to clear or neutralize all mines and obstacles from a route or area. (JP 3-15)

close air support — Air action by fixed- and rotary-wing aircraft against hostile targets that are in close proximity to friendly forces and that require detailed integration of each air mission with the fire and movement of those forces. Also called **CAS.** See also **air interdiction; immediate mission request; preplanned mission request.** (JP 3-0)

close support — That action of the supporting force against targets or objectives which are sufficiently near the supported force as to require detailed integration or coordination of the supporting action with the fire, movement, or other actions of the supported force. See also **direct support; general support; mutual support; support.** (JP 3-31)

close support area — Those parts of the ocean operating areas nearest to, but not necessarily in, the objective area. They are assigned to naval support carrier strike groups, surface action groups, surface action units, and certain logistic combat service support elements. (JP 3-02)

closure — In transportation, the process of a unit's arriving at a specified location. (JP 4-01.5)

coalition — An arrangement between two or more nations for common action. See also **alliance; multinational.** (JP 5-0)

coastal sea control — The employment of forces to ensure the unimpeded use of an offshore coastal area by friendly forces and, as appropriate, to deny the use of the area to enemy forces. (JP 3-10)

code word — 1. A word that has been assigned a classification and a classified meaning to safeguard intentions and information regarding a classified plan or operation. 2. A cryptonym used to identify sensitive intelligence data. (JP 3-50)

collateral damage — Unintentional or incidental injury or damage to persons or objects that would not be lawful military targets in the circumstances ruling at the time. (JP 3-60)

collection — In intelligence usage, the acquisition of information and the provision of this information to processing elements. See also **intelligence process.** (JP 2-01)

collection agency — Any individual, organization, or unit that has access to sources of information and the capability of collecting information from them. See also **agency.** (JP 2-01)

collection asset — A collection system, platform, or capability that is supporting, assigned, or attached to a particular commander. See also **collection.** (JP 2-01)

collection management — In intelligence usage, the process of converting intelligence requirements into collection requirements, establishing priorities, tasking or coordinating with appropriate collection sources or agencies, monitoring results, and retasking, as required. See also **collection; collection requirement; collection requirements management; intelligence; intelligence process.** (JP 2-0)

collection management authority — Within the Department of Defense, collection management authority constitutes the authority to establish, prioritize, and validate theater collection requirements, establish sensor tasking guidance, and develop theater-wide collection policies. Also called **CMA.** See also **collection manager; collection plan; collection requirement.** (JP 2-01.2)

collection manager — An individual with responsibility for the timely and efficient tasking of organic collection resources and the development of requirements for theater and national assets that could satisfy specific information needs in support of the mission. Also called **CM.** See also **collection; collection management authority.** (JP 2-01)

collection operations management — The authoritative direction, scheduling, and control of specific collection operations and associated processing, exploitation, and reporting resources. Also called **COM.** See also **collection management; collection requirements management.** (JP 2-0)

collection plan — A systematic scheme to optimize the employment of all available collection capabilities and associated processing, exploitation, and dissemination resources to satisfy specific information requirements. See also **information requirements; intelligence process.** (JP 2-0)

collection planning — A continuous process that coordinates and integrates the efforts of all collection units and agencies. See also **collection.** (JP 2-0)

collection point — A point designated for the assembly of personnel casualties, stragglers, disabled materiel, salvage, etc., for further movement to collecting stations or rear installations. Also called **CP.** (JP 4-06)

collection posture — The current status of collection assets and resources to satisfy identified information requirements. (JP 2-0)

collection requirement — A valid need to close a specific gap in intelligence holdings in direct response to a request for information. (JP 2-0)

collection requirements management — The authoritative development and control of collection, processing, exploitation, and/or reporting requirements that normally result in either the direct tasking of requirements to units over which the commander has authority, or the generation of tasking requests to collection management authorities at a higher, lower, or lateral echelon to accomplish the collection mission. Also called

CRM. See also **collection; collection management; collection operations management.** (JP 2-0)

collection resource — A collection system, platform, or capability that is not assigned or attached to a specific unit or echelon which must be requested and coordinated through the chain of command. See also **collection management.** (JP 2-01)

collection strategy — An analytical approach used by collection managers to determine which intelligence disciplines can be applied to satisfy information requirements. (JP 2-0)

collective protection — The protection provided to a group of individuals that permits relaxation of individual chemical, biological, radiological, and nuclear protection. Also called **COLPRO.** (JP 3-11)

colored beach — That portion of usable coastline sufficient for the assault landing of a regimental landing team or similar sized unit. In the event that the landing force consists of a single battalion landing team, a colored beach will be used and no further subdivision of the beach is required. See also **numbered beach.** (JP 3-02)

combat air patrol — An aircraft patrol provided over an objective area, the force protected, the critical area of a combat zone, or in an air defense area, for the purpose of intercepting and destroying hostile aircraft before they reach their targets. Also called **CAP.** See also **airborne alert; barrier combat air patrol; rescue combat air patrol.** (JP 3-01)

combat and operational stress — The expected and predictable emotional, intellectual, physical, and/or behavioral reactions of an individual who has been exposed to stressful events in war or stability operations. (JP 4-02)

combat and operational stress control — Programs developed and actions taken by military leadership to prevent, identify, and manage adverse combat and operational stress reactions in units; optimize mission performance; conserve fighting strength; prevent or minimize adverse effects of combat and operational stress on members' physical, psychological, intellectual and social health; and to return the unit or Service member to duty expeditiously. (JP 4-02)

combatant command — A unified or specified command with a broad continuing mission under a single commander established and so designated by the President, through the Secretary of Defense and with the advice and assistance of the Chairman of the Joint Chiefs of Staff. Also called **CCMD.** See also **specified combatant command; unified command.** (JP 1)

combatant command chaplain — The senior chaplain assigned to the staff of, or designated by, the combatant commander to provide advice on religion, ethical, and moral issues, and morale of assigned personnel and to coordinate religious ministries within the combatant commander's area of responsibility. See also **command chaplain; religious support; religious support team.** (JP 1-05)

combatant command (command authority) — Nontransferable command authority, which cannot be delegated, of a combatant commander to perform those functions of command over assigned forces involving organizing and employing commands and forces; assigning tasks; designating objectives; and giving authoritative direction over all aspects of military operations, joint training, and logistics necessary to accomplish the missions assigned to the command. Also called **COCOM.** See also **combatant command; combatant commander; operational control; tactical control.** (JP 1)

combatant commander — A commander of one of the unified or specified combatant commands established by the President. Also called **CCDR.** See also **combatant command; specified combatant command; unified combatant command.** (JP 3-0)

combatant commander logistic procurement support board — A combatant commander-level joint board established to ensure that contracting support and other sources of support are properly synchronized across the entire area of responsibility. Also called **CLPSB.** See also **joint acquisition review board; joint contracting support board.** (JP 4-10)

combatant command support agent — The Secretary of a Military Department to whom the Secretary of Defense or the Deputy Secretary of Defense has assigned administrative and logistical support of the headquarters of a combatant command, United States Element, North American Aerospace Defense Command, or subordinate unified command. The nature and scope of the combatant command support agent responsibilities, functions, and authorities shall be prescribed at the time of assignment or in keeping with existing agreements and practices, and they shall remain in effect until the Secretary of Defense or the Deputy Secretary of Defense revokes, supersedes, or modifies them. Also called **CCSA.** (DODD 5100.03)

combat assessment — The determination of the overall effectiveness of force employment during military operations. Combat assessment is composed of three major components: (a) battle damage assessment; (b) munitions effectiveness assessment; and (c) reattack recommendation. Also called **CA.** See also **battle damage assessment; munitions effectiveness assessment; reattack recommendation.** (JP 3-60)

combat camera — The acquisition and utilization of still and motion imagery in support of operational and planning requirements across the range of military operations and during joint exercises. Also called **COMCAM.** See also **visual information.** (JP 3-61)

combat cargo officer — An embarkation officer assigned to major amphibious ships or naval staffs, functioning primarily as an adviser to and representative of the naval commander in matters pertaining to embarkation and debarkation of troops and their supplies and equipment. Also called **CCO.** See also **embarkation officer.** (JP 3-02.1)

combat chart — A special naval chart, at a scale of 1:50,000, designed for naval surface fire support and close air support during coastal or amphibious operations and showing detailed hydrography and topography in the coastal belt. (JP 2-03)

combat control team — A task-organized team of special operations forces who are certified air traffic controllers that are trained and equipped to deploy into hostile environments to establish and control assault zones and airfields. Also called **CCT.** (JP 3-17)

combat engineering — Engineering capabilities and activities that closely support the maneuver of land combat forces consisting of three types: mobility, countermobility, and survivability. (JP 3-34)

combat identification — The process of attaining an accurate characterization of detected objects in the operational environment sufficient to support an engagement decision. Also called **CID.** (JP 3-09)

combat information — Unevaluated data, gathered by or provided directly to the tactical commander which, due to its highly perishable nature or the criticality of the situation, cannot be processed into tactical intelligence in time to satisfy the user's tactical intelligence requirements. (JP 2-01)

combat information center — The agency in a ship or aircraft manned and equipped to collect, display, evaluate, and disseminate tactical information for the use of the embarked flag officer, commanding officer, and certain control agencies. Also called **CIC.** (JP 3-04)

combating terrorism — Actions, including antiterrorism and counterterrorism, taken to oppose terrorism throughout the entire threat spectrum. Also called **CbT.** See also **antiterrorism; counterterrorism.** (JP 3-26)

combat loading — The arrangement of personnel and the stowage of equipment and supplies in a manner designed to conform to the anticipated tactical operation of the organization embarked. Each individual item is stowed so that it can be unloaded at the required time. (JP 3-02)

combat organizational loading — A method of loading by which a unit with its equipment and initial supplies is loaded into a single ship, together with other units, in such a manner as to be available for unloading in a predetermined order. (JP 3-02.1)

combat power — The total means of destructive and/or disruptive force which a military unit/formation can apply against the opponent at a given time. (JP 3-0)

combat readiness — Synonymous with operational readiness, with respect to missions or functions performed in combat. (JP 1-0)

combat search and rescue — The tactics, techniques, and procedures performed by forces to effect the recovery of isolated personnel during combat. Also called **CSAR.** See also **search and rescue.** (JP 3-50)

combat search and rescue task force — All forces committed to a specific combat search and rescue operation to locate, identify, support, and recover isolated personnel during combat. Also called **CSARTF.** See also **combat search and rescue; search; search and rescue.** (JP 3-50)

combat service support — The essential capabilities, functions, activities, and tasks necessary to sustain all elements of all operating forces in theater at all levels of war. Also called **CSS.** See also **combat support.** (JP 4-0)

combat service support area — An area ashore that is organized to contain the necessary supplies, equipment, installations, and elements to provide the landing force with combat service support throughout the operation. Also called **CSSA.** (JP 3-02)

combat service support elements — Those elements whose primary missions are to provide service support to combat forces and which are a part, or prepared to become a part, of a theater, command, or task force formed for combat operations. See also **service troops.**

combat spread loading — A method of combat loading by which some of the troops, equipment, and initial supplies of a unit are loaded in one ship and the remainder are loaded in one or more others. This method is commonly used for troop units with heavy equipment. (JP 3-02.1)

combat support — Fire support and operational assistance provided to combat elements. Also called **CS.** See also **combat service support.** (JP 4-0)

combat support agency — A Department of Defense agency so designated by Congress or the Secretary of Defense that supports military combat operations. Also called **CSA.** (JP 5-0)

combat surveillance — A continuous, all-weather, day-and-night, systematic watch over the battle area in order to provide timely information for tactical combat operations. (JP 3-01)

combat survival — Those measures to be taken by Service personnel when involuntarily separated from friendly forces in combat, including procedures relating to individual survival, evasion, escape, and conduct after capture. (JP 3-50)

combat unit loading — A method of loading by which all or a part of a combat unit, such as an assault battalion landing team, is completely loaded in a single ship, with essential combat equipment and supplies, in such a manner as to be immediately available to support the tactical plan upon debarkation, and to provide a maximum of flexibility to meet possible changes in the tactical plan. (JP 3-02.1)

combined — A term identifying two or more forces or agencies of two or more allies operating together. See also **joint.** (JP 3-16)

combined arms team — The full integration and application of two or more arms or elements of one Service into an operation. (JP 3-18)

command — 1. The authority that a commander in the armed forces lawfully exercises over subordinates by virtue of rank or assignment. 2. An order given by a commander; that is, the will of the commander expressed for the purpose of bringing about a particular action. 3. A unit or units, an organization, or an area under the command of one individual. Also called **CMD.** See also **area command; combatant command; combatant command (command authority).** (JP 1)

command and control — The exercise of authority and direction by a properly designated commander over assigned and attached forces in the accomplishment of the mission. Also called **C2.** (JP 1)

command and control system — The facilities, equipment, communications, procedures, and personnel essential to a commander for planning, directing, and controlling operations of assigned and attached forces pursuant to the missions assigned. (JP 6-0)

command chaplain — The senior chaplain assigned to or designated by a commander of a staff, command, or unit. See also **combatant command chaplain; religious support.** (JP 1-05)

command element — The core element of a Marine air-ground task force that is the headquarters. The command element is composed of the commander, general or executive and special staff sections, headquarters section, and requisite communications support, intelligence, and reconnaissance forces necessary to accomplish the mission. The command element provides command and control, intelligence, and other support essential for effective planning and execution of operations by the other elements of the Marine air-ground task force. The command element varies in size and composition. Also called **CE.** (JP 3-02)

commander, amphibious task force — The Navy officer designated in the initiating directive as the commander of the amphibious task force. Also called **CATF.** See also **amphibious operation; amphibious task force; commander, landing force.** (JP 3-02)

commander, landing force — The officer designated in the initiating directive as the commander of the landing force for an amphibious operation. Also called **CLF.** See also **amphibious operation; commander, amphibious task force; landing force.** (JP 3-02)

commander's critical information requirement — An information requirement identified by the commander as being critical to facilitating timely decision making. Also called **CCIR.** See also **information requirements; intelligence; priority intelligence requirement.** (JP 3-0)

commander's estimate — A developed course of action designed to provide the Secretary of Defense with military options to meet a potential contingency. (JP 5-0)

commander's intent — A clear and concise expression of the purpose of the operation and the desired military end state that supports mission command, provides focus to the staff, and helps subordinate and supporting commanders act to achieve the commander's desired results without further orders, even when the operation does not unfold as planned. See also **assessment; end state.** (JP 3-0)

commander's required delivery date — The original date relative to C-day, specified by the combatant commander for arrival of forces or cargo at the destination; shown in the time-phased force and deployment data to assess the impact of later arrival. (JP 5-0)

command information — Communication by a military organization directed to the internal audience that creates an awareness of the organization's goals, informs them of significant developments affecting them and the organization, increases their effectiveness as ambassadors of the organization, and keeps them informed about what is going on in the organization. Also called **internal information.** See also **command; public affairs.** (JP 3-61)

commanding officer of troops — On a ship that has embarked units, a designated officer (usually the senior embarking unit commander) who is responsible for the administration, discipline, and training of all embarked units. Also called **COT.** (JP 3-02)

command net — (*) A communications network which connects an echelon of command with some or all of its subordinate echelons for the purpose of command and control.

command post exercise — An exercise in which the forces are simulated, involving the commander, the staff, and communications within and between headquarters. Also called **CPX.** See also **exercise; maneuver.** (JP 3-0)

command relationships — The interrelated responsibilities between commanders, as well as the operational authority exercised by commanders in the chain of command; defined further as combatant command (command authority), operational control, tactical control, or support. See also **chain of command; combatant command (command authority); command; operational control; support; tactical control.** (JP 1)

command-sponsored dependent — A dependent entitled to travel to overseas commands at government expense and endorsed by the appropriate military commander to be present in a dependent's status. (JP 3-68)

commercial items — Articles of supply readily available from established commercial distribution sources which the Department of Defense or inventory managers in the Military Services have designated to be obtained directly or indirectly from such sources. (JP 4-06)

commercial loading — See **administrative loading.**

commercial vehicle — A vehicle that has evolved in the commercial market to meet civilian requirements and which is selected from existing production lines for military use. (JP 4-06)

commit — The process of assigning one or more aircraft or surface-to-air missile units to prepare to engage an entity, prior to authorizing such engagement. (JP 3-01)

commodity loading — A method of loading in which various types of cargoes are loaded together, such as ammunition, rations, or boxed vehicles, in order that each commodity can be discharged without disturbing the others. See also **combat loading.** (JP 3-02.1)

commodity manager — An individual within the organization of an inventory control point or other such organization assigned management responsibility for homogeneous grouping of materiel items.

commonality — A quality that applies to materiel or systems: a. possessing like and interchangeable characteristics enabling each to be utilized, or operated and maintained, by personnel trained on the others without additional specialized training; b. having interchangeable repair parts and/or components; and c. applying to consumable items interchangeably equivalent without adjustment. (JP 6-0)

common item — 1. Any item of materiel that is required for use by more than one activity. 2. Sometimes loosely used to denote any consumable item except repair parts or other technical items. 3. Any item of materiel that is procured for, owned by (Service stock), or used by any Military Department of the Department of Defense and is also required to be furnished to a recipient country under the grant-aid Military Assistance Program. 4. Readily available commercial items. 5. Items used by two or more Military Services of similar manufacture or fabrication that may vary between the Services as to color or shape (as vehicles or clothing). 6. Any part or component that is required in the assembly of two or more complete end-items. (JP 4-01.5)

common operating environment — Automation services that support the development of the common reusable software modules that enable interoperability across multiple combat support applications. Also called **COE.** (JP 4-01.2)

common operational picture — A single identical display of relevant information shared by more than one command that facilitates collaborative planning and assists all echelons to achieve situational awareness. Also called **COP.** (JP 3-0)

common servicing — Functions performed by one Service in support of another for which reimbursement is not required. (JP 3-34)

common tactical picture — An accurate and complete display of relevant tactical data that integrates tactical information from the multi-tactical data link network, ground network, intelligence network, and sensor networks. Also called **CTP.** (JP 3-01)

common use — Services, materiel, or facilities provided by a Department of Defense agency or a Military Department on a common basis for two or more Department of Defense agencies, elements, or other organizations as directed. (JP 4-01.5)

common-use container — Any Department of Defense-owned, -leased, or -controlled 20- or 40-foot International Organization for Standardization container managed by US Transportation Command as an element of the Department of Defense common-use container system. See also **component- owned container; Service-unique container.** (JP 4-09)

common-user airlift service — The airlift service provided on a common basis for all Department of Defense agencies and, as authorized, for other agencies of the United States Government. (JP 3-17)

common-user item — An item of an interchangeable nature that is in common use by two or more nations or Services of a nation. (JP 4-0)

common-user land transportation — Point-to-point land transportation service operated by a single Service for common use by two or more Services. Also called **CULT.** (JP 4-01.5)

common-user logistics — Materiel or service support shared with or provided by two or more Services, Department of Defense agencies, or multinational partners to another Service, Department of Defense agency, non-Department of Defense agency, and/or multinational partner in an operation. Common-user logistics is usually restricted to a particular type of supply and/or service and may be further restricted to specific unit(s) or types of units, specific times, missions, and/or geographic areas. Also called **CUL.** See also **common use.** (JP 4-09)

common-user network — A system of circuits or channels allocated to furnish communication paths between switching centers to provide communication service on a common basis to all connected stations or subscribers. (JP 3-33)

common-user ocean terminals — A military installation, part of a military installation, or a commercial facility operated under contract or arrangement by the Surface Deployment and Distribution Command which regularly provides for two or more Services terminal functions of receipt, transit storage or staging, processing, and loading and unloading of passengers or cargo aboard ships. (JP 4-01.2)

common-user sealift — The sealift services provided by the Military Sealift Command on a common basis for all Department of Defense agencies and, as authorized, for other departments and agencies of the United States Government. See also **Military Sealift Command; transportation component command.** (JP 3-35)

common-user transportation — Transportation and transportation services provided on a common basis for two or more Department of Defense agencies and, as authorized, non-Department of Defense agencies. See also **common use.** (JP 4-01.2)

communicate — To use any means or method to convey information of any kind from one person or place to another. (JP 6-0)

communications intelligence — Technical information and intelligence derived from foreign communications by other than the intended recipients. Also called **COMINT.** (JP 2-0)

communications network — An organization of stations capable of intercommunications, but not necessarily on the same channel. (JP 6-0)

communications satellite — An orbiting vehicle, which relays signals between communications stations. There are two types: a. **active communications satellite** — A satellite that receives, regenerates, and retransmits signals between stations; b. **passive communications satellite** — A satellite which reflects communications signals between stations. Also called **COMSAT.** (JP 6-0)

communications security — The protection resulting from all measures designed to deny unauthorized persons information of value that might be derived from the possession and study of telecommunications, or to mislead unauthorized persons in their interpretation of the results of such possession and study. Also called **COMSEC.** (JP 6-0)

communications security material — All documents, devices, equipment, apparatus, and cryptomaterial used in establishing or maintaining secure communications. (JP 4-01.6)

community engagement — Those public affairs activities that support the relationship between military and civilian communities. (JP 3-61)

compartmentation — 1. Establishment and management of an organization so that information about the personnel, internal organization, or activities of one component is made available to any other component only to the extent required for the performance of assigned duties. 2. Effects of relief and drainage upon avenues of approach so as to produce areas bounded on at least two sides by terrain features such as woods, ridges, or ravines that limit observation or observed fire into the area from points outside the area. (JP 3-05.1)

completeness — The joint operation plan review criterion for assessing whether operation plans incorporate major operations and tasks to be accomplished and to what degree they include forces required, deployment concept, employment concept, sustainment concept, time estimates for achieving objectives, description of the end state, mission success criteria, and mission termination criteria. (JP 5-0)

complex catastrophe — Any natural or man-made incident, including cyberspace attack, power grid failure, and terrorism, which results in cascading failures of multiple, interdependent, critical, life-sustaining infrastructure sectors and caused extraordinary levels of mass casualties, damage, or disruption severely affecting the population, environment, economy, public health, national morale, response efforts, and/or government functions. (DepSecDef Memo OSD001185-13)

component — 1. One of the subordinate organizations that constitute a joint force. (JP 1) 2. In logistics, a part or combination of parts having a specific function, which can be installed or replaced only as an entity. (JP 4-0) Also called **COMP.** See also **functional component command; Service component command.**

component-owned container — A 20- or 40-foot International Organization for Standardization container procured and owned by a single Department of Defense component. May be either on an individual unit property book or contained within a component pool (e.g., Marine Corps maritime pre-positioning force containers). May be temporarily assigned to the Department of Defense common-use container system. Also called **Service-unique container.** See also **common-use container.** (JP 4-09)

composite warfare commander — An officer to whom the officer in tactical command of a naval task organization may delegate authority to conduct some or all of the offensive and defensive functions of the force. Also called **CWC.** (JP 3-32)

compromise — The known or suspected exposure of clandestine personnel, installations, or other assets or of classified information or material, to an unauthorized person. (JP 2-01.2)

compromised — A term applied to classified matter, knowledge of which has, in whole or in part, passed to an unauthorized person or persons, or which has been subject to risk of such passing. (JP 2-01.2)

computer security — The protection resulting from all measures to deny unauthorized access and exploitation of friendly computer systems. Also called **COMPUSEC.** See also **communications security.** (JP 6-0)

concept of fires — A verbal or graphic statement that clearly and concisely expresses how lethal and nonlethal fires will be synchronized and integrated to support the commander's operational objectives. (JP 3-09)

concept of intelligence operations — Within the Department of Defense, a verbal or graphic statement, in broad outline, of an intelligence directorate's assumptions or intent in regard to intelligence support of an operation or series of operations. See also **concept of operations.** (JP 2-0)

concept of logistic support — A verbal or graphic statement, in a broad outline, of how a commander intends to support and integrate with a concept of operations in an operation or campaign. Also called **COLS.** (JP 4-0)

concept of operations — A verbal or graphic statement that clearly and concisely expresses what the joint force commander intends to accomplish and how it will be done using available resources. Also called **CONOPS.** (JP 5-0)

concept plan — In the context of joint operation planning level 3 planning detail, an operation plan in an abbreviated format that may require considerable expansion or alteration to convert it into a complete operation plan or operation order. Also called **CONPLAN.** See also **operation plan.** (JP 5-0)

condition — 1. Those variables of an operational environment or situation in which a unit, system, or individual is expected to operate and may affect performance. 2. A physical or behavioral state of a system that is required for the achievement of an objective. See also **joint mission-essential tasks.** (JP 3-0)

conduits — Within military deception, conduits are information or intelligence gateways to the deception target. Examples of conduits include: foreign intelligence and security services, intelligence collection platforms, open-source intelligence, news media—foreign and domestic. (JP 3-13.4)

confidential — Security classification that shall be applied to information, the unauthorized disclosure of which reasonably could be expected to cause damage to the national security that the original classification authority is able to identify or describe. (EO 13526)

configuration management — A discipline applying technical and administrative direction and surveillance to: (1) identify and document the functional and physical characteristics of a configuration item; (2) control changes to those characteristics; and (3) record and report changes to processing and implementation status. (JP 6-0)

conflict prevention — A peace operation employing complementary diplomatic, civil, and, when necessary, military means, to monitor and identify the causes of conflict, and take timely action to prevent the occurrence, escalation, or resumption of hostilities. (JP 3-07.3)

constellation — A system consisting of a number of like satellites acting in concert to perform a specific mission. See also **Global Positioning System.** (JP 3-14)

constraint — In the context of joint operation planning, a requirement placed on the command by a higher command that dictates an action, thus restricting freedom of action. See also **operational limitation; restraint.** (JP 5-0)

consumer — Person or agency that uses information or intelligence produced by either its own staff or other agencies. (JP 2-01)

consumption rate — The average quantity of an item consumed or expended during a given time interval, expressed in quantities by the most appropriate unit of measurement per applicable stated basis. (JP 4-05)

contact mine — A mine detonated by physical contact. See also **mine**. (JP 3-15)

contact point — 1. In land warfare, a point on the terrain, easily identifiable, where two or more units are required to make contact. (JP 3-50) 2. In air operations, the position at which a mission leader makes radio contact with an air control agency. (JP 3-09.3) 3. In personnel recovery, a location where isolated personnel can establish contact with recovery forces. Also called **CP.** See also **control point; coordinating point.** (JP 3-50)

contact procedure — Those predesignated actions taken by isolated personnel and recovery forces that permit link-up between the two parties in hostile territory and facilitate the return of isolated personnel to friendly control. See also **evader; recovery force.** (JP 3-50)

container — An article of transport equipment that meets American National Standards Institute/International Organization for Standardization standards that is designed to facilitate and optimize the carriage of goods by one or more modes of transportation without intermediate handling of the contents. (JP 4-01)

container control officer — A designated official (E6 or above or civilian equivalent) within a command, installation, or activity who is responsible for control, reporting, use, and maintenance of all Department of Defense-owned and controlled intermodal containers and equipment. This officer has custodial responsibility for containers from time received until dispatched. (JP 4-09)

container-handling equipment — Items of materials-handling equipment required to specifically receive, maneuver, and dispatch International Organization for Standardization containers. Also called **CHE.** (JP 4-09)

containership — A ship specially constructed and equipped to carry only containers without associated equipment, in all available cargo spaces, either below or above deck.

Containerships are usually non-self-sustaining, do not have built-in capability to load or off-load containers, and require port crane service. A containership with shipboard-installed cranes capable of loading and off-loading containers without assistance of port crane service is considered self-sustaining. (JP 4-09)

contaminated remains — Remains of personnel which have absorbed or upon which have been deposited radioactive material, or biological or chemical agents. See also **mortuary affairs.** (JP 4-06)

contamination — 1. The deposit, absorption, or adsorption of radioactive material, or of biological or chemical agents on or by structures, areas, personnel, or objects. Also called **fallout radiation**. 2. Food and/or water made unfit for consumption by humans or animals because of the presence of environmental chemicals, radioactive elements, bacteria or organisms, the byproduct of the growth of bacteria or organisms, the decomposing material or waste in the food or water. (JP 3-11)

contamination avoidance — Individual and/or unit measures taken to reduce the effects of chemical, biological, radiological, and nuclear hazards. (JP 3-11)

contamination control — A combination of preparatory and responsive measures designed to limit the vulnerability of forces to chemical, biological, radiological, nuclear, and toxic industrial hazards and to avoid, contain, control exposure to, and, where possible, neutralize them. See also **biological agent; chemical agent; contamination.** (JP 3-11)

contamination mitigation — The planning and actions taken to prepare for, respond to, and recover from contamination associated with all chemical, biological, radiological, and nuclear threats and hazards in order to continue military operations. (JP 3-11)

contiguous zone — 1. A maritime zone adjacent to the territorial sea that may not extend beyond 24 nautical miles from the baselines from which the breadth of the territorial sea is measured. 2. The zone of the ocean extending 3-12 nautical miles from the United States coastline. (JP 3-32)

continental United States — United States territory, including the adjacent territorial waters, located within North America between Canada and Mexico. Also called **CONUS.** (JP 1)

contingency — A situation requiring military operations in response to natural disasters, terrorists, subversives, or as otherwise directed by appropriate authority to protect US interests. See also **contingency contracting.** (JP 5-0)

contingency basing — The life-cycle process of planning, designing, constructing, operating, managing, and transitioning or closing a non-enduring location supporting a combatant commander's requirements. (DODD 3000.10).

contingency contract — A legally binding agreement for supplies, services, and construction let by government contracting officers in the operational area as well as other contracts that have a prescribed area of performance within a designated operational area. See also **external support contract; systems support contract; theater support contract.** (JP 4-10)

contingency contracting — The process of obtaining goods, services, and construction via contracting means in support of contingency operations. See also contingency; contingency contract. (JP 4-10)

contingency engineering management organization — An organization formed by the combatant commander, or subordinate joint force commander to augment their staffs with additional Service engineering expertise for planning and construction management. See also **combat engineering; contingency; crisis action planning; geospatial engineering.** (JP 3-34)

contingency operation — A military operation that is either designated by the Secretary of Defense as a contingency operation or becomes a contingency operation as a matter of law (Title 10, United States Code, Section 101[a][13]). See also **contingency; operation.** (JP 1)

contingency plan — A plan for major contingencies that can reasonably be anticipated in the principal geographic subareas of the command. (JP 5-0)

Contingency Planning Guidance — Secretary of Defense written guidance, approved by the President, for the Chairman of the Joint Chiefs of Staff, which focuses the guidance given in the national security strategy and Defense Planning Guidance, and is the principal source document for the Joint Strategic Capabilities Plan. Also called **CPG.** (JP 1)

contingency response program — Fast reaction transportation procedures intended to provide for priority use of land transportation assets by Department of Defense when required. Also called **CORE.** (JP 4-01)

contingency ZIP Code — A ZIP Code consisting of a five-digit base with a four-digit add-on to assist in routing and sorting assigned by Military Postal Service Agency to a contingency post office for the tactical use of the Armed Forces on a temporary basis. (JP 1-0)

continuity of operations — The degree or state of being continuous in the conduct of functions, tasks, or duties necessary to accomplish a military action or mission in carrying out the national military strategy. Also called **COOP.** (JP 3-0)

contour flight — See **terrain flight.**

contract administration — A subset of contracting that includes efforts to ensure that supplies, services, and construction are delivered in accordance with the terms and conditions of the contract. (JP 4-10)

contracted logistic support — Support in which maintenance operations for a particular military system are performed exclusively by contract support personnel. Also called **CLS.** See also **logistic support; support.** (JP 4-07)

contracting officer — The Service member or Department of Defense civilian with the legal authority to enter into, administer, modify, and/or terminate contracts. (JP 4-10)

contracting officer representative — A Service member or Department of Defense civilian appointed in writing and trained by a contracting officer, responsible for monitoring contract performance and performing other duties specified by their appointment letter. Also called **COR.** (JP 4-10)

contractor management — The oversight and integration of contractor personnel and associated equipment providing support to the joint force in a designated operational area. (JP 4-10)

contractors authorized to accompany the force — Contingency contractor employees and all tiers of subcontractor employees who are specifically authorized through their contract to accompany the force and have protected status in accordance with international conventions. Also called **CAAF.** (JP 4-10)

contractors not authorized to accompany the force — Contingency contractor employees and all tiers of subcontractor employees who are not authorized through their contract to accompany the force and do not have protected status in accordance with international conventions. Also called **non-CAAF.** (JP 4-10)

contract support integration — The coordination and synchronization of contracted support executed in a designated operational area in support of the joint force. (JP 4-10)

contract termination — Defense procurement: the cessation or cancellation, in whole or in part, of work under a prime contract or a subcontract thereunder for the convenience of, or at the option of, the government, or due to failure of the contractor to perform in accordance with the terms of the contract (default). (JP 4-10)

control — 1. Authority that may be less than full command exercised by a commander over part of the activities of subordinate or other organizations. (JP 1) 2. In mapping, charting, and photogrammetry, a collective term for a system of marks or objects on the Earth or on a map or a photograph, whose positions or elevations (or both) have been or will be determined. (JP 2-03) 3. Physical or psychological pressures exerted with the intent to assure that an agent or group will respond as directed. (JP 3-0) 4. An indicator governing the distribution and use of documents, information, or material. Such indicators are the subject of intelligence community agreement and are specifically defined in appropriate regulations. See also **administrative control; operational control; tactical control.** (JP 2-01)

control area — A controlled airspace extending upwards from a specified limit above the Earth. See also **controlled airspace; control zone; terminal control area.** (JP 3-04)

control group — Personnel, ships, and craft designated to control the waterborne ship-to-shore movement. (JP 3-02)

controlled airspace — An airspace of defined dimensions within which civilian air traffic control services are provided to control flights. (JP 3-52)

controlled firing area — An area in which ordnance firing is conducted under controlled conditions so as to eliminate hazard to aircraft in flight. See also **restricted area.**

controlled information — 1. Information conveyed to an adversary in a deception operation to evoke desired appreciations. 2. Information and indicators deliberately conveyed or denied to foreign targets to evoke invalid official estimates that result in foreign official actions advantageous to US interests and objectives. (JP 2-01.2)

controlled shipping — Shipping that is controlled by the Military Sealift Command. Included in this category are Military Sealift Command ships (United States naval ships), government-owned ships operated under a general agency agreement, and commercial ships under charter to the Military Sealift Command. See also **Military Sealift Command; United States naval ship.** (JP 3-02.1)

controlled source — In counterintelligence use, a person employed by or under the control of an intelligence activity and responding to intelligence tasking. (JP 2-01.2)

controlled substance — A drug or other substance, or immediate precursor included in Schedule I, II, III, IV, or V of the Controlled Substances Act. (JP 3-07.4)

controlled technical services — The controlled use of technology to enhance counterintelligence and human intelligence activities. Also called **CTS**. (JP 2-01.2)

control point — 1. A position along a route of march at which men are stationed to give information and instructions for the regulation of supply or traffic. 2. A position marked by coordinates (latitude, longitude), a buoy, boat, aircraft, electronic device, conspicuous terrain feature, or other identifiable object which is given a name or number and used as an aid to navigation or control of ships, boats, or aircraft. 3. In marking mosaics, a point located by ground survey with which a corresponding point on a photograph is matched as a check. (JP 3-09.3)

control zone — A controlled airspace extending upwards from the surface of the Earth to a specified upper limit. See also **control area; controlled airspace; terminal control area.** (JP 3-52)

conventional forces — 1. Those forces capable of conducting operations using nonnuclear weapons. 2. Those forces other than designated special operations forces. Also called **CF.** (JP 3-05)

conventional mines — Land mines, other than nuclear or chemical, that are not designed to self-destruct; are designed to be emplaced by hand or mechanical means; and can be buried or surface emplaced. See also **mine.** (JP 3-15)

convoy — 1. A number of merchant ships and/or naval auxiliaries usually escorted by warships and/or aircraft — or a single merchant ship or naval auxiliary under surface escort — assembled and organized for the purpose of passage together. 2. A group of vehicles organized for the purpose of control and orderly movement with or without escort protection that moves over the same route at the same time and under one commander. (JP 3-02.1)

convoy escort — 1. A naval ship(s) or aircraft in company with a convoy and responsible for its protection. 2. An escort to protect a convoy of vehicles from being scattered, destroyed, or captured. See also **escort.** (JP 4-01.5)

cooperative security location — A facility located outside the United States and US territories with little or no permanent US presence, maintained with periodic Service, contractor, or host-nation support. Cooperative security locations provide contingency access, logistic support, and rotational use by operating forces and are a focal point for security cooperation activities. Also called **CSL.** See also **forward operating site; main operating base.** (CJCS CM-0007-05)

coordinated fire line — A line beyond which conventional and indirect surface fire support means may fire at any time within the boundaries of the establishing headquarters without additional coordination. The purpose of the coordinated fire line is to expedite the surface-to-surface attack of targets beyond the coordinated fire line without coordination with the ground commander in whose area the targets are located. Also called **CFL.** See also **fire support.** (JP 3-09)

coordinating agency — An agency that supports the incident management mission by providing the leadership, staff, expertise, and authorities to implement critical and specific aspects of the response. (JP 3-28)

coordinating altitude — An airspace coordinating measure that uses altitude to separate users and as the transition between different airspace coordinating entities. Also called **CA.** (JP 3-52)

coordinating authority — A commander or individual who has the authority to require consultation between the specific functions or activities involving forces of two or more Services, joint force components, or forces of the same Service or agencies, but does not have the authority to compel agreement. (JP 1)

coordinating point — (*) Designated point at which, in all types of combat, adjacent units/formations must make contact for purposes of control and coordination.

coordinating review authority — An agency appointed by a Service or combatant command to coordinate with and assist the primary review authority in joint doctrine development and maintenance. Also called **CRA.** See also **joint doctrine; joint publication; lead agent; primary review authority.** (CJCSM 5120.01)

cost-plus award fee contract — A type of contract that provides for a payment consisting of a base amount fixed at inception of the contract along with an award amount that is based upon a judgmental evaluation by the United States Government. (JP 4-10)

cost-type contract — A contract that provides for payment to the contractor of allowable cost, to the extent prescribed in the contract, incurred in performance of the contract. (JP 4-10)

counterair — A mission that integrates offensive and defensive operations to attain and maintain a desired degree of air superiority and protection by neutralizing or destroying enemy aircraft and missiles, both before and after launch. See also **air superiority; mission; offensive counterair.** (JP 3-01)

counterdeception — Efforts to negate, neutralize, diminish the effects of, or gain advantage from a foreign deception operation. Counterdeception does not include the intelligence function of identifying foreign deception operations. (JP 3-13.4)

counterdrug — Those active measures taken to detect, monitor, and counter the production, trafficking, and use of illegal drugs. Also called **CD.** (JP 3-07.4)

counterdrug activities — Those measures taken to detect, interdict, disrupt, or curtail any activity that is reasonably related to illicit drug trafficking. (JP 3-07.4)

counterdrug operational support — Support to host nations and drug law enforcement agencies involving military personnel and their associated equipment, provided by the geographic combatant commanders from forces assigned to them or made available to them by the Services for this purpose. See also **counterdrug operations.** (JP 3-07.4)

counterdrug operations — Civil or military actions taken to reduce or eliminate illicit drug trafficking. See also **counterdrug; counterdrug operational support.** (JP 3-07.4)

counterespionage — That aspect of counterintelligence designed to detect, destroy, neutralize, exploit, or prevent espionage activities through identification, penetration, manipulation, deception, and repression of individuals, groups, or organizations conducting or suspected of conducting espionage activities. (JP 2-01.2)

counterfire — Fire intended to destroy or neutralize enemy weapons. Includes counterbattery and countermortar fire. (JP 3-09)

counterforce — The employment of strategic air and missile forces in an effort to destroy, or render impotent, selected military capabilities of an enemy force under any of the circumstances by which hostilities may be initiated.

counterguerrilla operations — Operations and activities conducted by armed forces, paramilitary forces, or nonmilitary agencies against guerrillas. (JP 3-24)

counter-improvised explosive device operations — The organization, integration, and synchronization of capabilities that enable offensive, defensive, stability, and support operations across all phases of operations or campaigns in order to defeat improvised explosive devices as operational and strategic weapons of influence. Also called **C-IED operations.** (JP 3-15.1)

counterinsurgency — Comprehensive civilian and military efforts designed to simultaneously defeat and contain insurgency and address its root causes. Also called **COIN.** (JP 3-24)

counterintelligence — Information gathered and activities conducted to identify, deceive, exploit, disrupt, or protect against espionage, other intelligence activities, sabotage, or assassinations conducted for or on behalf of foreign powers, organizations or persons or their agents, or international terrorist organizations or activities. Also called **CI**. See also **counterespionage; security.** (JP 2-01.2)

counterintelligence activities — One or more of the five functions of counterintelligence: operations, investigations, collection, analysis and production, and functional services. See also **analysis and production; collection; counterintelligence; operation.** (JP 2-01.2)

counterintelligence collection — The systematic acquisition of information (through investigations, operations, or liaison) concerning espionage, sabotage, terrorism, other intelligence activities or assassinations conducted by or on behalf of foreign governments or elements thereof, foreign organizations, or foreign persons that are directed against or threaten Department of Defense interests. See also **counterintelligence.** (JP 2-01.2)

counterintelligence insider threat — A person, known or suspected, who uses their authorized access to Department of Defense facilities, systems, equipment, information or infrastructure to damage, disrupt operations, commit espionage on behalf of a foreign intelligence entity or support international terrorist organizations. (JP 2-01.2)

counterintelligence investigation — An official, systematic search for facts to determine whether a person(s) is engaged in activities that may be injurious to US national security or advantageous to a foreign power. See also **counterintelligence.** (JP 2-01.2)

counterintelligence operational tasking authority — The levying of counterintelligence requirements specific to joint military activities and operations. Counterintelligence operational tasking authority is exercised through supporting components. Also called **CIOTA.** See also **counterintelligence.** (JP 2-01.2)

counterintelligence operations — Proactive activities designed to identify, exploit, neutralize, or deter foreign intelligence collection and terrorist activities directed against the United States. See also **counterintelligence; operation.** (JP 2-01.2)

counterintelligence production — The process of analyzing all-source information concerning espionage or other multidiscipline intelligence collection threats, sabotage, terrorism, and other related threats to US military commanders, the Department of Defense, and the US Intelligence Community and developing it into a final product that is disseminated. Counterintelligence production is used in formulating security policy, plans, and operations. See also **counterintelligence.** (JP 2-01.2)

counterintelligence support — Conducting counterintelligence activities to protect against espionage and other foreign intelligence activities, sabotage, international terrorist activities, or assassinations conducted for or on behalf of foreign powers, organizations, or persons. See also **counterintelligence.** (JP 2-01.2)

countermeasures — That form of military science that, by the employment of devices and/or techniques, has as its objective the impairment of the operational effectiveness of enemy activity. See also **electronic warfare.** (JP 3-13.1)

countermobility operations — The construction of obstacles and emplacement of minefields to delay, disrupt, and destroy the enemy by reinforcement of the terrain. See also **minefield; operation; target acquisition.** (JP 3-34)

counterproliferation — Those actions taken to defeat the threat and/or use of weapons of mass destruction against the United States, our forces, friends, allies, and partners. Also called **CP.** See also **nonproliferation.** (JP 3-40)

countersurveillance — All measures, active or passive, taken to counteract hostile surveillance. See also **surveillance.** (JP 3-07.2)

counterterrorism — Actions taken directly against terrorist networks and indirectly to influence and render global and regional environments inhospitable to terrorist networks. Also called **CT.** See also **antiterrorism; combating terrorism; terrorism.** (JP 3-26)

country team — The senior, in-country, United States coordinating and supervising body, headed by the chief of the United States diplomatic mission, and composed of the senior member of each represented United States department or agency, as desired by the chief of the United States diplomatic mission. Also called **CT.** (JP 3-07.4)

coup de main — An offensive operation that capitalizes on surprise and simultaneous execution of supporting operations to achieve success in one swift stroke. (JP 3-0)

courier — A messenger (usually a commissioned or warrant officer) responsible for the secure physical transmission and delivery of documents and material. (JP 2-01)

course of action — 1. Any sequence of activities that an individual or unit may follow. 2. A scheme developed to accomplish a mission. 3. A product of the course-of-action development step of the joint operation planning process. Also called **COA.** (JP 5-0)

cover — In intelligence usage, those measures necessary to give protection to a person, plan, operation, formation, or installation from the enemy intelligence effort and leakage of information. (JP 2-01.2)

covering fire — 1. Fire used to protect troops when they are within range of enemy small arms. 2. In amphibious usage, fire delivered prior to the landing to cover preparatory operations such as underwater demolition or mine countermeasures. (JP 3-02)

covering force — 1. A force operating apart from the main force for the purpose of intercepting, engaging, delaying, disorganizing, and deceiving the enemy before the enemy can attack the force covered. 2. Any body or detachment of troops which provides security for a larger force by observation, reconnaissance, attack, or defense, or by any combination of these methods. (JP 3-18)

covert operation — An operation that is so planned and executed as to conceal the identity of or permit plausible denial by the sponsor. See also **clandestine operation; overt operation.** (JP 3-05)

crash rescue and fire suppression — Extraction of aircrew members from crashed or burning aircraft and the control of related fires. (JP 3-34)

crisis — An incident or situation involving a threat to the United States, its citizens, military forces, or vital interests that develops rapidly and creates a condition of such diplomatic, economic, or military importance that commitment of military forces and resources is contemplated to achieve national objectives. (JP 3-0)

crisis action planning — The Adaptive Planning and Execution system process involving the time-sensitive development of joint operation plans and operation orders for the deployment, employment, and sustainment of assigned and allocated forces and resources in response to an imminent crisis. Also called **CAP.** See also **joint operation planning; Joint Operation Planning and Execution System.** (JP 5-0)

crisis management — Measures, normally executed under federal law, to identify, acquire, and plan the use of resources needed to anticipate, prevent, and/or resolve a threat or an act of terrorism. Also called **CrM.** (JP 3-28)

critical asset — A specific entity that is of such extraordinary importance that its incapacitation or destruction would have a very serious, debilitating effect on the ability of a nation to continue to function effectively. (JP 3-07.2)

critical asset list — A prioritized list of assets or areas, normally identified by phase of the operation and approved by the joint force commander, that should be defended against air and missile threats. Also called **CAL.** (JP 3-01)

critical capability — A means that is considered a crucial enabler for a center of gravity to function as such and is essential to the accomplishment of the specified or assumed objective(s). (JP 5-0)

critical element — 1. An element of an entity or object that enables it to perform its primary function. 2. An element of a target, which if effectively engaged, will serve to support the achievement of an operational objective and/or mission task. Also called **CE.** (JP 3-60)

critical information — Specific facts about friendly intentions, capabilities, and activities needed by adversaries for them to plan and act effectively so as to guarantee failure or unacceptable consequences for friendly mission accomplishment. Also called **CRITIC.** (JP 2-0)

critical infrastructure and key resources — The infrastructure and assets vital to a nation's security, governance, public health and safety, economy, and public confidence. Also called **CI/KR.** (JP 3-27)

critical infrastructure protection — Actions taken to prevent, remediate, or mitigate the risks resulting from vulnerabilities of critical infrastructure assets. Also called **CIP.** See also **defense critical infrastructure.** (JP 3-28)

critical intelligence — Intelligence that is crucial and requires the immediate attention of the commander. (JP 2-0)

critical item — An essential item which is in short supply or expected to be in short supply for an extended period. (JP 4-01.5)

critical item list — Prioritized list, compiled from a subordinate commander's composite critical item lists, identifying supply items and weapon systems that assist Service and Defense Logistics Agency's selection of supply items and systems for production surge planning. Also may be used in operational situations by the combatant commander and/or subordinate joint force commander (within combatant commander directives) to cross-level critical supply items between Service components. Also called **CIL.** See also **critical item.** (JP 4-07)

criticality assessment — An assessment that identifies key assets and infrastructure that support Department of Defense missions, units, or activities and are deemed mission critical by military commanders or civilian agency managers. It addresses the impact of temporary or permanent loss of key assets or infrastructures to the installation or a unit's ability to perform its mission. It examines costs of recovery and reconstitution including time, dollars, capability, and infrastructure support. (JP 3-07.2)

critical joint duty assignment billet — A joint duty assignment position for which, considering the duties and responsibilities of the position, it is highly important that the assigned officer be particularly trained in, and oriented toward, joint matters. (JP 1-0)

critical occupational specialty — A military occupational specialty selected from among the combat arms in the Army or equivalent military specialties in the Navy, Air Force, or Marine Corps. Equivalent military specialties are those engaged in operational art in order to attain strategic goals in an operational area through the design, organization, and conduct of campaigns and major operations. Critical occupational specialties are designated by the Secretary of Defense. Also called **COS.**

critical point — 1. A key geographical point or position important to the success of an operation. 2. In point of time, a crisis or a turning point in an operation. 3. A selected point along a line of march used for reference in giving instructions. 4. A point where there is a change of direction or change in slope in a ridge or stream. 5. Any point along a route of march where interference with a troop movement may occur.

critical requirement — An essential condition, resource, and means for a critical capability to be fully operational. (JP 5-0)

critical vulnerability — An aspect of a critical requirement which is deficient or vulnerable to direct or indirect attack that will create decisive or significant effects. (JP 5-0)

cross-leveling —At the theater strategic and operational levels, it is the process of diverting en route or in-theater materiel from one military element to meet the higher priority of another within the combatant commander's directive authority for logistics. (JP 4-0)

cross-loading — The distribution of leaders, key weapons, personnel, and key equipment among the aircraft, vessels, or vehicles of a formation to aid rapid assembly of units at the drop zone or landing zone or preclude the total loss of command and control or unit effectiveness if an aircraft, vessel, or vehicle is lost. (JP 3-17)

cross-servicing — A subset of common-user logistics in which a function is performed by one Military Service in support of another Service and for which reimbursement is required from the Service receiving support. See also **acquisition and cross-servicing agreement; common-user logistics.** (JP 4-08)

cruise missile — Guided missile, the major portion of whose flight path to its target is conducted at approximately constant velocity; depends on the dynamic reaction of air for lift and upon propulsion forces to balance drag. (JP 3-01)

culminating point — The point at which a force no longer has the capability to continue its form of operations, offense or defense. (JP 5-0)

current force — The actual force structure and/or manning available to meet present contingencies. See also **force.** (JP 5-0)

custody — 1. The responsibility for the control of, transfer and movement of, and access to, weapons and components. Custody also includes the maintenance of accountability for weapons and components. 2. Temporary restraint of a person.

customer direct — A materiel acquisition and distribution method that requires vendor delivery directly to the customer. Also called **CD.** (JP 4-09)

customer wait time — The total elapsed time between issuance of a customer order and satisfaction of that order. Also called **CWT.** (JP 4-09)

cyber counterintelligence — Measures to identify, penetrate, or neutralize foreign operations that use cyber means as the primary tradecraft methodology, as well as foreign intelligence service collection efforts that use traditional methods to gauge cyber capabilities and intentions. See also **counterintelligence.** (JP 2-01.2)

cyberspace — A global domain within the information environment consisting of the interdependent network of information technology infrastructures and resident data, including the Internet, telecommunications networks, computer systems, and embedded processors and controllers. (JP 3-12)

cyberspace operations — The employment of cyberspace capabilities where the primary purpose is to achieve objectives in or through cyberspace. (JP 3-0)

cyberspace superiority — The degree of dominance in cyberspace by one force that permits the secure, reliable conduct of operations by that force, and its related land, air, maritime, and space forces at a given time and place without prohibitive interference by an adversary. (JP 3-12)

D

damage assessment — 1. The determination of the effect of attacks on targets. 2. A determination of the effect of a compromise of classified information on national security. (JP 3-60)

damage criteria — The critical levels of various weapons effects required to create specified levels of damage. (JP 3-60)

damage estimation — A preliminary appraisal of the potential effects of an attack. See also **attack assessment.** (JP 3-60)

danger close — In close air support, artillery, mortar, and naval gunfire support fires, it is the term included in the method of engagement segment of a call for fire which indicates that friendly forces are within close proximity of the target. The close proximity distance is determined by the weapon and munition fired. See also **final protective fire.** (JP 3-09.3)

dangerous cargo — Cargo which, because of its dangerous properties, is subject to special regulations for its transport. (JP 4-01.5)

data element — 1. A basic unit of information built on standard structures having a unique meaning and distinct units or values. 2. In electronic recordkeeping, a combination of characters or bytes referring to one separate item of information, such as name, address, or age. (JP 1-0)

date-time group — The date and time, expressed as six digits followed by the time zone suffix at which the message was prepared for transmission (first pair of digits denotes the date, second pair the hours, third pair the minutes, followed by a three-letter month abbreviation and two-digit year abbreviation.). Also called **DTG.** (JP 5-0)

datum (geodetic) — 1. A reference surface consisting of five quantities: the latitude and longitude of an initial point, the azimuth of a line from that point, and the parameters of the reference ellipsoid. 2. The mathematical model of the earth used to calculate the coordinates on any map. Different nations use different datum for printing coordinates on their maps. (JP 2-03)

D-day — See **times.** (JP 3-02)

de-arming — An operation in which a weapon is changed from a state of readiness for initiation to a safe condition. Also called **safing.** (JP 3-04)

debarkation — The unloading of troops, equipment, or supplies from a ship or aircraft. (JP 3-02.1)

debarkation schedule —A schedule that provides for the timely and orderly debarkation of troops and equipment and emergency supplies for the waterborne ship-to-shore movement. (JP 3-02.1)

decedent effects — Personal effects found on human remains. Also called **DE.** (JP 4-06)

decentralized control — In air defense, the normal mode whereby a higher echelon monitors unit actions, making direct target assignments to units only when necessary to ensure proper fire distribution or to prevent engagement of friendly aircraft. See also **centralized control.** (JP 3-01)

decentralized execution — Delegation of execution authority to subordinate commanders. (JP 3-30)

deception action — A collection of related deception events that form a major component of a deception operation. (JP 3-13.4)

deception concept — The deception course of action forwarded to the Chairman of the Joint Chiefs of Staff for review as part of the combatant commander's strategic concept. (JP 3-13.4)

deception event — A deception means executed at a specific time and location in support of a deception operation. (JP 3-13.4)

deception means — Methods, resources, and techniques that can be used to convey information to the deception target. There are three categories of deception means: a. **physical means.** Activities and resources used to convey or deny selected information to a foreign power. b. **technical means.** Military material resources and their associated operating techniques used to convey or deny selected information to a foreign power. c. **administrative means.** Resources, methods, and techniques to convey or deny oral, pictorial, documentary, or other physical evidence to a foreign power. (JP 3-13.4)

deception objective — The desired result of a deception operation expressed in terms of what the adversary is to do or not to do at the critical time and/or location. (JP 3-13.4)

deception story — A scenario that outlines the friendly actions that will be portrayed to cause the deception target to adopt the desired perception. (JP 3-13.4)

deception target — The adversary decision maker with the authority to make the decision that will achieve the deception objective. (JP 3-13.4)

decision — In an estimate of the situation, a clear and concise statement of the line of action intended to be followed by the commander as the one most favorable to the successful accomplishment of the assigned mission. (JP 5-0)

decision point — A point in space and time when the commander or staff anticipates making a key decision concerning a specific course of action. See also **course of action; decision support template; target area of interest.** (JP 5-0)

decision support template — A combined intelligence and operations graphic based on the results of wargaming. The decision support template depicts decision points, timelines associated with movement of forces and the flow of the operation, and other key items of information required to execute a specific friendly course of action. See also **course of action; decision point.** (JP 2-01.3)

decisive point — A geographic place, specific key event, critical factor, or function that, when acted upon, allows commanders to gain a marked advantage over an adversary or contribute materially to achieving success. See also **center of gravity.** (JP 5-0)

deck status light — A three-colored light (red, amber, green) controlled from the primary flight control. Navy — The light displays the status of the ship to support flight operations. United States Coast Guard — The light displays clearance for a helicopter to conduct a given evolution. (JP 3-04)

decompression — In personnel recovery, the process of normalizing psychological and behavioral reactions that recovered isolated personnel experienced or are currently experiencing as a result of their isolation and recovery. (JP 3-50)

decontamination — The process of making any person, object, or area safe by absorbing, destroying, neutralizing, making harmless, or removing chemical or biological agents, or by removing radioactive material clinging to or around it. (JP 3-11)

decoy — An imitation in any sense of a person, object, or phenomenon which is intended to deceive enemy surveillance devices or mislead enemy evaluation. Also called **dummy.** (JP 3-13.4)

defended asset list — A listing of those assets from the critical asset list prioritized by the joint force commander to be defended with the resources available. Also called **DAL.** (JP 3-01)

defense coordinating element — A staff and military liaison officers who assist the defense coordinating officer in facilitating coordination and support to activated emergency support functions. Also called **DCE.** (JP 3-28)

defense coordinating officer — Department of Defense single point of contact for domestic emergencies who is assigned to a joint field office to process requirements for military support, forward mission assignments through proper channels to the appropriate military organizations, and assign military liaisons, as appropriate, to activated emergency support functions. Also called **DCO.** (JP 3-28)

defense critical infrastructure — Department of Defense and non-Department of Defense networked assets and facilities essential to project, support, and sustain military forces and operations worldwide. Also called **DCI.** (JP 3-27)

defense human intelligence executor — The senior Department of Defense intelligence official as designated by the head of each of the Department of Defense components who are authorized to conduct human intelligence and related intelligence activities. Also called **DHE.** (JP 2-01.2)

defense industrial base — The Department of Defense, government, and private sector worldwide industrial complex with capabilities to perform research and development, design, produce, and maintain military weapon systems, subsystems, components, or parts to meet military requirements. Also called **DIB.** (JP 3-27)

Defense Information Systems Network — Integrated network, centrally managed and configured to provide long-haul information transfer services for all Department of Defense activities. It is an information transfer utility designed to provide dedicated point-to-point, switched voice and data, imagery, and video teleconferencing services. Also called **DISN.** (JP 6-0)

defense message system — Consists of all hardware, software, procedures, standards, facilities, and personnel used to exchange messages electronically.

Defense Satellite Communications System — Geosynchronous military communications satellites that provide high data rate communications for military forces, diplomatic corps, and the White House. Also called **DSCS.** (JP 3-14)

defense sexual assault incident database — A Department of Defense database that captures and serves as the reporting source for all sexual assault data collected by the Services. Also called **DSAID.** (JP 1-0)

defense support of civil authorities — Support provided by US Federal military forces, Department of Defense civilians, Department of Defense contract personnel, Department of Defense component assets, and National Guard forces (when the Secretary of Defense, in coordination with the governors of the affected states, elects and requests to use those forces in Title 32, United States Code, status) in response to requests for assistance from civil authorities for domestic emergencies, law enforcement support, and other domestic activities, or from qualifying entities for special events. Also called **DSCA.** Also known as **civil support**. (DODD 3025.18)

Defense Support Program — Satellites that provide early warning of missile launches. Also called **DSP.** (JP 3-14)

Defense Switched Network — Component of the Defense Communications System that handles Department of Defense voice, data, and video communications. Also called **DSN.** (JP 6-0)

Defense Transportation System — That portion of the worldwide transportation infrastructure that supports Department of Defense transportation needs in peace and war. Also called **DTS**. See also **common-user transportation; transportation system.** (JP 4-01)

defensive counterair — All defensive measures designed to neutralize or destroy enemy forces attempting to penetrate or attack through friendly airspace. Also called **DCA.** See also **counterair; offensive counterair.** (JP 3-01)

defensive cyberspace operation response action — Deliberate, authorized defensive measures or activities taken outside of the defended network to protect and defend Department of Defense cyberspace capabilities or other designated systems. Also called **DCO-RA.** (JP 3-12)

defensive cyberspace operations — Passive and active cyberspace operations intended to preserve the ability to utilize friendly cyberspace capabilities and protect data, networks, net-centric capabilities, and other designated systems. Also called **DCO.** (JP 3-12)

defensive minefield — 1. In naval mine warfare, a minefield laid in international waters or international straits with the declared intention of controlling shipping in defense of sea communications. 2. In land mine warfare, a minefield laid in accordance with an established plan to prevent a penetration between positions and to strengthen the defense of the positions themselves. See also **minefield.** (JP 3-15)

defensive space control — Operations conducted to preserve the ability to exploit space capabilities via active and passive actions, while protecting friendly space capabilities from attack, interference, or unintentional hazards. (JP 3-14)

defilade — 1. Protection from hostile observation and fire provided by an obstacle such as a hill, ridge, or bank. 2. A vertical distance by which a position is concealed from enemy observation. 3. To shield from enemy fire or observation by using natural or artificial obstacles. (JP 3-09)

definitive care — Care rendered to conclusively manage a patient's condition, such as full range of preventive, curative acute, convalescent, restorative, and rehabilitative medical care. (JP 4-02)

degaussing — The process whereby a ship's magnetic field is reduced by the use of electromagnetic coils, permanent magnets, or other means. (JP 3-15)

delayed entry program — A program under which an individual may enlist in a Reserve Component of a military service and specify a future reporting date for entry on active duty that would coincide with availability of training spaces and with personal plans such as high school graduation. Also called **DEP.** See also **active duty.** (JP 4-05)

delaying operation — An operation in which a force under pressure trades space for time by slowing down the enemy's momentum and inflicting maximum damage on the enemy without, in principle, becoming decisively engaged. (JP 3-04)

delegation of authority — The action by which a commander assigns part of his or her authority, commensurate with the assigned task, to a subordinate commander. (JP 1)

deliberate planning — 1. The Adaptive Planning and Execution system process involving the development of joint operation plans for contingencies identified in joint strategic planning documents. 2. A planning process for the deployment and employment of apportioned forces and resources that occurs in response to a hypothetical situation. (JP 5-0)

demilitarized zone — A defined area in which the stationing or concentrating of military forces, or the retention or establishment of military installations of any description, is prohibited. (JP 3-07.3)

demobilization — The process of transitioning a conflict or wartime military establishment and defense-based civilian economy to a peacetime configuration while maintaining national security and economic vitality. See also **mobilization.** (JP 4-05)

demonstration — 1. An attack or show of force on a front where a decision is not sought, made with the aim of deceiving the enemy. See also **amphibious demonstration; diversion.** 2. In military deception, a show of force in an area where a decision is not sought that is made to deceive an adversary. It is similar to a feint but no actual contact with the adversary is intended. (JP 3-13.4)

denial measure — An action to hinder or deny the enemy the use of territory, personnel, or facilities to include destruction, removal, contamination, or erection of obstructions. (JP 3-15)

denied area — An area under enemy or unfriendly control in which friendly forces cannot expect to operate successfully within existing operational constraints and force capabilities. (JP 3-05)

Department of Defense civilian — A Federal civilian employee of the Department of Defense directly hired and paid from appropriated or nonappropriated funds, under permanent or temporary appointment. (JP 1-0)

Department of Defense components — The Office of the Secretary of Defense, the Military Departments, the Chairman of the Joint Chiefs of Staff and the Joint Staff, the combatant commands, the Office of the Inspector General of the Department of Defense, the Department of Defense agencies, Department of Defense field activities, and all other organizational entities in the Department of Defense. (JP 1)

Department of Defense construction agent — The Corps of Engineers, Naval Facilities Engineering Command, or other such approved Department of Defense activity, that is assigned design or execution responsibilities associated with military construction programs, facilities support, or civil engineering support to the combatant commanders in contingency operations. See also **contingency operation.** (JP 3-34)

Department of Defense container system — All Department of Defense owned, leased, and controlled 20- or 40-foot intermodal International Organization for Standardization containers and flatracks, supporting equipment such as generator sets and chassis, container handling equipment, information systems, and other infrastructure that supports Department of Defense transportation and logistic operations, including commercially provided transportation services. This also includes 463L pallets, nets, and tie down equipment as integral components of the Department of Defense container system. See also **container-handling equipment; International Organization for Standardization.** (JP 4-09)

Department of Defense information network operations — Operations to design, build, configure, secure, operate, maintain, and sustain Department of Defense networks to create and preserve information assurance on the Department of Defense information networks. (JP 3-12)

Department of Defense information networks — The globally interconnected, end-to-end set of information capabilities, and associated processes for collecting, processing, storing, disseminating, and managing information on-demand to warfighters, policy makers, and support personnel, including owned and leased communications and computing systems and services, software (including applications), data, and security. (JP 3-12)

Department of Defense Intelligence Information System — The combination of Department of Defense personnel, procedures, equipment, computer programs, and supporting communications that support the timely and comprehensive preparation and presentation of intelligence and information to military commanders and national-level decision makers. Also called **DODIIS.** (JP 2-0)

Department of Defense support to counterdrug operations — Support provided by the Department of Defense to law enforcement agencies to detect, monitor, and counter the production, trafficking, and use of illegal drugs. See also **counterdrug operations.** (JP 3-07.4)

Department of the Air Force — The executive part of the Department of the Air Force at the seat of government and all field headquarters, forces, Reserve Component, installations, activities, and functions under the control or supervision of the Secretary of the Air Force. Also called **DAF.** See also **Military Department.** (JP 1)

Department of the Army — The executive part of the Department of the Army at the seat of government and all field headquarters, forces, Reserve Component, installations, activities, and functions under the control or supervision of the Secretary of the Army. Also called **DA**. See also **Military Department**. (JP 1)

Department of the Navy — The executive part of the Department of the Navy at the seat of government; the headquarters, United States Marine Corps; the entire operating forces of the United States Navy and of the United States Marine Corps, including the Reserve Component of such forces; all field activities, headquarters, forces, bases, installations, activities, and functions under the control or supervision of the Secretary of the Navy; and the United States Coast Guard when operating as a part of the Navy pursuant to law. Also called **DON**. See also **Military Department**. (JP 1)

departure airfield — An airfield on which troops and/or materiel are enplaned for flight. See also **airfield**. (JP 3-17)

departure point — A navigational check point used by aircraft as a marker for setting course. (JP 3-17)

dependents/immediate family — An employee's spouse; children who are unmarried and under age 21 years or who, regardless of age, are physically or mentally incapable of self-support; dependent parents, including step and legally adoptive parents of the employee's spouse; and dependent brothers and sisters, including step and legally adoptive brothers and sisters of the employee's spouse who are unmarried and under 21 years of age or who, regardless of age, are physically or mentally incapable of self-support. (JP 3-68)

deployment — The rotation of forces into and out of an operational area. See also **deployment order; deployment planning; prepare to deploy order**. (JP 3-35)

deployment health surveillance — The regular or repeated collection, analysis, archiving, interpretation, and distribution of health-related data used for monitoring the health of a population or of individuals, and for intervening in a timely manner to prevent, treat, or control the occurrence of disease or injury, which includes occupational and environmental health surveillance and medical surveillance subcomponents. (JP 4-02)

deployment order — A planning directive from the Secretary of Defense, issued by the Chairman of the Joint Chiefs of Staff, that authorizes and directs the transfer of forces between combatant commands by reassignment or attachment. Also called **DEPORD**. See also **deployment; deployment planning; prepare to deploy order**. (JP 5-0)

deployment planning — Operational planning directed toward the movement of forces and sustainment resources from their original locations to a specific operational area for conducting the joint operations contemplated in a given plan. See also **deployment; deployment order; prepare to deploy order**. (JP 5-0)

depot — 1. **supply** — An activity for the receipt, classification, storage, accounting, issue, maintenance, procurement, manufacture, assembly, research, salvage, or disposal of material. 2. **personnel** — An activity for the reception, processing, training, assignment, and forwarding of personnel replacements. (JP 4-0)

design basis threat — The threat against which an asset must be protected and upon which the protective system's design is based. It is the baseline type and size of threat that buildings or other structures are designed to withstand. The design basis threat includes the tactics aggressors will use against the asset and the tools, weapons, and explosives employed in these tactics. Also called **DBT**. (JP 3-07.2)

desired perception — In military deception, what the deception target must believe for it to make the decision that will achieve the deception objective. (JP 3-13.4)

desired point of impact — A precise point, associated with a target and assigned as the impact point for a single unitary weapon to create a desired effect. Also called **DPI**. See also **aimpoint**. (JP 3-60)

destroyed — A condition of a target so damaged that it can neither function as intended nor be restored to a usable condition. In the case of a building, all vertical supports and spanning members are damaged to such an extent that nothing is salvageable. In the case of bridges, all spans must have dropped and all piers must require replacement.

destruction fire — Fire delivered for the sole purpose of destroying material objects.

detainee — A term used to refer to any person captured or otherwise detained by an armed force. (JP 3-63)

detainee collecting point — A facility or other location where detainees are assembled for subsequent movement to a detainee processing station.

detainee processing station — A facility or other location where detainees are administratively processed and provided custodial care pending disposition and subsequent release, transfer, or movement to a prisoner-of-war or civilian internee camp.

detection — 1. In tactical operations, the perception of an object of possible military interest but unconfirmed by recognition. 2. In surveillance, the determination and transmission by a surveillance system that an event has occurred. 3. In arms control, the first step in the process of ascertaining the occurrence of a violation of an arms control agreement. 4. In chemical, biological, radiological, and nuclear environments, the act of locating chemical, biological, radiological, and nuclear hazards by use of chemical, biological, radiological, and nuclear detectors or monitoring and/or survey teams. See also **hazard**. (JP 3-11)

deterrence — The prevention of action by the existence of a credible threat of unacceptable counteraction and/or belief that the cost of action outweighs the perceived benefits. (JP 3-0)

deterrent options — A course of action, developed on the best economic, diplomatic, and military judgment, designed to dissuade an adversary from a current course of action or contemplated operations. (JP 5-0)

development assistance. Programs, projects, and activities carried out by the United States Agency for International Development that improve the lives of the citizens of developing countries while furthering United States foreign policy interests in expanding democracy and promoting free market economic growth. (JP 3-08)

diplomatic authorization — Authority for overflight or landing obtained at government-to-government level through diplomatic channels. (JP 3-50)

direct action — Short-duration strikes and other small-scale offensive actions conducted as a special operation in hostile, denied, or diplomatically sensitive environments and which employ specialized military capabilities to seize, destroy, capture, exploit, recover, or damage designated targets. Also called **DA.** See also **special operations; special operations forces.** (JP 3-05)

direct air support center — The principal air control agency of the US Marine air command and control system responsible for the direction and control of air operations directly supporting the ground combat element. It processes and coordinates requests for immediate air support and coordinates air missions requiring integration with ground forces and other supporting arms. It normally collocates with the senior fire support coordination center within the ground combat element and is subordinate to the tactical air command center. Also called **DASC.** See also **Marine air command and control system; tactical air operations center.** (JP 3-09.3)

direct air support center (airborne) — An airborne aircraft equipped with the necessary staff personnel, communications, and operations facilities to function as a direct air support center. Also called **DASC(A).** See also **direct air support center.** (JP 3-09.3)

directed energy — An umbrella term covering technologies that relate to the production of a beam of concentrated electromagnetic energy or atomic or subatomic particles. Also called **DE.** See also **directed-energy device; directed-energy weapon.** (JP 3-13.1)

directed-energy device — A system using directed energy primarily for a purpose other than as a weapon. See also **directed energy; directed-energy weapon.** (JP 3-13.1)

directed-energy warfare — Military action involving the use of directed-energy weapons, devices, and countermeasures. Also called **DEW.** See also **directed energy; directed-energy device; directed-energy weapon; electromagnetic spectrum; electronic warfare.** (JP 3-13.1)

directed-energy weapon — A weapon or system that uses directed energy to incapacitate, damage, or destroy enemy equipment, facilities, and/or personnel. See also **directed energy; directed-energy device.** (JP 3-13.1)

direct fire — Fire delivered on a target using the target itself as a point of aim for either the weapon or the director. (JP 3-09.3)

direction finding — A procedure for obtaining bearings of radio frequency emitters by using a highly directional antenna and a display unit on an intercept receiver or ancillary equipment. Also called **DF.** (JP 3-13.1)

directive authority for logistics — Combatant commander authority to issue directives to subordinate commanders to ensure the effective execution of approved operation plans, optimize the use or reallocation of available resources, and prevent or eliminate redundant facilities and/or overlapping functions among the Service component commands. Also called **DAFL.** See also **combatant command (command authority); logistics.** (JP 1)

direct liaison authorized — That authority granted by a commander (any level) to a subordinate to directly consult or coordinate an action with a command or agency within or outside of the granting command. Also called **DIRLAUTH.** (JP 1)

director of mobility forces — The designated agent for all air mobility issues in the area of responsibility or joint operations area, exercising coordinating authority between the air operations center (or appropriate theater command and control node), the 618 Air Operations Center (Tanker Airlift Control Center), and the joint deployment and distribution operation center or joint movement center, in order to expedite the resolution of air mobility issues. Also called **DIRMOBFOR.** See also **air and space operations center; coordinating authority.** (JP 3-17)

direct support — A mission requiring a force to support another specific force and authorizing it to answer directly to the supported force's request for assistance. Also called **DS.** See also **close support; general support; mission; mutual support; support.** (JP 3-09.3)

disabling fire — The firing of ordnance by ships or aircraft at the steering or propulsion system of a vessel. The intent is to disable with minimum injury to personnel or damage to vessel.

disarmament — The reduction of a military establishment to some level set by international agreement. See also **arms control agreement.** (JP 3-0)

disaster assistance response team — A team of specialists, trained in a variety of disaster relief skills, rapidly deployed to assist US embassies and United States Agency for International Development missions with the management of US Government response to disasters. Also called **DART.** See also **foreign disaster; foreign disaster relief.** (JP 3-08)

disease and nonbattle injury — All illnesses and injuries not resulting from enemy or terrorist action or caused by conflict. Also called **DNBI.** (JP 4-02)

disembarkation schedule — See **debarkation schedule.**

disengagement — The act of geographically separating the forces of disputing parties. (JP 3-07.3)

dislocated civilian — A broad term primarily used by the Department of Defense that includes a displaced person, an evacuee, an internally displaced person, a migrant, a refugee, or a stateless person. Also called **DC.** See also **displaced person; evacuee; internally displaced person; migrant; refugee; stateless person.** (JP 3-29)

dispersal — Relocation of forces for the purpose of increasing survivability. (JP 3-01)

dispersal airfield — An airfield, military or civil, to which aircraft might move before H-hour on either a temporary duty or permanent change of station basis and be able to conduct operations. See also **airfield.** (JP 3-01)

dispersion — 1. The spreading or separating of troops, materiel, establishments, or activities, which are usually concentrated in limited areas to reduce vulnerability. (JP 5-0) 2. In chemical and biological operations, the dissemination of agents in liquid or aerosol form. (JP 3-41) 3. In airdrop operations, the scatter of personnel and/or cargo on the drop zone. (JP 3-17) 4. In naval control of shipping, the reberthing of a ship in the periphery of the port area or in the vicinity of the port for its own protection in order to minimize the risk of damage from attack. (JP 4-01.2)

displaced person — A broad term used to refer to internally and externally displaced persons collectively. See also **evacuee; refugee.** (JP 3-29)

display — In military deception, a static portrayal of an activity, force, or equipment intended to deceive the adversary's visual observation. (JP 3-13.4)

dissemination and integration — In intelligence usage, the delivery of intelligence to users in a suitable form and the application of the intelligence to appropriate missions, tasks, and functions. See also **intelligence process.** (JP 2-01)

distant retirement area — In amphibious operations, the sea area located to seaward of the landing area. This area is divided into a number of operating areas to which assault ships may retire and operate in the event of adverse weather or to prevent concentration of ships in the landing area. See also **amphibious operation; landing area.** (JP 3-02)

distressed person — An individual who requires search and rescue assistance to remove he or she from life-threatening or isolating circumstances in a permissive environment. (JP 3-50)

distribution — 1. The arrangement of troops for any purpose, such as a battle, march, or maneuver. 2. A planned pattern of projectiles about a point. 3. A planned spread of fire to cover a desired frontage or depth. 4. An official delivery of anything, such as orders or supplies. 5. The operational process of synchronizing all elements of the logistic system to deliver the "right things" to the "right place" at the "right time" to support the geographic combatant commander. 6. The process of assigning military personnel to activities, units, or billets. (JP 4-0)

distribution manager — The executive agent for managing distribution with the combatant commander's area of responsibility. See also **area of responsibility; distribution.** (JP 4-09)

distribution pipeline — Continuum or channel through which the Department of Defense conducts distribution operations. The distribution pipeline represents the end-to-end flow of resources from supplier to consumer and, in some cases, back to the supplier in retrograde activities. See also **distribution.** (JP 4-09)

distribution plan — A reporting system comprising reports, updates, and information systems feeds that articulate the requirements of the theater distribution system to the strategic and operational resources assigned responsibility for support to the theater. It portrays the interface of the physical, financial, information and communications networks for gaining visibility of the theater distribution system and communicates control activities necessary for optimizing capacity of the system. It depicts, and is continually updated to reflect changes in, infrastructure, support relationships, and customer locations to all elements of the distribution system (strategic operational, and tactical). See also **distribution; distribution system; theater distribution; theater distribution system.** (JP 4-09)

distribution point — A point at which supplies and/or ammunition, obtained from supporting supply points by a division or other unit, are broken down for distribution to subordinate units. Distribution points usually carry no stocks; items drawn are issued completely as soon as possible. (JP 4-09)

distribution system — That complex of facilities, installations, methods, and procedures designed to receive, store, maintain, distribute, and control the flow of military materiel between the point of receipt into the military system and the point of issue to using activities and units. (JP 4-09)

ditching — Controlled landing of a distressed aircraft on water. (JP 3-50)

diversion — 1. The act of drawing the attention and forces of an enemy from the point of the principal operation; an attack, alarm, or feint that diverts attention. 2. A change made in a prescribed route for operational or tactical reasons that does not constitute a change of destination. 3. A rerouting of cargo or passengers to a new transshipment point or destination or on a different mode of transportation prior to arrival at ultimate destination. 4. In naval mine warfare, a route or channel bypassing a dangerous area by connecting one channel to another or it may branch from a channel and rejoin it on the other side of the danger. See also **demonstration.** (JP 3-03)

dock landing ship — A ship designed to transport and launch loaded amphibious craft and/or amphibian vehicles with their crews and embarked personnel and/or equipment and to render limited docking and repair services to small ships and craft. Also called **LSD.** (JP 3-02)

doctrine — Fundamental principles by which the military forces or elements thereof guide their actions in support of national objectives. It is authoritative but requires judgment in application. See also **multinational doctrine; joint doctrine.**

domestic emergencies — Civil defense emergencies, civil disturbances, major disasters, or natural disasters affecting the public welfare and occurring within the United States and its territories. See also **natural disaster.** (JP 3-27)

domestic intelligence — Intelligence relating to activities or conditions within the United States that threaten internal security and that might require the employment of troops; and intelligence relating to activities of individuals or agencies potentially or actually dangerous to the security of the Department of Defense. (JP 3-08)

dominant user — The Service or multinational partner who is the principal consumer of a particular common-user logistic supply or service within a joint or multinational operation and will normally act as the lead Service to provide this particular common-user logistic supply or service to other Service components, multinational partners, other governmental agencies, or nongovernmental agencies as directed by the combatant commander. See also **common-user logistics; lead Service or agency for common-user logistics.** (JP 4-0)

double agent — Agent in contact with two opposing intelligence services, only one of which is aware of the double contact or quasi-intelligence services. Also called **DA.** (JP 2-01.2)

downgrade — To determine that classified information requires, in the interests of national security, a lower degree of protection against unauthorized disclosure than currently provided, coupled with a changing of the classification designation to reflect such a lower degree. (JP 3-08)

downloading — An operation that removes airborne weapons or stores from an aircraft. (JP 3-04)

drop altitude — The altitude above mean sea level at which airdrop is executed. (JP 3-17)

drop zone — A specific area upon which airborne troops, equipment, or supplies are airdropped. Also called **DZ.** (JP 3-17)

drug interdiction — A continuum of events focused on interrupting illegal drugs smuggled by air, sea, or land. See also **counterdrug operations.** (JP 3-07.4)

dry deck shelter — A shelter module that attaches to the hull of a specially configured submarine to provide the submarine with the capability to launch and recover special operations personnel, vehicles, and equipment while submerged. The dry deck shelter provides a working environment at one atmosphere for the special operations element during transit and has structural integrity to the collapse depth of the host submarine. Also called **DDS.** (JP 3-05.1)

dual-capable aircraft — Allied and US fighter aircraft tasked and configured to perform either conventional or theater nuclear missions. Also called **DCA.**

dual-role tanker — An aircraft that can carry support personnel, supplies, and equipment for the deploying force while escorting and/or refueling combat aircraft to the area of responsibility. See also **air refueling.** (JP 3-17)

dummy — See **decoy**. (JP 3-13.4)

dwell time — 1. The length of time a target is expected to remain in one location. (JP 3-60) 2. The period of time between the release from involuntary active and the reporting date for a subsequent tour of active duty pursuant to Title 10, United States Code, Section 12302. Such time includes any voluntary active duty performed between two periods of involuntary active duty pursuant to Title 10, United States Code, Section 12302. (DODD 1235.10)

dynamic targeting — Targeting that prosecutes targets identified too late, or not selected for action in time to be included in deliberate targeting. (JP 3-60)

dynamic threat assessment — An intelligence assessment developed by the Defense Intelligence Agency that details the threat, capabilities, and intentions of adversaries in each of the priority plans in the Joint Strategic Capabilities Plan. Also called **DTA.** (JP 2-0)

Intentionally Blank

E

earliest anticipated launch time — The earliest time expected for a special operations tactical element and its supporting platform to depart the staging or marshalling area together en route to the operations area. Also called **EALT**. (JP 3-05.1)

earliest arrival date — A day, relative to C-day, that is specified as the earliest date when a unit, a resupply shipment, or replacement personnel can be accepted at a port of debarkation during a deployment. Also called **EAD**. See also **latest arrival date**. (JP 5-0)

early warning — Early notification of the launch or approach of unknown weapons or weapons carriers. Also called **EW**. See also **attack assessment; tactical warning.** (JP 3-01)

economy of force — The judicious employment and distribution of forces so as to expend the minimum essential combat power on secondary efforts in order to allocate the maximum possible combat power on primary efforts. (JP 3-0)

E-day — See **times.** (JP 3-02.1)

effect — 1. The physical or behavioral state of a system that results from an action, a set of actions, or another effect. 2. The result, outcome, or consequence of an action. 3. A change to a condition, behavior, or degree of freedom. (JP 3-0)

effective United States controlled ships — United States-owned foreign flag ships that can be tasked by the Maritime Administration to support Department of Defense requirements when necessary. Also called **EUSCS**. (JP 4-01.2)

electro-explosive device — An explosive or pyrotechnic component that initiates an explosive, burning, electrical, or mechanical train and is activated by the application of electrical energy. Also called **EED**. (JP 3-04)

electromagnetic battle management — The dynamic monitoring, assessing, planning, and directing of joint electromagnetic spectrum operations in support of the commander's scheme of maneuver. Also called **EMBM**. (JP 3-13.1)

electromagnetic compatibility — The ability of systems, equipment, and devices that use the electromagnetic spectrum to operate in their intended environments without causing or suffering unacceptable or unintentional degradation because of electromagnetic radiation or response. Also called **EMC**. See also **electromagnetic spectrum; electromagnetic spectrum management; electronic warfare.** (JP 3-13.1)

electromagnetic environment — The resulting product of the power and time distribution, in various frequency ranges, of the radiated or conducted electromagnetic emission levels encountered by a military force, system, or platform when performing its assigned mission in its intended operational environment. Also called **EME.** (JP 3-13.1)

electromagnetic environmental effects — The impact of the electromagnetic environment upon the operational capability of military forces, equipment, systems, and platforms. Also called **E3.** (JP 3-13.1)

electromagnetic hardening — Action taken to protect personnel, facilities, and/or equipment by blanking, filtering, attenuating, grounding, bonding, and/or shielding against undesirable effects of electromagnetic energy. See also **electronic warfare.** (JP 3-13.1)

electromagnetic interference — Any electromagnetic disturbance, induced intentionally or unintentionally, that interrupts, obstructs, or otherwise degrades or limits the effective performance of electronics and electrical equipment. Also called **EMI.** (JP 3-13.1)

electromagnetic intrusion — The intentional insertion of electromagnetic energy into transmission paths in any manner, with the objective of deceiving operators or of causing confusion. See also **electronic warfare.** (JP 3-13.1)

electromagnetic jamming — The deliberate radiation, reradiation, or reflection of electromagnetic energy for the purpose of preventing or reducing an enemy's effective use of the electromagnetic spectrum, and with the intent of degrading or neutralizing the enemy's combat capability. See also **electromagnetic spectrum; electromagnetic spectrum management; electronic warfare.** (JP 3-13.1)

electromagnetic operational environment — The background electromagnetic environment and the friendly, neutral, and adversarial electromagnetic order of battle within the electromagnetic area of influence associated with a given operational area. Also called **EMOE.** (JP 6-01)

electromagnetic pulse — The electromagnetic radiation from a strong electronic pulse, most commonly caused by a nuclear explosion that may couple with electrical or electronic systems to produce damaging current and voltage surges. Also called **EMP.** See also **electromagnetic radiation.** (JP 3-13.1)

electromagnetic radiation — Radiation made up of oscillating electric and magnetic fields and propagated with the speed of light. (JP 6-01)

electromagnetic radiation hazards —Transmitter or antenna installation that generates or increases electromagnetic radiation in the vicinity of ordnance, personnel, or fueling operations in excess of established safe levels. Also called **EMR hazards or RADHAZ.** (JP 3-13.1)

electromagnetic spectrum — The range of frequencies of electromagnetic radiation from zero to infinity. It is divided into 26 alphabetically designated bands. See also **electronic warfare.** (JP 3-13.1)

electromagnetic spectrum control — The coordinated execution of joint electromagnetic spectrum operations with other lethal and nonlethal operations that enable freedom of action in the electromagnetic operational environment. Also called **EMSC.** (JP 3-13.1)

electromagnetic spectrum management — Planning, coordinating, and managing use of the electromagnetic spectrum through operational, engineering, and administrative procedures. See also **electromagnetic spectrum.** (JP 6-01)

electromagnetic vulnerability — The characteristics of a system that cause it to suffer a definite degradation (incapability to perform the designated mission) as a result of having been subjected to a certain level of electromagnetic environmental effects. Also called **EMV.** (JP 3-13.1)

electronic attack — Division of electronic warfare involving the use of electromagnetic energy, directed energy, or antiradiation weapons to attack personnel, facilities, or equipment with the intent of degrading, neutralizing, or destroying enemy combat capability and is considered a form of fires. Also called **EA.** See also **electronic protection; electronic warfare; electronic warfare support.** (JP 3-13.1)

electronic intelligence — Technical and geolocation intelligence derived from foreign noncommunications electromagnetic radiations emanating from other than nuclear detonations or radioactive sources. Also called **ELINT.** See also **electronic warfare; foreign instrumentation signals intelligence; intelligence; signals intelligence.** (JP 3-13.1)

electronic masking —The controlled radiation of electromagnetic energy on friendly frequencies in a manner to protect the emissions of friendly communications and electronic systems against enemy electronic warfare support measures/signals intelligence without significantly degrading the operation of friendly systems. (JP 3-13.1)

electronic probing — Intentional radiation designed to be introduced into the devices or systems of potential enemies for the purpose of learning the functions and operational capabilities of the devices or systems. (JP 3-13.1)

electronic protection — Division of electronic warfare involving actions taken to protect personnel, facilities, and equipment from any effects of friendly or enemy use of the electromagnetic spectrum that degrade, neutralize, or destroy friendly combat capability. Also called **EP.** See also **electronic attack, electronic warfare; electronic warfare support.** (JP 3-13.1)

electronic reconnaissance — The detection, location, identification, and evaluation of foreign electromagnetic radiations. See also **electromagnetic radiation; reconnaissance.** (JP 3-13.1)

electronics security — The protection resulting from all measures designed to deny unauthorized persons information of value that might be derived from their interception and study of noncommunications electromagnetic radiations, e.g., radar. (JP 3-13.1)

electronic warfare — Military action involving the use of electromagnetic and directed energy to control the electromagnetic spectrum or to attack the enemy. Also called **EW.** See also **directed energy; electromagnetic spectrum; electronic attack; electronic protection; electronic warfare support.** (JP 3-13.1)

electronic warfare frequency deconfliction — Actions taken to integrate those frequencies used by electronic warfare systems into the overall frequency deconfliction process. See also **electronic warfare.** (JP 3-13.1)

electronic warfare reprogramming — The deliberate alteration or modification of electronic warfare or target sensing systems, or the tactics and procedures that employ them, in response to validated changes in equipment, tactics, or the electromagnetic environment. See also **electronic warfare.** (JP 3-13.1)

electronic warfare support — Division of electronic warfare involving actions tasked by, or under direct control of, an operational commander to search for, intercept, identify, and locate or localize sources of intentional and unintentional radiated electromagnetic energy for the purpose of immediate threat recognition, targeting, planning and conduct of future operations. Also called **ES.** See also **electronic attack; electronic protection; electronic warfare.** (JP 3-13.1)

electro-optical-infrared countermeasure — A device or technique employing electro-optical-infrared materials or technology that is intended to impair the effectiveness of enemy activity, particularly with respect to precision guided weapons and sensor systems. Also called **EO-IR CM.** (JP 3-13.1)

element — An organization formed around a specific function within a designated directorate of a joint force commander's headquarters. (JP 3-33)

elevated causeway system — An elevated causeway pier that provides a means of delivering containers, certain vehicles, and bulk cargo ashore without the lighterage contending with the surf zone. Also called **ELCAS.** See also **causeway.** (JP 4-01.6)

elicitation — In intelligence usage, the acquisition of information from a person or group in a manner that does not disclose the intent of the interview or conversation. (JP 2-0)

embarkation — The process of putting personnel and/or vehicles and their associated stores and equipment into ships and/or aircraft. (JP 3-02.1)

embarkation and tonnage table — A consolidated table showing personnel and cargo, by troop or naval units, loaded aboard a combat-loaded ship.

embarkation area — An area ashore, including a group of embarkation points, in which final preparations for embarkation are completed and through which assigned personnel and loads for craft and ships are called forward to embark. See also **mounting area.** (JP 3-02.1)

embarkation element — A temporary administrative formation of personnel with supplies and equipment embarking or to be embarked (combat loaded) aboard the ships of one transport element. It is dissolved upon completion of the embarkation. An embarkation element normally consists of two or more embarkation teams. (JP 3-02.1)

embarkation group — A temporary administrative formation of personnel with supplies and equipment embarking or to be embarked (combat loaded) aboard the ships of one transport element group. It is dissolved upon completion of the embarkation. An embarkation group normally consists of two or more embarkation units. (JP 3-02.1)

embarkation officer — An officer on the staff of units of the landing force who advises the commander thereof on matters pertaining to embarkation planning and loading ships. See also **combat cargo officer.** (JP 3-02.1)

embarkation order — An order specifying dates, times, routes, loading diagrams, and methods of movement to shipside or aircraft for troops and their equipment. (JP 3-02.1)

embarkation organization — A temporary administrative formation of personnel with supplies and equipment embarking or to be embarked (combat loaded) aboard amphibious shipping. See also **embarkation team**. (JP 3-02.1)

embarkation phase — In amphibious operations, the phase that encompasses the orderly assembly of personnel and materiel and their subsequent loading aboard ships and/or aircraft in a sequence designed to meet the requirements of the landing force concept of operations ashore. (JP 3-02.1)

embarkation plans — The plans prepared by the landing force and appropriate subordinate commanders containing instructions and information concerning the organization for embarkation, assignment to shipping, supplies and equipment to be embarked, location and assignment of embarkation areas, control and communication arrangements, movement schedules and embarkation sequence, and additional pertinent instructions relating to the embarkation of the landing force. (JP 3-02)

embarkation team — A temporary administrative formation of all personnel with supplies and equipment embarking or to be embarked (combat loaded) aboard one ship. See also **embarkation organization**. (JP 3-02.1)

embarkation unit — A temporary administrative formation of personnel with supplies and equipment embarking or to be embarked (combat loaded) aboard the ships of one transport unit. It is dissolved upon completion of the embarkation. An embarkation unit normally consists of two or more embarkation elements. JP 3-02.1)

emergency action committee — An organization established at a foreign service post by the chief of mission or principal officer for the purpose of directing and coordinating the post's response to contingencies. It consists of consular representatives and members of other local US Government agencies in a foreign country who assist in the implementation of a Department of State emergency action plan. Also called **EAC.** (JP 3-68)

emergency authority — A Federal military commander's authority, in extraordinary emergency circumstances where prior authorization by the President is impossible and duly constituted local authorities are unable to control the situation, to engage temporarily in activities that are necessary to quell large-scale, unexpected civil disturbances because (1) such activities are necessary to prevent significant loss of life or wanton destruction of property and are necessary to restore governmental function and public order or (2) duly constituted Federal, state, or local authorities are unable or decline to provide adequate protection for Federal property or Federal governmental functions. (DODD 3025.18)

emergency-essential employee — A Department of Defense civilian employee whose assigned duties and responsibilities must be accomplished following the evacuation of non-essential personnel (including dependents) during a declared emergency or outbreak of war. See also **evacuation.** (JP 1-0)

emergency locator beacon — A generic term for all radio beacons used for emergency locating purposes. See also **personal locator beacon.** (JP 3-50)

emergency operations center — A temporary or permanent facility where the coordination of information and resources to support domestic incident management activities normally takes place. Also called **EOC.** (JP 3-41)

emergency preparedness — Measures taken in advance of an emergency to reduce the loss of life and property and to protect a nation's institutions from all types of hazards through a comprehensive emergency management program of preparedness, mitigation, response, and recovery. Also called **EP.** (JP 3-28)

emergency preparedness liaison officer — A senior reserve officer who represents their Service at the appropriate joint field office conducting planning and coordination responsibilities in support of civil authorities. Also called **EPLO.** (JP 3-28)

emergency repair — The least amount of immediate repair to damaged facilities necessary for the facilities to support the mission. See also **facility substitutes.** (JP 3-34)

emergency resupply — A resupply mission that occurs based on a predetermined set of circumstances and time interval should radio contact not be established or, once established, is lost between a special operations tactical element and its base. See also **on-call resupply.** (JP 3-05.1)

emergency support functions — A grouping of government and certain private-sector capabilities into an organizational structure to provide the support, resources, program implementation, and services that are most likely to be needed to save lives, protect property and the environment, restore essential services and critical infrastructure, and help victims and communities return to normal, when feasible, following domestic incidents. Also called **ESFs.** (JP 3-28)

emission control — The selective and controlled use of electromagnetic, acoustic, or other emitters to optimize command and control capabilities while minimizing, for operations security: a. detection by enemy sensors; b. mutual interference among friendly systems; and/or c. enemy interference with the ability to execute a military deception plan. Also called **EMCON.** See also **electronic warfare.** (JP 3-13.1)

emission security — The component of communications security that results from all measures taken to deny unauthorized persons information of value that might be derived from intercept and analysis of compromising emanations from crypto-equipment and telecommunications systems. See also **communications security.** (JP 6-0)

employment — The strategic, operational, or tactical use of forces. (JP 5-0)

end evening civil twilight — The time period when the sun has dropped 6 degrees beneath the western horizon; it is the instant at which there is no longer sufficient light to see objects with the unaided eye. Light intensification devices are recommended from this time until begin morning civil twilight. Also called **EECT.**

end item — A final combination of end products, component parts, and/or materials that is ready for its intended use. (JP 4-02)

end of evening nautical twilight — Occurs when the sun has dropped 12 degrees below the western horizon, and is the instant of last available daylight for the visual control of limited ground operations. At end of evening nautical twilight there is no further sunlight available. (JP 2-01.3)

end state — The set of required conditions that defines achievement of the commander's objectives. (JP 3-0)

end-to-end — Joint distribution operations boundaries begin at the point of origin and terminate at the combatant commander's designated point of need within a desired operational area, including the return of forces and materiel. (JP 4-09)

enemy capabilities — Those courses of action of which the enemy is physically capable and that, if adopted, will affect accomplishment of the friendly mission. The term "capabilities" includes not only the general courses of action open to the enemy, such as attack, defense, reinforcement, or withdrawal, but also all the particular courses of action possible under each general course of action. "Enemy capabilities" are considered in the light of all known factors affecting military operations, including time, space, weather, terrain, and the strength and disposition of enemy forces. In strategic thinking, the capabilities of a nation represent the courses of action within the power of the nation for accomplishing its national objectives throughout the range of military operations. See also **course of action; mission.** (JP 2-01.3)

enemy combatant — In general, a person engaged in hostilities against the United States or its coalition partners during an armed conflict. Also called **EC.** (DODD 2310.01E)

engage — 1. In air defense, a fire control order used to direct or authorize units and/or weapon systems to fire on a designated target. (JP 3-01) See also **cease engagement; hold fire.** 2. To bring the enemy under fire. (JP 3-09.3)

engagement — 1. In air defense, an attack with guns or air-to-air missiles by an interceptor aircraft, or the launch of an air defense missile by air defense artillery and the missile's subsequent travel to intercept. (JP 3-01) 2. A tactical conflict, usually between opposing lower echelons maneuver forces. (JP 3-0) See also **battle; campaign.**

engagement authority — An authority vested with a joint force commander that may be delegated to a subordinate commander, that permits an engagement decision. (JP 3-01)

engineer support plan — An appendix to the logistics annex or separate annex of an operation plan that identifies the minimum essential engineering services and construction requirements required to support the commitment of military forces. Also called **ESP.** See also **operation plan.** (JP 3-34)

en route care — Continuation of the provision of care during movement (evacuation) between the health service support capabilities in the roles of care, without clinically compromising the patient's condition. See also **evacuation.** (JP 4-02)

entity — Within the context of targeting, a term used to describe facilities, organizations, individuals, equipment, or virtual (nontangible) things. (JP 3-60)

environmental baseline survey — A multi-disciplinary site survey conducted prior to or in the initial stage of a joint operational deployment. Also called **EBS.** See also **general engineering.** (JP 3-34)

environmental considerations — The spectrum of environmental media, resources, or programs that may affect the planning and execution of military operations. (JP 3-34)

equipment — In logistics, all nonexpendable items needed to outfit or equip an individual or organization. See also **component; supplies.** (JP 4-0)

escapee — Any person who has been physically captured by the enemy and succeeds in getting free. (JP 3-50)

escort — A member of the Armed Forces assigned to accompany, assist, or guide an individual or group, e.g., an escort officer. (JP 4-06)

espionage — The act of obtaining, delivering, transmitting, communicating, or receiving information about the national defense with an intent, or reason to believe, that the information may be used to the injury of the United States or to the advantage of any foreign nation. Espionage is a violation of Title 18 United States Code, Sections 792-798 and Article 106, *Uniform Code of Military Justice.* See also **counterintelligence.** (JP 2-01.2)

essential care — Medical treatment provided to manage the casualty throughout the roles of care, which includes all care and treatment to either return the patient to duty (within the theater evacuation policy), or begin initial treatment required for optimization of outcome, and/or stabilization to ensure the patient can tolerate evacuation. See also **en route care; first responder; forward resuscitative care; theater.** (JP 4-02)

essential elements of friendly information — Key questions likely to be asked by adversary officials and intelligence systems about specific friendly intentions, capabilities, and activities, so they can obtain answers critical to their operational effectiveness. Also called **EEFI.** (JP 2-01)

essential elements of information — The most critical information requirements regarding the adversary and the environment needed by the commander by a particular time to relate with other available information and intelligence in order to assist in reaching a logical decision. Also called **EEIs.** (JP 2-0)

essential task — A specified or implied task that an organization must perform to accomplish the mission that is typically included in the mission statement. See also **implied task; specified task.** (JP 5-0)

establishing directive — An order issued to specify the purpose of the support relationship. (JP 3-02)

estimate — 1. An analysis of a foreign situation, development, or trend that identifies its major elements, interprets the significance, and appraises the future possibilities and the prospective results of the various actions that might be taken. 2. An appraisal of the capabilities, vulnerabilities, and potential courses of action of a foreign nation or combination of nations in consequence of a specific national plan, policy, decision, or contemplated course of action. 3. An analysis of an actual or contemplated clandestine operation in relation to the situation in which it is or would be conducted in order to identify and appraise such factors as available as well as needed assets and potential obstacles, accomplishments, and consequences. See also **intelligence estimate.** (JP 2-01)

estimative intelligence — Intelligence that identifies, describes, and forecasts adversary capabilities and the implications for planning and executing military operations. (JP 2-0)

evacuation — 1. Removal of a patient by any of a variety of transport means from a theater of military operation, or between health service support capabilities, for the purpose of preventing further illness or injury, providing additional care, or providing disposition of patients from the military health care system. (JP 4-02) 2. The clearance of personnel, animals, or materiel from a given locality. (JP 3-68) 3. The controlled process of collecting, classifying, and shipping unserviceable or abandoned materiel, United States or foreign, to appropriate reclamation, maintenance, technical intelligence, or disposal facilities. (JP 4-09) 4. The ordered or authorized departure of noncombatants from a specific area by Department of State, Department of Defense, or appropriate military commander. This refers to the movement from one area to another in the same or different countries. The evacuation is caused by unusual or emergency circumstances and applies equally to command or non-command sponsored family members. See also **evacuee; noncombatant evacuation operations.** (JP 3-68)

evacuee — A civilian removed from a place of residence by military direction for reasons of personal security or the requirements of the military situation. See also **displaced person; refugee.** (JP 3-57)

evader — Any person isolated in hostile or unfriendly territory who eludes capture. (JP 3-50)

evaluation — In intelligence usage, appraisal of an item of information in terms of credibility, reliability, pertinence, and accuracy. (JP 2-01)

evaluation agent — The command or agency designated in the evaluation directive to be responsible for the planning, coordination, and conduct of the required evaluation of a joint test publication. Also called **EA.** See also **joint doctrine; joint test publication.** (CJCSM 5120.01)

evaluation and feedback — In intelligence usage, continuous assessment of intelligence operations throughout the intelligence process to ensure that the commander's intelligence requirements are being met. See **intelligence process.** (JP 2-01)

evasion — The process whereby isolated personnel avoid capture with the goal of successfully returning to areas under friendly control. (JP 3-50)

evasion aid — In personnel recovery, any piece of information or equipment designed to assist an individual in avoiding capture. See also **blood chit; evasion; evasion chart; pointee-talkee; recovery; recovery operations.** (JP 3-50)

evasion chart — A special map or chart designed as an evasion aid. Also called **EVC**. See also **evasion; evasion aid.** (JP 3-50)

evasion plan of action — A course of action, developed prior to executing a combat mission, that is intended to improve a potential isolated person's chances of successful evasion and recovery by providing the recovery forces with an additional source of information that can increase the predictability of the evader's action and movement. Also called **EPA**. See also **course of action; evader; evasion; recovery force.** (JP 3-50)

event matrix — A description of the indicators and activity expected to occur in each named area of interest. It normally cross-references each named area of interest and indicator with the times they are expected to occur and the courses of action they will confirm or deny. There is no prescribed format. See also **activity; area of interest; indicator.** (JP 2-01.3)

event template — A guide for collection planning. The event template depicts the named areas of interest where activity, or its lack of activity, will indicate which course of action the adversary has adopted. See also **activity; area of interest; collection planning; course of action.** (JP 2-01.3)

exclusion zone — A zone established by a sanctioning body to prohibit specific activities in a specific geographic area in order to persuade nations or groups to modify their behavior to meet the desires of the sanctioning body or face continued imposition of sanctions, or use or threat of force. (JP 3-0)

exclusive economic zone — A maritime zone adjacent to the territorial sea that may not extend beyond 200 nautical miles from the baselines from which the breadth of the territorial sea is measured. Also called **EEZ**. (JP 3-15)

execute order — 1. An order issued by the Chairman of the Joint Chiefs of Staff, at the direction of the Secretary of Defense, to implement a decision by the President to initiate military operations. 2. An order to initiate military operations as directed. Also called **EXORD**. (JP 5-0)

execution planning — The Adaptive Planning and Execution system translation of an approved course of action into an executable plan of action through the preparation of a complete operation plan or operation order. Also called **EP**. See also **Joint Operation Planning and Execution System**. (JP 5-0)

executive agent — A term used to indicate a delegation of authority by the Secretary of Defense or Deputy Secretary of Defense to a subordinate to act on behalf of the Secretary of Defense. Also called **EA.** (JP 1)

exercise — A military maneuver or simulated wartime operation involving planning, preparation, and execution that is carried out for the purpose of training and evaluation. See also **command post exercise; maneuver.** (JP 3-34)

exfiltration — The removal of personnel or units from areas under enemy control by stealth, deception, surprise, or clandestine means. See also **special operations; unconventional warfare.** (JP 3-50)

expeditionary force — An armed force organized to accomplish a specific objective in a foreign country. (JP 3-0)

expendable supplies — Supplies that are consumed in use, such as ammunition, paint, fuel, cleaning and preserving materials, surgical dressings, drugs, medicines, etc., or that lose their identity, such as spare parts, etc., and may be dropped from stock record accounts when it is issued or used. (JP 4-02)

exploitation — 1. Taking full advantage of success in military operations, following up initial gains, and making permanent the temporary effects already achieved. 2. Taking full advantage of any information that has come to hand for tactical, operational, or strategic purposes. 3. An offensive operation that usually follows a successful attack and is designed to disorganize the enemy in depth. See also **attack.** (JP 2-01.3)

explosive cargo — Cargo such as artillery ammunition, bombs, depth charges, demolition material, rockets, and missiles. (JP 4-01.5)

explosive hazard — Any hazard containing an explosive component to include unexploded explosive ordnance (including land mines), booby traps (some booby traps are nonexplosive), improvised explosive devices (which are an improvised type of booby trap), captured enemy ammunition, and bulk explosives. Also called **EH.** (JP 3-15)

explosive hazard incident — The suspected or detected presence of unexploded or damaged explosive ordnance that constitutes a hazard to operations, installations, personnel, or material. Not included in this definition are the accidental arming or other conditions that develop during the manufacture of high explosive material, technical service assembly operations, or the laying of mines and demolition charges. (JP 3-15.1)

explosive ordnance — All munitions containing explosives, nuclear fission or fusion materials, and biological and chemical agents. (JP 3-34)

explosive ordnance disposal — The detection, identification, on-site evaluation, rendering safe, recovery, and final disposal of unexploded explosive ordnance. Also called **EOD.** (JP 3-34)

explosive ordnance disposal unit — Personnel with special training and equipment who render explosive ordnance safe, make intelligence reports on such ordnance, and supervise the safe removal thereof. (JP 3-34)

exposure dose — The amount of radiation, as measured in roentgen, at a given point in relation to its ability to produce ionization. (JP 3-41)

external audience — All people who are not US military members, Department of Defense civilian employees, and their immediate families. See also **internal audience; public.** (JP 3-61)

external support contract — Contract awarded by contracting organizations whose contracting authority does not derive directly from the theater support contracting head(s) of contracting activity or from systems support contracting authorities. See also **systems support contract; theater support contract.** (JP 4-10)

Intentionally Blank

F

fabricator — An individual or group who, usually without genuine resources, invents or inflates information for personal or political gain or political purposes. (JP 2-01.2)

facility — A real property entity consisting of one or more of the following: a building, a structure, a utility system, pavement, and underlying land. (JP 3-34)

facility substitutes — Items such as tents and prepackaged structures requisitioned through the supply system that may be used to substitute for constructed facilities. (JP 3-34)

feasibility — The joint operation plan review criterion for assessing whether the assigned mission can be accomplished using available resources within the time contemplated by the plan. See also **acceptability; adequacy.** (JP 5-0)

feasibility assessment — A basic target analysis that provides an initial determination of the viability of a proposed target for special operations forces employment. Also called **FA.** (JP 3-05.1)

federal service — A term applied to National Guard members and units when called to active duty to serve the United States Government under Article I, Section 8 and Article II, Section 2 of the Constitution and the Title 10, United States Code, Sections 12401 to 12408. See also **active duty; Reserve Component.** (JP 4-05)

feint — In military deception, an offensive action involving contact with the adversary conducted for the purpose of deceiving the adversary as to the location and/or time of the actual main offensive action. (JP 3-13.4)

F-hour — See **times.**

field artillery — Equipment, supplies, ammunition, and personnel involved in the use of cannon, rocket, or surface-to-surface missile launchers. Field artillery cannons are classified according to caliber as follows.
Light — 120mm and less.
Medium — 121-160mm.
Heavy — 161-210mm.
Very heavy — greater than 210mm.
Also called **FA.** (JP 3-09)

field ordering officer — A Service member or Department of Defense civilian, who is appointed in writing and trained by a contracting officer and authorized to execute micropurchases in support of forces and/or designated civil-military operations. Also called **FOO.** (JP 4-10)

fighter engagement zone — In air defense, that airspace of defined dimensions within which the responsibility for engagement of air threats normally rests with fighter aircraft. Also called **FEZ.** (JP 3-01)

fighter escort — An offensive counterair operation providing dedicated protection sorties by air-to-air capable fighters in support of other offensive air and air support missions over enemy territory, or in a defensive counterair role to protect high value airborne assets. (JP 3-01)

fighter sweep — An offensive mission by fighter aircraft to seek out and destroy enemy aircraft or targets of opportunity in a designated area. (JP 3-01)

final governing standards — A comprehensive set of country-specific substantive environmental provisions, typically technical limitations on effluent, discharges, etc., or a specific management practice. (JP 3-34)

final protective fire — (*) An immediately available prearranged barrier of fire designed to impede enemy movement across defensive lines or areas.

finance support — A financial management function to provide financial advice and recommendations, pay support, disbursing support, establishment of local depository accounts, essential accounting support, and support of the procurement process. See also **financial management.** (JP 1-06)

financial management — The combination of the two core functions of resource management and finance support. Also called **FM.** See also **finance support; resource management.** (JP 1-06)

fire direction center — That element of a command post, consisting of gunnery and communications personnel and equipment, by means of which the commander exercises fire direction and/or fire control. The fire direction center receives target intelligence and requests for fire, and translates them into appropriate fire direction. The fire direction center provides timely and effective tactical and technical fire control in support of current operations. Also called **FDC.** (JP 3-09.3)

fires — The use of weapon systems to create specific lethal or nonlethal effects on a target. (JP 3-09)

fire support — Fires that directly support land, maritime, amphibious, and special operations forces to engage enemy forces, combat formations, and facilities in pursuit of tactical and operational objectives. See also **fires.** (JP 3-09)

fire support area — An appropriate maneuver area assigned to fire support ships by the naval force commander from which they can deliver gunfire support to an amphibious operation. Also called **FSA.** See also **amphibious operation; fire support.** (JP 3-09)

fire support coordination — The planning and executing of fire so that targets are adequately covered by a suitable weapon or group of weapons. (JP 3-09)

fire support coordination center — A single location in which are centralized communications facilities and personnel incident to the coordination of all forms of fire support. Also called **FSCC**. See also **fire support; fire support coordination; support; supporting arms coordination center.** (JP 3-09)

fire support coordination line — A fire support coordination measure that is established and adjusted by appropriate land or amphibious force commanders within their boundaries in consultation with superior, subordinate, supporting, and affected commanders. Fire support coordination lines facilitate the expeditious attack of surface targets of opportunity beyond the coordinating measure. A fire support coordination line does not divide an area of operations by defining a boundary between close and deep operations or a zone for close air support. The fire support coordination line applies to all fires of air, land, and sea-based weapon systems using any type of ammunition. Forces attacking targets beyond a fire support coordination line must inform all affected commanders in sufficient time to allow necessary reaction to avoid fratricide. Supporting elements attacking targets beyond the fire support coordination line must ensure that the attack will not produce adverse effects on, or to the rear of, the line. Short of a fire support coordination line, all air-to-ground and surface-to-surface attack operations are controlled by the appropriate land or amphibious force commander. The fire support coordination line should follow well-defined terrain features. Coordination of attacks beyond the fire support coordination line is especially critical to commanders of air, land, and special operations forces. In exceptional circumstances, the inability to conduct this coordination will not preclude the attack of targets beyond the fire support coordination line. However, failure to do so may increase the risk of fratricide and could waste limited resources. Also called **FSCL.** See also **fires; fire support.** (JP 3-09)

fire support coordination measure — A measure employed by commanders to facilitate the rapid engagement of targets and simultaneously provide safeguards for friendly forces. Also called **FSCM**. See also **fire support coordination.** (JP 3-0)

fire support element — That portion of the force tactical operations center at every echelon above company or troop (to corps) that is responsible for targeting coordination and for integrating fires delivered on surface targets by fire-support means under the control, or in support, of the force. Also called **FSE**. See also **fire support; force; support.** (JP 3-09)

fire support officer — Senior field artillery officer assigned to Army maneuver battalions and brigades. Advises commander on fire-support matters. Also called **FSO**. See also **field artillery; fire support; support.** (JP 3-09)

fire support station — An exact location at sea within a fire support area from which a fire support ship delivers fire. Also called **FSS**. (JP 3-02)

fire support team — A field artillery team provided for each maneuver company/troop and selected units to plan and coordinate all supporting fires available to the unit, including mortars, field artillery, naval surface fire support, and close air support integration. Also called **FIST**. See also **close air support; field artillery; fire support; support.** (JP 3-09.3)

first light — The beginning of morning nautical twilight; i.e., when the center of the morning sun is 12 degrees below the horizon.

first responder — A primary health care provider who provides immediate clinical care and stabilization in preparation for evacuation to the next health service support capability in the roles of care, and treats Service members for common acute minor illnesses. See also **essential care; evacuation.** (JP 4-02)

first responder care — The health care capability that provides immediate clinical care and stabilization to the patient in preparation for evacuation to the next health service support capability in the continuum of care. (JP 4-02)

fixed port — Terminals with an improved network of cargo-handling facilities designed for the transfer of freight. See also **maritime terminal.** (JP 4-01.5)

fixed price type contract — A type of contract that generally provides for a firm price or, under appropriate circumstances, may provide for an adjustable price for the supplies or services being procured. Fixed price contracts are of several types so designed as to facilitate proper pricing under varying circumstances. (JP 4-10)

flame field expedients — Simple, handmade devices used to produce flame or illumination. Also called **FFE.** (JP 3-15)

flash burn — A burn caused by excessive exposure (of bare skin) to thermal radiation. (JP 3-41)

flatrack — Portable, open-topped, open-sided units that fit into existing below-deck container cell guides and provide a capability for container ships to carry oversized cargo and wheeled and tracked vehicles. (JP 4-09)

flatted cargo — Cargo placed in the bottom of the holds, covered with planks and dunnage, and held for future use. Flatted cargo usually has room left above it for the loading of vehicles that may be moved without interfering with the flatted cargo. Frequently, flatted cargo serves in lieu of ballast. Sometimes called understowed cargo.

fleet — An organization of ships, aircraft, Marine forces, and shore-based fleet activities all under the command of a commander who may exercise operational as well as administrative control. See also **numbered fleet.** (JP 3-02.1)

Fleet Marine Force — A balanced force of combined arms comprising land, air, and service elements of the United States Marine Corps, which is an integral part of a United States fleet and has the status of a type command. Also called **FMF.** (JP 4-02)

flexible deterrent option — A planning construct intended to facilitate early decision making by developing a wide range of interrelated responses that begin with deterrent-oriented actions carefully tailored to produce a desired effect. Also called **FDO.** See also **deterrent options.** (JP 5-0)

flexible response — The capability of military forces for effective reaction to any enemy threat or attack with actions appropriate and adaptable to the circumstances existing. (JP 5-0)

flight — 1. In Navy and Marine Corps usage, a specified group of aircraft usually engaged in a common mission. 2. The basic tactical unit in the Air Force, consisting of four or more aircraft in two or more elements. 3. A single aircraft airborne on a nonoperational mission. (JP 3-30)

flight deck — 1. In certain airplanes, an elevated compartment occupied by the crew for operating the airplane in flight. 2. The upper deck of an aircraft carrier that serves as a runway. The deck of an air-capable ship, amphibious aviation assault ship, or aircraft carrier used to launch and recover aircraft. (JP 3-04)

flight deck officer — Officer responsible for the safe movement of aircraft on or about the flight deck of an aviation-capable ship. Also called **FDO.** (JP 3-04)

flight information region — An airspace of defined dimensions within which flight information service and alerting service are provided. Also called **FIR.** (JP 3-52)

flight information service — A service provided for the purpose of giving advice and information useful for the safe and efficient conduct of flights. Also called **FIS.** JP 3-52)

flight quarters — A ship configuration that assigns and stations personnel at critical positions to conduct safe flight operations. (JP 3-04)

floating craft company — A company-sized unit made up of various watercraft teams such as tugs, barges, and barge cranes. (JP 4-01.6)

floating dump — Emergency supplies preloaded in landing craft, amphibious vehicles, or in landing ships. Floating dumps are located in the vicinity of the appropriate control officer, who directs their landing as requested by the troop commander concerned. (JP 3-02)

follow-up — In amphibious operations, the reinforcements and stores carried on transport ships and aircraft (not originally part of the amphibious force) that are offloaded after the assault and assault follow-on echelons have been landed. See also **amphibious operation; assault; assault follow-on echelon.** (JP 3-02)

follow-up shipping — Ships not originally a part of the amphibious task force but which deliver troops and supplies to the objective area after the action phase has begun. (JP 3-02)

footprint — 1. The area on the surface of the earth within a satellite's transmitter or sensor field of view. 2. The amount of personnel, spares, resources, and capabilities physically present and occupying space at a deployed location. (JP 4-01.5)

force — 1. An aggregation of military personnel, weapon systems, equipment, and necessary support, or combination thereof. 2. A major subdivision of a fleet. (JP 1)

force/activity designator — Number used in conjunction with urgency of need designators to establish a matrix of priorities used for supply requisitions. Defines the relative importance of the unit to accomplish the objectives of the Department of Defense. Also called **F/AD.** See also **force.** (JP 4-09)

force beddown — The provision of expedient facilities for troop support to provide a platform for the projection of force. See also **facility substitutes.** (JP 3-34)

force closure — The point in time when a supported joint force commander determines that sufficient personnel and equipment resources are in the assigned operational area to carry out assigned tasks. See also **closure; force.** (JP 3-35)

force health protection — Measures to promote, improve, or conserve the behavioral and physical well-being of Service members to enable a healthy and fit force, prevent injury and illness, and protect the force from health hazards. Also called **FHP.** See also **force; protection.** (JP 4-02)

force module — A grouping of combat, combat support, and combat service support forces, with their accompanying supplies and the required nonunit resupply and personnel necessary to sustain forces for a minimum of 30 days. Also called **FM.** (JP 4-01.5)

force multiplier — A capability that, when added to and employed by a combat force, significantly increases the combat potential of that force and thus enhances the probability of successful mission accomplishment. (JP 3-05.1)

force planning — 1. Planning associated with the creation and maintenance of military capabilities by the Military Departments, Services, and US Special Operations Command. 2. In the Joint Operation Planning and Execution System, the planning conducted by the supported combatant command and its components to determine required force capabilities to accomplish an assigned mission. (JP 5-0)

force projection — The ability to project the military instrument of national power from the United States or another theater, in response to requirements for military operations. See also **force.** (JP 3-0)

force protection — Preventive measures taken to mitigate hostile actions against Department of Defense personnel (to include family members), resources, facilities, and critical information. Also called **FP.** See also **force; force protection condition; protection.** (JP 3-0)

force protection condition — A Chairman of the Joint Chiefs of Staff-approved standard for identification of and recommended responses to terrorist threats against US personnel and facilities. Also called **FPCON.** See also **antiterrorism; force protection.** (JP 3-07.2)

force protection detachment — A counterintelligence element that provides counterintelligence support to transiting and assigned ships, personnel, and aircraft in regions of elevated threat. Also called **FPD.** (JP 2-01.2)

force protection working group — Cross-functional working group whose purpose is to conduct risk assessment and risk management and to recommend mitigating measures to the commander. Also called **FPWG.** (JP 3-10)

force requirement number — An alphanumeric code used to uniquely identify force entries in a given operation plan time-phased force and deployment data. Also called **FRN.** (JP 3-35)

force sequencing — The phased introduction of forces into and out of the operational area. (JP 3-68)

force sourcing — The identification of the actual units, their origins, ports of embarkation, and movement characteristics to satisfy the time-phased force requirements of a supported commander. (JP 5-0)

force tracking — The process of gathering and maintaining information on the location, status, and predicted movement of each element of a unit including the unit's command element, personnel, and unit-related supplies and equipment while in transit to the specified operational area. (JP 3-35)

force visibility — The current and accurate status of forces; their current mission; future missions; location; mission priority; and readiness status. (JP 3-35)

forcible entry — Seizing and holding of a military lodgment in the face of armed opposition. See also **lodgment.** (JP 3-18)

foreign assistance — Assistance to foreign nations ranging from the sale of military equipment to donations of food and medical supplies to aid survivors of natural and man-made disasters. US foreign assistance takes three forms: development assistance, humanitarian assistance, and security assistance. See also **domestic emergencies; foreign disaster; foreign humanitarian assistance; security assistance.** (JP 3-29)

foreign consequence management — United States Government activity that assists friends and allies in responding to the effects from an intentional or accidental chemical, biological, radiological, or nuclear incident on foreign territory in order to maximize preservation of life. Also called **FCM.** (JP 3-41)

foreign disaster — An act of nature (such as a flood, drought, fire, hurricane, earthquake, volcanic eruption, or epidemic), or an act of man (such as a riot, violence, civil strife, explosion, fire, or epidemic), which is or threatens to be of sufficient severity and magnitude to warrant United States foreign disaster relief to a foreign country, foreign persons, or to an intergovernmental organization. See also **foreign disaster relief.** (JP 3-29)

foreign disaster relief — Prompt aid that can be used to alleviate the suffering of foreign disaster victims. Normally it includes humanitarian services and transportation; the provision of food, clothing, medicine, beds, and bedding; temporary shelter and housing; the furnishing of medical materiel and medical and technical personnel; and making repairs to essential services. See also **foreign disaster.** (JP 3-29)

foreign humanitarian assistance — Department of Defense activities, normally in support of the United States Agency for International Development or Department of State, conducted outside the United States, its territories, and possessions to relieve or reduce human suffering, disease, hunger, or privation. Also called **FHA.** See also **foreign assistance.** (JP 3-29)

foreign instrumentation signals intelligence — A subcategory of signals intelligence, consisting of technical information and intelligence derived from the intercept of foreign electromagnetic emissions associated with the testing and operational deployment of non-US aerospace, surface, and subsurface systems. Foreign instrumentation signals include but are not limited to telemetry, beaconry, electronic interrogators, and video data links. Also called **FISINT.** See also **signals intelligence.** (JP 2-01)

foreign intelligence — Information relating to capabilities, intentions, and activities of foreign governments or elements thereof, foreign organizations, or foreign persons, or international terrorist activities. Also called **FI.** See also **intelligence.** (JP 2-0)

foreign intelligence entity — Any known or suspected foreign organization, person, or group (public, private, or governmental) that conducts intelligence activities to acquire US information, block or impair US intelligence collection, influence US policy, or disrupts US systems and programs. The term includes foreign intelligence and security services and international terrorists. Also called **FIE.** (JP 2-01.2)

foreign internal defense — Participation by civilian and military agencies of a government in any of the action programs taken by another government or other designated organization to free and protect its society from subversion, lawlessness, insurgency, terrorism, and other threats to its security. Also called **FID.** (JP 3-22)

foreign military intelligence collection activities — Entails the overt debriefing, by trained human intelligence personnel, of all US persons employed by the Department of Defense who have access to information of potential national security value. Also called **FORMICA.** (JP 2-01.2)

foreign military sales — That portion of United States security assistance authorized by the Foreign Assistance Act of 1961, as amended, and the Arms Export Control Act of 1976, as amended. This assistance differs from the Military Assistance Program and the International Military Education and Training Program in that the recipient provides reimbursement for defense articles and services transferred. Also called **FMS.** (JP 4-08)

foreign national — Any person other than a US citizen, US permanent or temporary legal resident alien, or person in US custody. (JP 1-0)

foreign nation support — Civil and/or military assistance rendered to a nation when operating outside its national boundaries during military operations based on agreements mutually concluded between nations or on behalf of intergovernmental organizations. Also called **FNS.** See also **host-nation support.** (JP 1-06)

foreign object damage — Rags, pieces of paper, line, articles of clothing, nuts, bolts, or tools that, when misplaced or caught by air currents normally found around aircraft operations (jet blast, rotor or prop wash, engine intake), cause damage to aircraft systems or weapons or injury to personnel. Also called **FOD.** (JP 3-04)

foreign service national — Foreign nationals who provide clerical, administrative, technical, fiscal, and other support at foreign service posts abroad and are not citizens of the United States. The term includes third country nationals who are individuals employed by a United States mission abroad and are neither a citizen of the United States nor of the country to which assigned for duty. Also called **FSN.** (JP 3-68)

forensic-enabled intelligence — The intelligence resulting from the integration of scientifically examined materials and other information to establish full characterization, attribution, and the linkage of events, locations, items, signatures, nefarious intent, and persons of interest. Also called **FEI.** (JP 2-0)

formerly restricted data — Information removed from the restricted data category upon a joint determination by the Department of Energy (or antecedent agencies) and Department of Defense that such information relates primarily to the military utilization of atomic weapons and that such information can be adequately safeguarded as classified defense information. (Section 142d, Atomic Energy Act of 1954, as amended.) (JP 2-01)

forward air controller — An officer (aviator/pilot) member of the tactical air control party who, from a forward ground or airborne position, controls aircraft in close air support of ground troops. Also called **FAC.** See also **close air support.** (JP 3-09.3)

forward air controller (airborne) — A specifically trained and qualified aviation officer who exercises control from the air of aircraft engaged in close air support of ground troops. The forward air controller (airborne) is normally an airborne extension of the tactical air control party. A qualified and current forward air controller (airborne) will be recognized across the Department of Defense as capable and authorized to perform terminal attack control. Also called **FAC(A).** (JP 3-09.3)

forward area — An area in proximity to combat. (JP 4-02)

forward arming and refueling point — A temporary facility — organized, equipped, and deployed by an aviation commander, and normally located in the main battle area closer to the area where operations are being conducted than the aviation unit's combat service area — to provide fuel and ammunition necessary for the employment of aviation maneuver units in combat. The forward arming and refueling point permits combat aircraft to rapidly refuel and rearm simultaneously. Also called **FARP.** (JP 3-09.3)

forward aviation combat engineering — A mobility operation in which engineers perform tasks in support of forward aviation ground facilities. Also called **FACE.** See also **combat engineering; reconnaissance.** (JP 3-34)

forward edge of the battle area — The foremost limits of a series of areas in which ground combat units are deployed, excluding the areas in which the covering or screening forces are operating, designated to coordinate fire support, the positioning of forces, or the maneuver of units. Also called **FEBA.** (JP 3-09.3)

forward line of own troops — A line that indicates the most forward positions of friendly forces in any kind of military operation at a specific time. Also called **FLOT.** (JP 3-03)

forward-looking infrared — An airborne, electro-optical thermal imaging device that detects far-infrared energy, converts the energy into an electronic signal, and provides a visible image for day or night viewing. Also called **FLIR.** (JP 3-09.3)

forward observer — An observer operating with front line troops and trained to adjust ground or naval gunfire and pass back battlefield information. In the absence of a forward air controller, the observer may control close air support strikes. Also called **FO.** See also **forward air controller; spotter.** (JP 3-09)

forward operating base — An airfield used to support tactical operations without establishing full support facilities. The base may be used for an extended time period. Support by a main operating base will be required to provide backup support for a forward operating base. Also called **FOB.** (JP 3-09.3)

forward operating site — A scaleable location outside the United States and US territories intended for rotational use by operating forces. Such expandable "warm facilities" may be maintained with a limited US military support presence and possibly pre-positioned equipment. Forward operating sites support rotational rather than permanently stationed forces and are a focus for bilateral and regional training. Also called **FOS.** See also **cooperative security location; main operating base.** (CJCS CM-0007-05)

forward operations base — In special operations, a base usually located in friendly territory or afloat that is established to extend command and control or communications or to provide support for training and tactical operations. Facilities may be established for temporary or longer duration operations and may include an airfield or an unimproved airstrip, an anchorage, or a pier. A forward operations base may be the location of special operations component headquarters or a smaller unit that is controlled and/or supported by a main operations base. Also called **FOB.** See also **advanced operations base; main operations base.** (JP 3-05.1)

forward presence — Maintaining forward-deployed or stationed forces overseas to demonstrate national resolve, strengthen alliances, dissuade potential adversaries, and enhance the ability to respond quickly to contingencies. (JP 3-32)

forward resuscitative care — Care provided as close to the point of injury as possible based on current operational requirements to attain stabilization, achieve the most efficient use of life-and-limb saving medical treatment, and provide essential care so the patient can tolerate evacuation, which is known as Role 2 care in the North Atlantic Treaty Organization doctrine. Also called **FRC.** See **also essential care; evacuation; medical treatment facility.** (JP 4-02)

foundation geospatial-intelligence data — The base underlying data to provide context and a framework for display and visualization of the environment to support analysis operations and intelligence, which consists of: features; elevation; controlled imagery; geodetic sciences; geographic names and boundaries; aeronautical, maritime and human geography. (JP 2-03)

463L system — Aircraft pallets, nets, tie down, and coupling devices, facilities, handling equipment, procedures, and other components designed to interface with military and civilian aircraft cargo restraint systems. Though designed for airlift, system components may have to move intermodally via surface to support geographic combatant commander objectives. (JP 4-09)

fragmentary order — An abbreviated form of an operation order issued as needed after an operation order to change or modify that order or to execute a branch or sequel to that order. Also called **FRAGORD.** (JP 5-0)

freedom of navigation operations — Operations conducted to demonstrate US navigation, overflight, and related interests on, or under, and over the seas. (JP 3-0)

free drop — The dropping of equipment or supplies from an aircraft without the use of parachutes. See also **airdrop; air movement; free fall; high velocity drop; low velocity drop.** (JP 3-17)

free fall — A parachute maneuver in which the parachute is manually activated at the discretion of the jumper or automatically at a preset altitude. See also **airdrop; air movement; free drop; high velocity drop; low velocity drop.** (JP 3-17)

free-fire area — A specific area into which any weapon system may fire without additional coordination with the establishing headquarters. Also called **FFA.** (JP 3-09)

free mail — Correspondence of a personal nature that weighs less than 11 ounces, to include audio and video recording tapes, from a member of the Armed Forces or designated civilian, mailed postage free from a Secretary of Defense approved free mail zone. (JP 1-0)

frequency deconfliction — A systematic management procedure to coordinate the use of the electromagnetic spectrum for operations, communications, and intelligence functions. Frequency deconfliction is one element of electromagnetic spectrum management. See also **electromagnetic spectrum; electromagnetic spectrum management; electronic warfare.** (JP 3-13.1)

frequency management — The requesting, recording, deconfliction of and issuance of authorization to use frequencies (operate electromagnetic spectrum dependent systems) coupled with monitoring and interference resolution processes. (JP 6-0)

friendly — A contact positively identified as friendly. (JP 3-01)

friendly force information requirement — Information the commander and staff need to understand the status of friendly force and supporting capabilities. Also called **FFIR.** (JP 3-0)

friendly force tracking — A system that provides commanders and forces with location information about friendly and hostile military forces. Also called **FFT.** (JP 3-14)

frustrated cargo — Any shipment of supplies and/or equipment which, while en route to destination, is stopped prior to receipt and for which further disposition instructions must be obtained. (JP 4-01.5)

full mobilization — See **mobilization.**

full-spectrum superiority — The cumulative effect of dominance in the air, land, maritime, and space domains and information environment (which includes cyberspace) that permits the conduct of joint operations without effective opposition or prohibitive interference. (JP 3-0)

function — The broad, general, and enduring role for which an organization is designed, equipped, and trained. (JP 1)

functional component command — A command normally, but not necessarily, composed of forces of two or more Military Departments which may be established across the range of military operations to perform particular operational missions that may be of short duration or may extend over a period of time. See also **component; Service component command.** (JP 1)

functional damage assessment — The estimate of the effect of military force to degrade or destroy the functional or operational capability of the target to perform its intended mission and on the level of success in achieving operational objectives established against the target. See also **damage assessment; target.** (JP 3-60)

fusion — In intelligence usage, the process of managing information to conduct all-source analysis and derive a complete assessment of activity. (JP 2-0)

Intentionally Blank

G

general agency agreement — A contract between the Maritime Administration and a steamship company which, as general agent, exercises administrative control over a government-owned ship for employment by the Military Sealift Command. Also called **GAA**. See also **Military Sealift Command.** (JP 3-02.1)

general cargo — Cargo that is susceptible for loading in general, nonspecialized stowage areas or standard shipping containers; e.g., boxes, barrels, bales, crates, packages, bundles, and pallets. (JP 4-09)

general engineering — Those engineering capabilities and activities, other than combat engineering, that modify, maintain, or protect the physical environment. Also called **GE.** (JP 3-34)

general military intelligence — Intelligence concerning the military capabilities of foreign countries or organizations, or topics affecting potential United States or multinational military operations. Also called **GMI.** See also **intelligence.** (JP 2-0)

general purchasing agents — Agents who have been appointed in the principal overseas areas to supervise, control, coordinate, negotiate, and develop the local procurement of supplies, services, and facilities by Armed Forces of the United States, in order that the most effective utilization may be made of local resources and production.

general support — 1. That support which is given to the supported force as a whole and not to any particular subdivision thereof. See also **close support; direct support; mutual support; support.** 2. A tactical artillery mission. Also called **GS.** See also **direct support; general support-reinforcing.** (JP 3-09.3)

general support-reinforcing — General support-reinforcing artillery has the mission of supporting the force as a whole and of providing reinforcing fires for other artillery units. Also called **GSR.**

general unloading period — In amphibious operations, that part of the ship-to-shore movement in which unloading is primarily logistic in character, and emphasizes speed and volume of unloading operations. It encompasses the unloading of units and cargo from the ships as rapidly as facilities on the beach permit. It proceeds without regard to class, type, or priority of cargo, as permitted by cargo handling facilities ashore. See also **initial unloading period.** (JP 3-02)

geographic coordinates — The quantities of latitude and longitude which define the position of a point on the surface of the Earth with respect to the reference spheroid. (JP 2-03)

geospatial engineering — Those engineering capabilities and activities that contribute to a clear understanding of the physical environment by providing geospatial information and services to commanders and staffs. See also **geospatial information and services.** (JP 3-34)

geospatial information — Information that identifies the geographic location and characteristics of natural or constructed features and boundaries on the Earth, including: statistical data and information derived from, among other things, remote sensing, mapping, and surveying technologies; and mapping, charting, geodetic data and related products. (JP 2-03)

geospatial information and services — The collection, information extraction, storage, dissemination, and exploitation of geodetic, geomagnetic, imagery, gravimetric, aeronautical, topographic, hydrographic, littoral, cultural, and toponymic data accurately referenced to a precise location on the Earth's surface. Also called **GI&S.** (JP 2-03)

geospatial intelligence — The exploitation and analysis of imagery and geospatial information to describe, assess, and visually depict physical features and geographically referenced activities on the Earth. Geospatial intelligence consists of imagery, imagery intelligence, and geospatial information. Also called **GEOINT.** (JP 2-03)

geospatial intelligence base for contingency operations — A mobile visualization tool available through National Geospatial-Intelligence Agency and the Defense Logistics Agency. Applications are broad, including the capability to become familiar with a foreign environment, develop a battle scene, plan and execute noncombatant evacuations, contingency operations, urban area missions, and provide access to geospatial data where networks or infrastructure have been damaged or do not exist. Also called **GIBCO.** (JP 3-68)

geospatial intelligence operations — The tasks, activities, and events to collect, manage, analyze, generate, visualize, and provide imagery, imagery intelligence, and geospatial information necessary to support national and defense missions and international arrangements. Also called **GEOINT operations.** (JP 2-03)

Global Air Transportation Execution System — The Air Mobility Command's aerial port operations and management information system designed to support automated cargo and passenger processing, the reporting of in-transit visibility data to the Global Transportation Network, and billing to Air Mobility Command's financial management directorate. Also called **GATES.** See also **Air Mobility Command.** (JP 3-17)

global ballistic missile defense. Defense against ballistic missile threats that cross one or more geographical combatant command boundaries and requires synchronization among the affected combatant commands. Also called **GBMD.** (JP 3-01)

Global Combat Support System-Joint — The primary information technology application used to provide automation support to the joint logistician. Also called **GCSS-J.** (JP 4-0)

Global Command and Control System — A deployable command and control system supporting forces for joint and multinational operations across the range of military operations with compatible, interoperable, and integrated communications systems. Also called **GCCS**. See also **command and control; command and control system.** (JP 6-0)

Global Decision Support System — The command and control system employed by mobility air forces that provides schedules, arrival and/or departure information, and status data to support in-transit visibility of mobility airlift and air refueling aircraft and aircrews. Also called **GDSS**. See also **Air Mobility Command; in-transit visibility.** (JP 3-17)

global distribution — The process that coordinates and synchronizes fulfillment of joint force requirements from point of origin to point of employment. See also **distribution.** (JP 4-09)

global distribution of materiel — The process of providing materiel from the source of supply to its point of consumption or use on a worldwide basis. See also **global distribution.** (JP 4-09)

global fleet station — A persistent sea base of operations from which to interact with partner nation military and civilian populations and the global maritime community. Also called **GFS**. (JP 3-32)

global force management — A process that provides near-term sourcing solutions while providing the integrating mechanism between force apportionment, allocation, and assignment. Also call **GFM**. (JP 3-35)

Global Information Grid — The globally interconnected, end-to-end set of information capabilities, and associated processes for collecting, processing, storing, disseminating, and managing information on demand to warfighters, policy makers, and support personnel. The Global Information Grid includes owned and leased communications and computing systems and services, software (including applications), data, security services, other associated services and National Security Systems. Also called **GIG**. (JP 6-0)

global maritime partnership — An approach to cooperation among maritime nations with a shared stake in international commerce, safety, security, and freedom of the seas. Also called **GMP**. (JP 3-32)

Global Network Operations Center — United States Strategic Command operational element responsible for providing global satellite communications system status; maintaining global situational awareness to include each combatant commander's planned and current operations as well as contingency plans; supporting radio frequency interference resolution management; supporting satellite anomaly resolution and management; facilitating satellite communications interface to the defense information infrastructure; and managing the regional satellite communications support centers. Also called **GNC.** (JP 6-0)

Global Patient Movement Requirements Center — A joint activity reporting directly to the Commander, United States Transportation Command, which provides medical regulating and aeromedical evacuation scheduling for the continental United States and intertheater operations, provides support to the theater patient movement requirements centers, and coordinates with supporting resource providers to identify available assets and communicates transport to bed plans to the appropriate transportation agency for execution. Also called **GPMRC.** See also **medical treatment facility.** (JP 4-02)

Global Positioning System — A satellite-based radio navigation system operated by the Department of Defense to provide all military, civil, and commercial users with precise positioning, navigation, and timing. Also called **GPS.** (JP 3-14)

global transportation management — The integrated process of satisfying transportation requirements using the Defense Transportation System to meet national security objectives. Also called **GTM.** See also **Defense Transportation System.** (JP 4-01)

go no-go — The condition or state of operability of a component or system: "go," functioning properly; or "no-go," not functioning properly. Alternatively, a critical point at which a decision to proceed or not must be made. (JP 3-02)

governance — The state's ability to serve the citizens through the rules, processes, and behavior by which interests are articulated, resources are managed, and power is exercised in a society, including the representative participatory decision-making processes typically guaranteed under inclusive, constitutional authority. (JP 3-24)

governing factors — In the context of joint operation planning, those aspects of the situation (or externally imposed factors) that the commander deems critical to the accomplishment of the mission. (JP 5-0)

grid coordinates — Coordinates of a grid coordinate system to which numbers and letters are assigned for use in designating a point on a gridded map, photograph, or chart. (JP 3-09)

ground alert — That status in which aircraft on the ground/deck are fully serviced and armed, with combat crews in readiness to take off within a specified period of time after receipt of a mission order. See also **airborne alert.** (JP 3-01)

ground-based interceptor — A fixed-based, surface-to-air missile for defense against long-range ballistic missiles using an exo-atmospheric hit-to-kill interception of the targeted reentry vehicle in the midcourse phase of flight. Also called **GBI.** (JP 3-01)

ground-based midcourse defense — A surface-to-air ballistic missile defense system for exo-atmospheric midcourse phase interception of long-range ballistic missiles using the ground-based interceptors. Also called **GMD.** (JP 3-01)

group — 1. A flexible administrative and tactical unit composed of either two or more battalions or two or more squadrons. 2. A number of ships and/or aircraft, normally a subdivision of a force, assigned for a specific purpose. 3. A long-standing functional organization that is formed to support a broad function within a joint force commander's headquarters. Also called **GP.** (JP 3-33)

guarded frequencies — A list of time-oriented, enemy frequencies that are currently being exploited for combat information and intelligence or jammed after the commander has weighed the potential operational gain against the loss of the technical information. See also **electronic warfare.** (JP 3-13.1)

guerrilla force — A group of irregular, predominantly indigenous personnel organized along military lines to conduct military and paramilitary operations in enemy-held, hostile, or denied territory. (JP 3-05)

guerrilla warfare — Military and paramilitary operations conducted in enemy-held or hostile territory by irregular, predominantly indigenous forces. Also called **GW.** See also **unconventional warfare.** (JP 3-05.1)

guided missile — An unmanned vehicle moving above the surface of the Earth whose trajectory or flight path is capable of being altered by an external or internal mechanism. See also **ballistic missile.** (JP 3-01)

gun-target line — (*) An imaginary straight line from gun to target. Also called **GTL.**

Intentionally Blank

half-life — The time required for the activity of a given radioactive species to decrease to half of its initial value due to radioactive decay. (JP 3-11)

hardstand — 1. A paved or stabilized area where vehicles are parked. 2. Open ground with a prepared surface used for the storage of materiel. (JP 3-34)

hasty breach — The creation of lanes through enemy minefields by expedient methods such as blasting with demolitions, pushing rollers or disabled vehicles through the minefields when the time factor does not permit detailed reconnaissance, deliberate breaching, or bypassing the obstacle. (JP 3-15)

hazard — A condition with the potential to cause injury, illness, or death of personnel; damage to or loss of equipment or property; or mission degradation. See also **injury; risk.** (JP 3-33)

hazardous cargo — Cargo that includes not only large bulk-type categories such as explosives, pyrotechnics, petroleum, oils, and lubricants, compressed gases, corrosives and batteries, but lesser quantity materials like super-tropical bleach (oxiderizer), pesticides, poisons, medicines, specialized medical chemicals and medical waste that can be loaded as cargo. (JP 3-02.1)

hazards of electromagnetic radiation to fuels — The potential hazard that is created when volatile combustibles, such as fuel, are exposed to electromagnetic fields of sufficient energy to cause ignition. Also called **HERF.** (JP 3-04)

hazards of electromagnetic radiation to ordnance — The danger of accidental actuation of electro-explosive devices or otherwise electrically activating ordnance because of radio frequency electromagnetic fields. Also called **HERO.** See also **electromagnetic radiation; HERO SAFE ordnance; HERO UNSAFE ordnance; ordnance.** (JP 3-04)

hazards of electromagnetic radiation to personnel — The potential hazard that exists when personnel are exposed to an electromagnetic field of sufficient intensity to heat the human body. Also called **HERP.** (JP 3-04)

head of contracting activity — The official who has overall responsibility for managing the contracting activity. Also called **HCA.** (JP 4-10)

head-up display — (*) A display of flight, navigation, attack, or other information superimposed upon the pilot's forward field of view. Also called **HUD.** See also **flight.**

health care provider — Any member of the Armed Forces, civilian employee of the Department of Defense, or personal services contract employee under Title 10, United

States Code, Section 1091 authorized by the Department of Defense to perform health care functions. Also called **DOD health care provider.** (JP 4-02)

health service support — All services performed, provided, or arranged to promote, improve, conserve, or restore the mental or physical well-being of personnel, which include, but are not limited to, the management of health services resources, such as manpower, monies, and facilities; preventive and curative health measures; evacuation of the wounded, injured, or sick; selection of the medically fit and disposition of the medically unfit; blood management; medical supply, equipment, and maintenance thereof; combat and operational stress control; and medical, dental, veterinary, laboratory, optometric, nutrition therapy, and medical intelligence services. Also called **HSS.** (JP 4-02)

health surveillance — The regular or repeated collection, analysis, and interpretation of health-related data and the dissemination of information to monitor the health of a population and to identify potential health risks, thereby enabling timely interventions to prevent, treat, reduce, or control disease and injury, which includes occupational and environmental health surveillance and medical surveillance subcomponents. (JP 4-02)

health threat — A composite of ongoing or potential enemy actions; adverse environmental, occupational, and geographic and meteorological conditions; endemic diseases; and employment of chemical, biological, radiological, and nuclear weapons (to include weapons of mass destruction) that have the potential to affect the short- or long-term health (including psychological impact) of personnel. (JP 4-02)

heavy-lift cargo — 1. Any single cargo lift, weighing over 5 long tons, and to be handled aboard ship. 2. In Marine Corps usage, individual units of cargo that exceed 800 pounds in weight or 100 cubic feet in volume. (JP 4-01.5)

heavy-lift ship — A ship specially designed and capable of loading and unloading heavy and bulky items and has booms of sufficient capacity to accommodate a single lift of 100 tons. (JP 4-01.2)

height of burst — The vertical distance from the Earth's surface or target to the point of burst. Also called **HOB.** (JP 3-41)

helicopter coordination section — The section within the Navy tactical air control center that coordinates rotary-wing air operations with all helicopter direction centers and air traffic control center(s) in the amphibious force. Also called **HCS.** (JP 3-02)

helicopter direction center — In amphibious operations, the primary direct control agency for the helicopter group/unit commander operating under the overall control of the tactical air control center. Also called **HDC.** (JP 3-02)

helicopter support team — A task organization formed and equipped for employment in a landing zone to facilitate the landing and movement of helicopter-borne troops, equipment, and supplies, and to evacuate selected casualties and enemy prisoners of war. Also called **HST.** (JP 3-50)

helicopter transport area — Areas to the seaward and on the flanks of the outer transport and landing ship areas, but preferably inside the area screen, used for launching and/or recovering helicopters. (JP 3-02)

helicopter wave — See **wave.**

HERO SAFE ordnance — Any ordnance item that is percussion initiated, sufficiently shielded or otherwise so protected that all electro-explosive devices contained by the item are immune to adverse effects (safety or reliability) when the item is employed in its expected radio frequency environments, provided that the general hazards of electromagnetic radiation to ordnance requirements defined in the hazards from electromagnetic radiation manual are observed. See also **electromagnetic radiation; hazards of electromagnetic radiation to ordnance; HERO SUSCEPTIBLE ordnance; HERO UNSAFE ordnance; ordnance.** (JP 3-04)

HERO SUSCEPTIBLE ordnance — Any ordnance item containing electro-explosive devices proven by test or analysis to be adversely affected by radio frequency energy to the point that the safety and/or reliability of the system is in jeopardy when the system is employed in its expected radio frequency environment. See also **electromagnetic radiation; hazards of electromagnetic radiation to ordnance; HERO SAFE ordnance; HERO UNSAFE ordnance; ordnance.** (JP 3-04)

HERO UNSAFE ordnance — Any ordnance item containing electro-explosive devices that has not been classified as HERO SAFE or HERO SUSCEPTIBLE ordnance as a result of a hazards of electromagnetic radiation to ordnance (HERO) analysis or test is considered HERO UNSAFE ordnance. Additionally, any ordnance item containing electro-explosive devices (including those previously classified as HERO SAFE or HERO SUSCEPTIBLE ordnance) that has its internal wiring exposed; when tests are being conducted on that item that result in additional electrical connections to the item; when electro-explosive devices having exposed wire leads are present and handled or loaded in any but the tested condition; when the item is being assembled or disassembled; or when such ordnance items are damaged causing exposure of internal wiring or components or destroying engineered HERO protective devices. See also **electromagnetic radiation; hazards of electromagnetic radiation to ordnance; HERO SAFE ordnance; HERO SUSCEPTIBLE ordnance; ordnance.** (JP 3-04)

H-hour — The specific hour on D-day at which a particular operation commences. (JP 5-0)

high altitude bombing — Horizontal bombing with the height of release over 15,000 feet. (JP 3-09.3)

high-altitude missile engagement zone — In air defense, that airspace of defined dimensions within which the responsibility for engagement of air threats normally rests with high-altitude surface-to-air missiles. Also called **HIMEZ**. (JP 3-01)

high-density airspace control zone — Airspace designated in an airspace control plan or airspace control order, in which there is a concentrated employment of numerous and varied weapons and airspace users. A high-density airspace control zone has defined dimensions which usually coincide with geographical features or navigational aids. Access to a high-density airspace control zone is normally controlled by the maneuver commander. The maneuver commander can also direct a more restrictive weapons status within the high-density airspace control zone. Also called **HIDACZ**. (JP 3-52)

high-payoff target — A target whose loss to the enemy will significantly contribute to the success of the friendly course of action. Also called **HPT**. See also **high-value target; target.** (JP 3-60)

high-risk personnel — Personnel who, by their grade, assignment, symbolic value, or relative isolation, are likely to be attractive or accessible terrorist targets. Also called **HRP**. See also **antiterrorism.** (JP 3-07.2)

high value airborne asset protection — A defensive counterair mission using fighter escorts that defends airborne national assets which are so important that the loss of even one could seriously impact United States warfighting capabilities or provide the enemy with significant propaganda value. Also called **HVAA protection**. See also **defensive counterair.** (JP 3-01)

high-value target — A target the enemy commander requires for the successful completion of the mission. Also called **HVT**. See also **high-payoff target; target.** (JP 3-60)

high velocity drop — A drop procedure in which the drop velocity is greater than 30 feet per second and lower than free drop velocity. See also **airdrop.** (JP 3-17)

holding point — A geographically or electronically defined location used in stationing aircraft in flight in a predetermined pattern in accordance with air traffic control clearance. (JP 3-50)

homeland — The physical region that includes the continental United States, Alaska, Hawaii, United States territories, and surrounding territorial waters and airspace. (JP 3-28)

homeland defense — The protection of United States sovereignty, territory, domestic population, and critical infrastructure against external threats and aggression or other threats as directed by the President. Also called **HD**. (JP 3-27)

homeland security — A concerted national effort to prevent terrorist attacks within the United States; reduce America's vulnerability to terrorism, major disasters, and other emergencies; and minimize the damage and recover from attacks, major disasters, and other emergencies that occur. Also called **HS.** (JP 3-27)

home station — The permanent location of active duty units and Reserve Component units (e.g., location of armory or reserve center). See also **active duty; Reserve Component.** (JP 4-05)

homing — The technique whereby a mobile station directs itself, or is directed, towards a source of primary or reflected energy, or to a specified point. (JP 3-50)

homing adaptor — A device which, when used with an aircraft radio receiver, produces aural and/or visual signals that indicate the direction of a transmitting radio station with respect to the heading of the aircraft. (JP 3-50)

honey pot — A trap set to detect, deflect, or in some manner counteract attempts at unauthorized use of information systems. Generally it consists of a computer, data, or a network site that appears to be part of a network, but is actually isolated, (un)protected, and monitored, and which seems to contain information or a resource of value to attackers. (JP 3-13.4)

horizontal stowage — The lateral distribution of unit equipment or categories of supplies so that they can be unloaded simultaneously from two or more holds. (JP 3-02.1)

hostage rescue — A personnel recovery method used to recover isolated personnel who are specifically designated as hostages. Also called **HR.** (JP 3-50)

host country — A nation which permits, either by written agreement or official invitation, government representatives and/or agencies of another nation to operate, under specified conditions, within its borders. (JP 2-01.2)

hostile act — An attack or other use of force against the United States, United States forces, or other designated persons or property to preclude or impede the mission and/or duties of United States forces, including the recovery of United States personnel or vital United States Government property. (JP 3-28)

hostile casualty — A person who is the victim of a terrorist activity or who becomes a casualty "in action." "In action" characterizes the casualty as having been the direct result of hostile action, sustained in combat or relating thereto, or sustained going to or returning from a combat mission provided that the occurrence was directly related to hostile action. Included are persons killed or wounded mistakenly or accidentally by friendly fire directed at a hostile force or what is thought to be a hostile force. However, not to be considered as sustained in action and not to be interpreted as hostile casualties are injuries or death due to the elements, self-inflicted wounds, combat fatigue, and except in unusual cases, wounds or death inflicted by a friendly force while the individual is in an absent-without-leave, deserter, or dropped-from-rolls status or is voluntarily absent from a place of duty. See also **casualty.**

hostile intent — The threat of imminent use of force against the United States, United States forces, or other designated persons or property. (JP 3-01)

host nation — A nation which receives the forces and/or supplies of allied nations and/or NATO organizations to be located on, to operate in, or to transit through its territory. Also called **HN.** (JP 3-57)

host-nation support — Civil and/or military assistance rendered by a nation to foreign forces within its territory during peacetime, crises or emergencies, or war based on agreements mutually concluded between nations. Also called **HNS.** See also **host nation.** (JP 4-0)

howitzer — 1. A cannon that combines certain characteristics of guns and mortars. The howitzer delivers projectiles with medium velocities, either by low or high trajectories. 2. Normally a cannon with a tube length of 20 to 30 calibers; however, the tube length can exceed 30 calibers and still be considered a howitzer when the high angle fire zoning solution permits range overlap between charges.

hub — An organization that sorts and distributes inbound cargo from wholesale supply sources (airlifted, sealifted, and ground transportable) and/or from within the theater. See also **hub and spoke distribution; spoke.** (JP 4-09)

hub and spoke distribution — A physical distribution system developed and modeled on industry standards to provide cargo management for a theater. It is based on a "hub" moving cargo to and between several "spokes". It is designed to increase transportation efficiencies and in-transit visibility and reduce order ship time. See also **distribution; distribution system; hub; in-transit visibility; spoke.** (JP 4-09)

human factors — The physical, cultural, psychological, and behavioral attributes of an individual or group that influence perceptions, understanding, and interactions. (JP 2-0)

human intelligence — A category of intelligence derived from information collected and provided by human sources. Also called **HUMINT.** (JP 2-0)

humanitarian and civic assistance — Assistance to the local populace provided by predominantly US forces in conjunction with military operations and exercises. This assistance is specifically authorized by Title 10, United States Code, Section 401, and funded under separate authorities. Also called **HCA.** See also **foreign humanitarian assistance.** (JP 3-29)

humanitarian assistance coordination center — A temporary center established by a geographic combatant commander to assist with interagency coordination and planning. A humanitarian assistance coordination center operates during the early planning and coordination stages of foreign humanitarian assistance operations by providing the link between the geographic combatant commander and other United States Government agencies, nongovernmental organizations, and international and regional organizations at the strategic level. Also called **HACC.** See also **foreign humanitarian assistance; interagency coordination.** (JP 3-29)

humanitarian demining assistance — The activities related to the furnishing of education, training, and technical assistance with respect to the detection and clearance of land mines and other explosive remnants of war. (JP 3-29)

humanitarian mine action — Activities that strive to reduce the social, economic, and environmental impact of land mines, unexploded ordnance and small arms ammunition - also characterized as explosive remnants of war. (JP 3-15)

humanitarian operations center — An international and interagency body that coordinates the overall relief strategy and unity of effort among all participants in a large foreign humanitarian assistance operation. It normally is established under the direction of the government of the affected country or the United Nations, or a US Government agency during a US unilateral operation. Because the humanitarian operations center operates at the national level, it will normally consist of senior representatives from the affected country, assisting countries, the United Nations, nongovernmental organizations, intergovernmental organizations, and other major organizations involved in the operation. Also called **HOC.** See also **operation.** (JP 3-29)

hung ordnance — Those weapons or stores on an aircraft that the pilot has attempted to drop or fire but could not because of a malfunction of the weapon, rack or launcher, or aircraft release and control system. (JP 3-04)

hydrographic reconnaissance — Reconnaissance of an area of water to determine depths, beach gradients, the nature of the bottom, and the location of coral reefs, rocks, shoals, and man-made obstacles. (JP 3-02)

hygiene services — The provision of personal hygiene facilities and waste collection; and the cleaning, repair, replacement, and return of individual clothing and equipment items in a deployed environment. (JP 4-0)

hyperspectral imagery — Term used to describe the imagery derived from subdividing the electromagnetic spectrum into very narrow bandwidths allowing images useful in precise terrain or target analysis to be formed. Also called **HSI**. (JP 2-03)

Intentionally Blank

I

identification — 1. The process of determining the friendly or hostile character of an unknown detected contact. 2. In arms control, the process of determining which nation is responsible for the detected violations of any arms control measure. 3. In ground combat operations, discrimination between recognizable objects as being friendly or enemy, or the name that belongs to the object as a member of a class. Also called **ID.** (JP 3-01)

identification, friend or foe — A device that emits a signal positively identifying it as a friendly. Also called **IFF.** See also **air defense.** (JP 3-52)

identification maneuver — A maneuver performed for identification purposes. (JP 3-52)

identity intelligence — The intelligence resulting from the processing of identity attributes concerning individuals, groups, networks, or populations of interest. Also called **I2.** (JP 2-0)

imagery — A likeness or presentation of any natural or man-made feature or related object or activity, and the positional data acquired at the same time the likeness or representation was acquired, including: products produced by space-based national intelligence reconnaissance systems; and likeness and presentations produced by satellites, airborne platforms, unmanned aerial vehicles, or other similar means (except that such term does not include handheld or clandestine photography taken by or on behalf of human intelligence collection organizations). (JP 2-03)

imagery exploitation — The cycle of processing, using, interpreting, mensuration and/or manipulating imagery, and any assembly or consolidation of the results for dissemination. (JP 2-03)

imagery intelligence — The technical, geographic, and intelligence information derived through the interpretation or analysis of imagery and collateral materials. Also called **IMINT.** See also **intelligence.** (JP 2-03)

immediate air support — Air support to meet specific requests which arise during the course of a battle and which by their nature cannot be planned in advance. (JP 3-09.3)

immediate decontamination — Decontamination carried out by individuals immediately upon becoming contaminated to save lives, minimize casualties, and limit the spread of contamination. Also called **emergency decontamination.** See also **contamination; decontamination.** (JP 3-11)

immediate message — A category of precedence reserved for messages relating to situations that gravely affect the security of national and multinational forces or populace and that require immediate delivery to the addressee(s).

---

immediate mission request — A request for an air strike on a target that, by its nature, could not be identified sufficiently in advance to permit detailed mission coordination and planning. See also **preplanned mission request.** (JP 3-09.3)

immediate response — Any form of immediate action taken in the United States and territories to save lives, prevent human suffering, or mitigate great property damage in response to a request for assistance from a civil authority, under imminently serious conditions when time does not permit approval from a higher authority. (JP 3-28)

immediate response authority — A Federal military commander's, Department of Defense component head's, and/or responsible Department of Defense civilian official's authority temporarily to employ resources under their control, subject to any supplemental direction provided by higher headquarters, and provide those resources to save lives, prevent human suffering, or mitigate great property damage in response to a request for assistance from a civil authority, under imminently serious conditions when time does not permit approval from a higher authority within the United States. Immediate response authority does not permit actions that would subject civilians to the use of military power that is regulatory, prescriptive, proscriptive, or compulsory. (DODD 3025.18)

implementation — Procedures governing the mobilization of the force and the deployment, employment, and sustainment of military operations in response to execution orders issued by the Secretary of Defense. Also called **IMP.** (JP 5-0)

implied task — In the context of joint operation planning, a task derived during mission analysis that an organization must perform or prepare to perform to accomplish a specified task or the mission, but which is not stated in the higher headquarters order. See also **essential task; specified task.** (JP 5-0)

imprest fund — A cash fund of a fixed amount established through an advance of funds, without appropriation change, to an authorized imprest fund cashier to effect immediate cash payments of relatively small amounts for authorized purchases of supplies and nonpersonal services. (JP 1-0)

improvised explosive device — A weapon that is fabricated or emplaced in an unconventional manner incorporating destructive, lethal, noxious, pyrotechnic, or incendiary chemicals designed to kill, destroy, incapacitate, harass, deny mobility, or distract. Also called **IED.** (JP 3-15.1)

inactive duty training — Authorized training performed by a member of a Reserve Component not on active duty or active duty for training and consisting of regularly scheduled unit training assemblies, additional training assemblies, periods of appropriate duty or equivalent training, and any special additional duties authorized for Reserve Component personnel by the Secretary concerned, and performed by them in connection with the prescribed activities of the organization in which they are assigned with or without pay. Also called **IDT.** See also **active duty for training.** (JP 1)

Inactive National Guard — Army National Guard personnel in an inactive status not in the Selected Reserve who are attached to a specific National Guard unit but do not participate in training activities. Upon mobilization, they will mobilize with their units. In order for these personnel to remain members of the Inactive National Guard, they must muster once a year with their assigned unit. Like the Individual Ready Reserve, all members of the Inactive National Guard have legal, contractual obligations. Members of the Inactive National Guard may not train for retirement credit or pay and are not eligible for promotion. Also called **ING.** See also **Individual Ready Reserve; Selected Reserve.** (JP 4-05)

inactive status — Status of reserve members on an inactive status list of a Reserve Component or assigned to the Inactive Army National Guard. Those in an inactive status may not train for points or pay, and may not be considered for promotion.

incapacitating agent — A chemical agent, which produces temporary disabling conditions that can be physical or mental and persist for hours or days after exposure to the agent has ceased. (JP 3-11)

incident — An occurrence, caused by either human action or natural phenomena, that requires action to prevent or minimize loss of life, or damage, loss of, or other risks to property, information, and/or natural resources. See also **information operations.** (JP 3-28)

incident awareness and assessment — The Secretary of Defense approved use of Department of Defense intelligence, surveillance, reconnaissance, and other intelligence capabilities for domestic non-intelligence support for defense support of civil authorities. Also called **IAA.** (JP 3-28)

incident command system — A standardized on-scene emergency management construct designed to aid in the management of resources during incidents. Also called **ICS.** (JP 3-28)

incident management — A national comprehensive approach to preventing, preparing for, responding to, and recovering from terrorist attacks, major disasters, and other emergencies. (JP 3-28)

incremental costs — Costs which are additional costs to the Service appropriations that would not have been incurred absent support of the contingency operation. See also **financial management.** (JP 1-06)

independent government estimate — The government's estimate of the resources and projected cost of the resources a contractor will incur in the performance of the contract. Also called **IGE.** (JP 4-10)

indications — In intelligence usage, information in various degrees of evaluation, all of which bear on the intention of a potential enemy to adopt or reject a course of action. (JP 2-0)

indicator — In intelligence usage, an item of information which reflects the intention or capability of an adversary to adopt or reject a course of action. (JP 2-0)

indigenous populations and institutions — The societal framework of an operational environment including citizens, legal and illegal immigrants, dislocated civilians, and governmental, tribal, ethnic, religious, commercial, and private organizations and entities. Also called **IPI.** (JP 3-57)

individual mobilization augmentee — An individual reservist attending drills who receives training and is preassigned to an Active Component organization, a Selective Service System, or a Federal Emergency Management Agency billet that must be filled on, or shortly after, mobilization. Individual mobilization augmentees train on a part-time basis with these organizations to prepare for mobilization. Inactive duty training for individual mobilization augmentees is decided by component policy and can vary from 0 to 48 drills a year. Also called **IMA.** (JP 4-05)

individual protective equipment — In chemical, biological, radiological, or nuclear operations, the personal clothing and equipment required to protect an individual from chemical, biological, and radiological hazards and some nuclear hazards. Also called **IPE.** (JP 3-11)

Individual Ready Reserve — A manpower pool consisting of individuals who have had some training or who have served previously in the Active Component or in the Selected Reserve, and may have some period of their military service obligation remaining. Members may voluntarily participate in training for retirement points and promotion with or without pay. Also called **IRR.** See also **Selected Reserve.** (JP 4-05)

industrial mobilization — The transformation of industry from its peacetime activity to the industrial program necessary to support the national military objectives. It includes the mobilization of materials, labor, capital, production facilities, and contributory items and services essential to the industrial program. See also **mobilization.** (JP 4-05)

industrial preparedness — The state of preparedness of industry to produce essential materiel to support the national military objectives. (JP 4-05)

industrial preparedness program — Plans, actions, or measures for the transformation of the industrial base, both government-owned and civilian-owned, from its peacetime activity to the emergency program necessary to support the national military objectives. It includes industrial preparedness measures such as modernization, expansion, and preservation of the production facilities and contributory items and services for planning with industry. Also called **IPP.** (JP 4-05)

inertial navigation system — A self-contained navigation system using inertial detectors, which automatically provides vehicle position, heading, and velocity. Also called **INS.** (JP 3-09)

infiltration — 1. The movement through or into an area or territory occupied by either friendly or enemy troops or organizations. The movement is made, either by small groups or by individuals, at extended or irregular intervals. When used in connection with the enemy, it implies that contact is avoided. 2. In intelligence usage, placing an agent or other person in a target area in hostile territory. Usually involves crossing a frontier or other guarded line. Methods of infiltration are: black (clandestine); grey (through legal crossing point but under false documentation); and white (legal). (JP 3-05.1)

influence mine — A mine actuated by the effect of a target on some physical condition in the vicinity of the mine or on radiations emanating from the mine. See also **mine.** (JP 3-15)

influence sweep — A sweep designed to produce an influence similar to that produced by a ship and thus actuate mines. (JP 3-15)

information assurance — Actions that protect and defend information systems by ensuring availability, integrity, authentication, confidentiality, and nonrepudiation. Also called **IA.** See also **information operations.** (JP 3-12)

information environment — The aggregate of individuals, organizations, and systems that collect, process, disseminate, or act on information. (JP 3-13)

information management — The function of managing an organization's information resources for the handling of data and information acquired by one or many different systems, individuals, and organizations in a way that optimizes access by all who have a share in that data or a right to that information. Also called **IM.** (JP 3-0)

information operations — The integrated employment, during military operations, of information-related capabilities in concert with other lines of operation to influence, disrupt, corrupt, or usurp the decision-making of adversaries and potential adversaries while protecting our own. Also called **IO.** See also **electronic warfare; military deception; operations security; military information support operations.** (JP 3-13)

information operations force — A force consisting of units, staff elements, individual military professionals in the Active and Reserve Components, and DOD civilian employees who conduct or directly support the integration of information-related capabilities against adversaries and potential adversaries during military operations as well as those who train these professionals. Also called **IO force.** (DODD 3600.01)

information operations intelligence integration — The integration of intelligence disciplines and analytic methods to characterize and forecast, identify vulnerabilities, determine effects, and assess the information environment. Also called **IOII**. (JP 3-13)

information-related capability — A tool, technique, or activity employed within a dimension of the information environment that can be used to create effects and operationally desirable conditions. Also called **IRC**. (JP 3-13)

information report — Report used to forward raw information collected to fulfill intelligence requirements. (JP 2-01)

information requirements — In intelligence usage, those items of information regarding the adversary and other relevant aspects of the operational environment that need to be collected and processed in order to meet the intelligence requirements of a commander. Also called **IR**. See also **priority intelligence requirement.** (JP 2-0)

information superiority — The operational advantage derived from the ability to collect, process, and disseminate an uninterrupted flow of information while exploiting or denying an adversary's ability to do the same. See also **information operations.** (JP 3-13)

infrared imagery — That imagery produced as a result of sensing electromagnetic radiations emitted or reflected from a given target surface in the infrared position of the electromagnetic spectrum (approximately 0.72 to 1,000 microns). (JP 2-03)

infrared pointer — A low power laser device operating in the near infrared light spectrum that is visible with light amplifying night vision devices. Also called **IR pointer.** (JP 3-09.3)

initial assessment — An assessment that provides a basic determination of the viability of the infiltration and exfiltration portion of a proposed special operations forces mission. Also called **IA.** (JP 3-05.1)

initial operational capability — The first attainment of the capability to employ effectively a weapon, item of equipment, or system of approved specific characteristics that is manned or operated by an adequately trained, equipped, and supported military unit or force. Also called **IOC.**

initial radiation — The radiation, essentially neutrons and gamma rays, resulting from a nuclear burst and emitted from the fireball within one minute after burst. See also **residual radiation.** (JP 3-11)

initial reception point — In personnel recovery, a secure area or facility under friendly control where initial reception of recovered isolated personnel can safely take place. (JP 3-50)

initial response force — The first unit, usually military police, on the scene of a terrorist incident. See also **antiterrorism.** (JP 3-07.2)

initial unloading period — In amphibious operations, that part of the ship-to-shore movement in which unloading is primarily tactical in character and must be instantly responsive to landing force requirements. All elements intended to land during this period are serialized. See also **general unloading period.** (JP 3-02)

initiating directive — An order to a subordinate commander to conduct military operations as directed. Also called **ID.** (JP 3-18)

injury — 1. A term comprising such conditions as fractures, wounds, sprains, strains, dislocations, concussions, and compressions. 2. Conditions resulting from extremes of temperature or prolonged exposure. 3. Acute poisonings (except those due to contaminated food) resulting from exposure to a toxic or poisonous substance. See also **casualty.** (JP 4-02)

inland petroleum distribution system —A multi-product system consisting of both commercially available and military standard petroleum equipment that can be assembled by military personnel and, when assembled into an integrated petroleum distribution system, provides the military with the capability required to support an operational force with bulk fuels. The inland petroleum distribution system is comprised of three primary subsystems: tactical petroleum terminal, pipeline segments, and pump stations. Also called **IPDS.** (JP 4-03)

inland search and rescue region — The inland areas of the continental United States, except waters under the jurisdiction of the United States. See also **search and rescue region.** (JP 3-50)

inner transport area — In amphibious operations, an area as close to the landing beach as depth of water, navigational hazards, boat traffic, and enemy action permit, to which transports may move to expedite unloading. See also **outer transport area; transport area.** (JP 3-02)

instrument approach procedure — A series of predetermined maneuvers for the orderly transfer of an aircraft under instrument flight conditions from the beginning of the initial approach to a landing or to a point from which a landing may be made visually or the missed approach procedure is initiated. (JP 3-04)

instrument meteorological conditions — Meteorological conditions expressed in terms of visibility, distance from cloud, and ceiling; less than minimums specified for visual meteorological conditions. Also called **IMC.** See also **visual meteorological conditions.** (JP 3-04)

instruments of national power — All of the means available to the government in its pursuit of national objectives. They are expressed as diplomatic, economic, informational and military. (JP 1)

in support of — Assisting or protecting another formation, unit, or organization while remaining under original control. (JP 1)

insurgency — The organized use of subversion and violence to seize, nullify, or challenge political control of a region. Insurgency can also refer to the group itself. (JP 3-24)

integrated air and missile defense — The integration of capabilities and overlapping operations to defend the homeland and United States national interests, protect the joint force, and enable freedom of action by negating an adversary's ability to create adverse effects from their air and missile capabilities. Also called **IAMD.** (JP 3-01)

integrated consumable item support — A decision support system that takes time-phased force and deployment data (i.e., Department of Defense deployment plans) and calculates the ability of the Defense Logistics Agency, the warehousing unit of the Department of Defense, to support those plans. Integrated consumable item support can calculate for the planned deployment supply/demand curves for over two million individual items stocked by the Defense Logistics Agency in support of deployment. Also called **ICIS.** (JP 4-03)

integrated financial operations — The integration, synchronization, prioritization, and targeting of fiscal resources and capabilities across United States departments and agencies, multinational partners, and nongovernmental organizations against an adversary and in support of the population. Also called **IFO.** (JP 1-06)

integrated logistic support — A composite of all the support considerations necessary to assure the effective and economical support of a system for its life cycle. Also called **ILS.** (JP 4-01.5)

integrated planning — In amphibious operations, the planning accomplished by commanders and staffs of corresponding echelons from parallel chains of command within the amphibious task force. See also **amphibious operation; amphibious task force.** (JP 3-02)

integrated priority list — A list of a combatant commander's highest priority requirements, prioritized across Service and functional lines, defining shortfalls in key programs that, in the judgment of the combatant commander, adversely affect the capability of the combatant commander's forces to accomplish their assigned mission. Also called **IPL.** (JP 1-04)

integrated staff — A staff in which one officer only is appointed to each post on the establishment of the headquarters, irrespective of nationality and Service. See also **multinational staff; joint staff.** (JP 3-16)

integration — 1. In force protection, the synchronized transfer of units into an operational commander's force prior to mission execution. 2. The arrangement of military forces and their actions to create a force that operates by engaging as a whole. 3. In photography, a process by which the average radar picture seen on several scans of the time base may be obtained on a print, or the process by which several photographic images are combined into a single image. See also **force protection.** (JP 1)

intelligence — 1. The product resulting from the collection, processing, integration, evaluation, analysis, and interpretation of available information concerning foreign nations, hostile or potentially hostile forces or elements, or areas of actual or potential operations. 2. The activities that result in the product. 3. The organizations engaged in such activities. See also **acoustic intelligence; all-source intelligence; communications intelligence; critical intelligence; domestic intelligence; electronic intelligence; foreign intelligence; foreign instrumentation signals intelligence; general military intelligence; imagery intelligence; joint intelligence; measurement and signature intelligence; medical intelligence; national intelligence; open-source intelligence; operational intelligence; scientific and technical intelligence; strategic intelligence; tactical intelligence; target intelligence; technical intelligence; terrain intelligence.** (JP 2-0)

intelligence annex — A supporting document of an operation plan or order that provides detailed information on the enemy situation, assignment of intelligence tasks, and intelligence administrative procedures. (JP 2-01)

intelligence asset — Any resource utilized by an intelligence organization for an operational support role. (JP 2-0)

intelligence community — All departments or agencies of a government that are concerned with intelligence activity, either in an oversight, managerial, support, or participatory role. Also called **IC.** (JP 2-0)

intelligence database — The sum of holdings of intelligence data and finished intelligence products at a given organization. (JP 2-01)

intelligence discipline — A well-defined area of intelligence planning, collection, processing, exploitation, analysis, and reporting using a specific category of technical or human resources. See also **counterintelligence; human intelligence; imagery intelligence; intelligence; measurement and signature intelligence; open-source intelligence; signals intelligence; technical intelligence.** (JP 2-0)

intelligence estimate — The appraisal, expressed in writing or orally, of available intelligence relating to a specific situation or condition with a view to determining the courses of action open to the enemy or adversary and the order of probability of their adoption. (JP 2-0)

intelligence federation — A formal agreement in which a combatant command joint intelligence center receives preplanned intelligence support from other joint intelligence centers, Service intelligence organizations, reserve organizations, and national agencies during crisis or contingency operations. (JP 2-01)

intelligence information report — The primary vehicle used to provide human intelligence information to the consumer. It utilizes a message format structure that supports automated data entry into intelligence community databases. Also called **IIR.** (JP 2-01.2)

intelligence interrogation — The systematic process of using approved interrogation approaches to question a captured or detained person to obtain reliable information to satisfy intelligence requirements, consistent with applicable law. (JP 2-01.2)

intelligence mission management — A systematic process by a joint intelligence staff to proactively and continuously formulate and revise command intelligence requirements, and track the resulting information through the processing, exploitation, and dissemination process to satisfy user requirements. Also called **IMM.** (JP 2-01)

intelligence operations — The variety of intelligence and counterintelligence tasks that are carried out by various intelligence organizations and activities within the intelligence process. See also **analysis and production; collection; dissemination and integration; evaluation and feedback; planning and direction; processing and exploitation.** (JP 2-01)

intelligence planning — The intelligence component of the Adaptive Planning and Execution system, which coordinates and integrates all available Defense Intelligence Enterprise capabilities to meet combatant commander intelligence requirements. Also called **IP.** (JP 2-0)

intelligence preparation of the battlespace — The analytical methodologies employed by the Services or joint force component commands to reduce uncertainties concerning the enemy, environment, time, and terrain. Intelligence preparation of the battlespace supports the individual operations of the joint force component commands. Also called **IPB.** See also **joint intelligence preparation of the operational environment.** (JP 2-01.3)

intelligence process — The process by which information is converted into intelligence and made available to users, consisting of the six interrelated intelligence operations: planning and direction, collection, processing and exploitation, analysis and production, dissemination and integration, and evaluation and feedback. See also **analysis and production; collection; dissemination and integration; evaluation and feedback; intelligence; planning and direction; processing and exploitation.** (JP 2-01)

intelligence production — The integration, evaluation, analysis, and interpretation of information from single or multiple sources into finished intelligence for known or anticipated military and related national security consumer requirements. (JP 2-0)

intelligence-related activities — Those activities outside the consolidated defense intelligence program that: respond to operational commanders' tasking for time-sensitive information on foreign entities; respond to national intelligence community tasking of systems whose primary mission is support to operating forces; train personnel for intelligence duties; provide an intelligence reserve; or are devoted to research and development of intelligence or related capabilities. (Specifically excluded are programs that are so closely integrated with a weapon system that their primary function is to provide immediate-use targeting data.) (JP 2-01)

intelligence report — A specific report of information, usually on a single item, made at any level of command in tactical operations and disseminated as rapidly as possible in keeping with the timeliness of the information. Also called **INTREP.** (JP 2-01)

intelligence reporting — The preparation and conveyance of information by any means. More commonly, the term is restricted to reports as they are prepared by the collector and as they are transmitted by the collector to the latter's headquarters and by this component of the intelligence structure to one or more intelligence-producing components. Thus, even in this limited sense, reporting embraces both collection and dissemination. The term is applied to normal and specialist intelligence reports. (JP 2-01.2)

intelligence requirement — 1. Any subject, general or specific, upon which there is a need for the collection of information, or the production of intelligence. 2. A requirement for intelligence to fill a gap in the command's knowledge or understanding of the operational environment or threat forces. Also called **IR.** See also **intelligence; priority intelligence requirement.** (JP 2-0)

intelligence source — The means or system that can be used to observe and record information relating to the condition, situation, or activities of a targeted location, organization, or individual. See also **intelligence; source.** (JP 2-0)

intelligence, surveillance, and reconnaissance — An activity that synchronizes and integrates the planning and operation of sensors, assets, and processing, exploitation, and dissemination systems in direct support of current and future operations. This is an integrated intelligence and operations function. Also called **ISR.** See also **intelligence; intelligence, surveillance, and reconnaissance visualization; reconnaissance; surveillance.** (JP 2-01)

intelligence, surveillance, and reconnaissance visualization — The capability to graphically display the current and future locations of intelligence, surveillance, and reconnaissance sensors, their projected platform tracks, vulnerability to threat capabilities and meteorological and oceanographic phenomena, fields of regard, tasked collection targets, and products to provide a basis for dynamic retasking and time-sensitive decision making. Also called **ISR visualization.** See also **intelligence; intelligence, surveillance, and reconnaissance; reconnaissance; surveillance.** (JP 2-01)

intelligence system — Any formal or informal system to manage data gathering, to obtain and process the data, to interpret the data, and to provide reasoned judgments to decision makers as a basis for action. (JP 2-01)

interagency — Of or pertaining to United States Government agencies and departments, including the Department of Defense. See also **interagency coordination.** (JP 3-08)

interagency coordination — Within the context of Department of Defense involvement, the coordination that occurs between elements of Department of Defense, and engaged US Government agencies and departments for the purpose of achieving an objective. (JP 3-0)

intercontinental ballistic missile — A land-based, long-range ballistic missile with a range capability greater than 3,000 nautical miles. Also called **ICBM.** (JP 3-01)

interdepartmental or agency support — Provision of logistic and/or administrative support in services or materiel by one or more Military Services to one or more departments or agencies of the United States Government (other than military) with or without reimbursement. See also **inter-Service support; support.**

interdiction — 1. An action to divert, disrupt, delay, or destroy the enemy's military surface capability before it can be used effectively against friendly forces, or to otherwise achieve objectives. 2. In support of law enforcement, activities conducted to divert, disrupt, delay, intercept, board, detain, or destroy, under lawful authority, vessels, vehicles, aircraft, people, cargo, and money. See also **air interdiction.** (JP 3-03)

intergovernmental organization — An organization created by a formal agreement between two or more governments on a global, regional, or functional basis to protect and promote national interests shared by member states. Also called **IGO.** (JP 3-08)

intermediate-range ballistic missile — A land-based ballistic missile with a range capability from 1,500 to 3,000 nautical miles. Also called **IRBM.** (JP 3-01)

intermediate staging base — A tailorable, temporary location used for staging forces, sustainment and/or extraction into and out of an operational area. Also called **ISB.** See also **base; staging base.** (JP 3-35)

intermodal — Type of international freight system that permits transshipping among sea, highway, rail, and air modes of transportation through use of American National Standards Institute and International Organization for Standardization containers, line-haul assets, and handling equipment. See also **International Organization for Standardization.** (JP 4-09)

internal audience — US military members and Department of Defense civilian employees and their immediate families. See also **external audience.** (JP 3-61)

internal defense and development — The full range of measures taken by a nation to promote its growth and to protect itself from subversion, lawlessness, insurgency, terrorism, and other threats to its security. Also called **IDAD.** See also **foreign internal defense.** (JP 3-22)

internal information — See **command information.**

internally displaced person — Any person who has been forced or obliged to flee or to leave their home or places of habitual residence, in particular as a result of or in order to avoid the effects of armed conflict, situations of generalized violence, violations of human rights or natural or human-made disasters, and who have not crossed an internationally recognized state border. (JP 3-29)

internal security — The state of law and order prevailing within a nation. (JP 3-08)

internal waters — All waters, other than lawfully claimed archipelagic waters, landward of the baseline from which the territorial sea is measured. Archipelagic states may also delimit internal waters consistent with the 1982 convention on the law of the sea. All states have complete sovereignty over their internal waters.

International Convention for Safe Containers — A convention held in Geneva, Switzerland, on 2 Dec 1972, which resulted in setting standard safety requirements for containers moving in international transport. These requirements were ratified by the United States on 3 January 1978. Also called **CSC.** (JP 4-09)

international military education and training — Formal or informal instruction provided to foreign military students, units, and forces on a nonreimbursable (grant) basis by offices or employees of the United States, contract technicians, and contractors. Instruction may include correspondence courses; technical, educational, or informational publications; and media of all kinds. Also called **IMET.** See also **United States Military Service funded foreign training.** (JP 3-22)

International Organization for Standardization — A worldwide federation of national standards bodies from some 100 countries, one from each country. The International Organization for Standardization is a nongovernmental organization, established to promote the development of standardization and related activities in the world with a view to facilitating the international exchange of goods and services, and to developing cooperation in the spheres of intellectual, scientific, technological, and economic activity. Also called **ISO.** (JP 4-09)

interoperability — 1. The ability to operate in synergy in the execution of assigned tasks. (JP 3-0)   2. The condition achieved among communications-electronics systems or items of communications-electronics equipment when information or services can be exchanged directly and satisfactorily between them and/or their users. The degree of interoperability should be defined when referring to specific cases. (JP 6-0)

interorganizational coordination — The interaction that occurs among elements of the Department of Defense; engaged United States Government agencies; state, territorial, local, and tribal agencies; foreign military forces and government agencies; intergovernmental organizations; nongovernmental organizations; and the private sector. (JP 3-08)

interpretation — A part of the analysis and production phase in the intelligence process in which the significance of information is judged in relation to the current body of knowledge. See also **intelligence process.** (JP 2-01)

interrogation — Systematic effort to procure information by direct questioning of a person under the control of the questioner. (JP 2-01.2)

inter-Service support — Action by one Service or element thereof to provide logistics and/or administrative support to another Service or element thereof. See also **interdepartmental or agency support; support.** (JP 4-0)

intertheater airlift — The common-user airlift linking theaters to the continental United States and to other theaters as well as the airlift within the continental United States. See also **intratheater airlift.** (JP 3-17)

intertheater patient movement — Moving patients between, into, and out of the different theaters of the geographic combatant commands and into the continental United States or another supporting theater. See also **en route care; evacuation; intratheater patient movement.** (JP 4-02)

in-transit visibility — The ability to track the identity, status, and location of Department of Defense units, and non-unit cargo (excluding bulk petroleum, oils, and lubricants) and passengers; patients; and personal property from origin to consignee or destination across the range of military operations. Also called **ITV.** (JP 4-01.2)

intratheater airlift — Airlift conducted within a theater with assets assigned to a geographic combatant commander or attached to a subordinate joint force commander. See also **intertheater airlift.** (JP 3-17)

intratheater patient movement — Moving patients within the theater of a combatant command or in the continental United States. See also **en route care; evacuation; intertheater patient movement.** (JP 4-02)

inventory control — That phase of military logistics that includes managing, cataloging, requirements determinations, procurement, distribution, overhaul, and disposal of materiel. Also called **inventory management; materiel control; materiel management; supply management.** (JP 4-09)

inventory control point — An organizational unit or activity within a Department of Defense supply system that is assigned the primary responsibility for the materiel management of a group of items either for a particular Service or for the Defense Department as a whole. Materiel inventory management includes cataloging direction, requirements computation, procurement direction, distribution management, disposal direction and, generally, rebuild direction. Also called **ICP.** (JP 4-09)

inventory management — See **inventory control.**

inventory managers — See **inventory control point.**

ionizing radiation — Particulate (alpha, beta, and neutron) and electromagnetic (X-ray and gamma) radiation of sufficient energy to displace electrons from atoms, producing ions. (JP 3-11)

IR pointer — See **infrared pointer.** (JP 3-09.3)

irregular warfare — A violent struggle among state and non-state actors for legitimacy and influence over the relevant population(s). Also called **IW.** (JP 1)

isolated personnel — US military, Department of Defense civilians and contractor personnel (and others designated by the President or Secretary of Defense) who are separated from their unit (as an individual or a group) while participating in a US sponsored military activity or mission and are, or may be, in a situation where they must survive, evade, resist, or escape. See also **combat search and rescue; search and rescue.** (JP 3-50)

isolated personnel report — A Department of Defense Form (DD 1833) containing information designed to facilitate the identification and authentication of an isolated person by a recovery force. Also called **ISOPREP.** See also **authentication; evader; recovery force.** (JP 3-50)

item manager — An individual within the organization of an inventory control point or other such organization assigned management responsibility for one or more specific items of materiel. (JP 4-09)

Intentionally Blank

J

J-2X — The staff element of the intelligence directorate of a joint staff that combines and represents the principal authority for counterintelligence and human intelligence support. See also **counterintelligence; human intelligence.** (JP 2-01.2)

joint — Connotes activities, operations, organizations, etc., in which elements of two or more Military Departments participate. (JP 1)

joint acquisition review board — A joint task force or subunified commander established board used to review and make recommendations for controlling critical common-user logistic supplies and services within the joint operational area and to recommend the proper sources of support for approved support requirements. Also called **JARB.** See also **combatant commander logistic procurement support board; joint contracting support board.** (JP 4-10)

joint after action report — A report consisting of summary joint universal lessons learned. It describes a real world operation or training exercise and identifies significant lessons learned. Also called **JAAR.**

joint air attack team — A combination of attack and/or scout rotary-wing aircraft and fixed-wing close air support aircraft operating together to locate and attack high-priority targets and other targets of opportunity. The joint air attack team normally operates as a coordinated effort supported by fire support, air defense artillery, naval surface fire support, intelligence, surveillance, and reconnaissance systems, electronic warfare systems, and ground maneuver forces against enemy forces. Joint terminal attack controllers may perform duties as directed by the air mission commander in support of the ground commander's scheme of maneuver. Also called **JAAT.** See also **close air support.** (JP 3-09.3)

joint air component coordination element — A general term for the liaison element that serves as the direct representative of the joint force air component commander for joint air operations. Also called **JACCE.** (JP 3-30)

joint air operations — Air operations performed with air capabilities/forces made available by components in support of the joint force commander's operation or campaign objectives, or in support of other components of the joint force. (JP 3-30)

joint air operations center — A jointly staffed facility established for planning, directing, and executing joint air operations in support of the joint force commander's operation or campaign objectives. Also called **JAOC.** See also **joint air operations.** (JP 3-30)

joint air operations plan — A plan for a connected series of joint air operations to achieve the joint force commander's objectives within a given time and joint operational area. Also called **JAOP.** See also **joint air operations.** (JP 3-30)

joint base — For purposes of base defense operations, a joint base is a locality from which operations of two or more of the Military Departments are projected or supported and which is manned by significant elements of two or more Military Departments or in which significant elements of two or more Military Departments are located. See also **base.** (JP 3-10)

joint captured materiel exploitation center — An element responsible for deriving intelligence information from captured enemy materiel. It is normally subordinate to the intelligence directorate of a joint staff. Also called **JCMEC.** (JP 2-01)

joint civil-military operations task force — A joint task force composed of civil-military operations units from more than one Service. Also called **JCMOTF.** See also **civil-military operations; joint task force.** (JP 3-57)

joint combined exchange training — A program conducted overseas to fulfill US forces training requirements and at the same time exchange the sharing of skills between US forces and host nation counterparts. Also called **JCET.** (JP 3-05)

joint communications network — The aggregation of all the joint communications systems in a theater. The joint communications network includes the joint multi-channel trunking and switching system and the joint command and control communications system(s). Also called **JCN.** (JP 6-0)

joint concept — Links strategic guidance to the development and employment of future joint force capabilities and serve as "engines for transformation" that may ultimately lead to doctrine, organization, training, materiel, leadership and education, personnel and facilities (DOTMLPF) and policy changes. (CJCSI 3010.02)

joint contracting support board — A joint task force or subunified commander established board to coordinate all contracting support and to determine specific contracting mechanisms to obtain commercially procured common logistic supplies and services within the joint operational area. Also called **JCSB.** See also **combatant commander logistic procurement support board; joint acquisition review board.** (JP 4-10)

joint counterintelligence unit — An organization composed of Service and Department of Defense agency counterintelligence personnel, formed under the authority of the Secretary of Defense and assigned to a combatant commander, which focuses on combatant command strategic and operational counterintelligence missions. Also called **JCIU.** (JP 2-01.2)

joint data network operations officer — The joint task force operations directorate officer responsible to the commander for integrating data from supporting components into a common database used to generate the common tactical picture. Also called **JDNO.** (JP 3-01)

joint deployable intelligence support system — A transportable workstation and communications suite that electronically extends a joint intelligence center to a joint task force or other tactical user. Also called **JDISS.** (JP 2-0)

joint deployment and distribution enterprise — The complex of equipment, procedures, doctrine, leaders, technical connectivity, information, shared knowledge, organizations, facilities, training, and materiel necessary to conduct joint distribution operations. Also called **JDDE.** (JP 4-0)

joint deployment and distribution operations center — A combatant command movement control organization designed to synchronize and optimize national and theater multimodal resources for deployment, distribution, and sustainment, Also called **JDDOC.** (JP 4-09)

joint desired point of impact — A unique, alpha-numeric coded precise aimpoint associated with a target to achieve an explicit weaponeering objective, and identified by a three dimensional (latitude, longitude, elevation) mensurated coordinate. Also called a **JDPI.** See also **aimpoint; desired point of impact.** (JP 3-60)

joint distribution — The operational process of synchronizing all elements of the joint logistic system using the Joint Deployment and Distribution Enterprise for end-to-end movement of forces and materiel from point of origin to the designated point of need. (JP 4-09)

joint doctrine — Fundamental principles that guide the employment of United States military forces in coordinated action toward a common objective and may include terms, tactics, techniques, and procedures. See also **Chairman of the Joint Chiefs of Staff instruction; Chairman of the Joint Chiefs of Staff manual; doctrine; joint publication; joint test publication; multinational doctrine.** (CJCSI 5120.02)

joint doctrine development community — The Chairman of the Joint Chiefs of Staff, the Services, the combatant commands, the Joint Staff, the combat support agencies, and the doctrine development agencies of the Services and the joint community. Also called **JDDC.** (CJCSI 5120.02)

Joint Doctrine Development System — The system of lead agents, Joint Staff doctrine sponsors, primary review authorities, coordinating review authorities, technical review authorities, assessment agents, evaluation agents, Joint Doctrine Planning Conference, procedures, and hierarchical framework designed to initiate, develop, approve, and maintain joint publications. Also called **JDDS.** (CJCSI 5120.02)

Joint Doctrine Planning Conference — A forum convened by the Joint Staff J-7 that meets semiannually to address and vote on project proposals; discuss key joint doctrinal and operational issues; discuss potential changes to the joint doctrine development process; keep up to date on the status of the joint publication projects and emerging publications; and keep abreast of other initiatives of interest to the members. Also called **JDPC.** (CJCSM 5120.01)

joint document exploitation center — An element, normally subordinate to the intelligence directorate of a joint staff, responsible for deriving intelligence information from captured adversary documents including all forms of electronic data and other forms of stored textual and graphic information. Also called **JDEC.** See also **intelligence.** (JP 2-01)

joint duty assignment — An assignment to a designated position in a multi-Service, joint or multinational command or activity that is involved in the integrated employment or support of the land, sea, and air forces of at least two of the three Military Departments. Such involvement includes, but is not limited to, matters relating to national military strategy, joint doctrine and policy, strategic planning, contingency planning, and command and control of combat operations under a unified or specified command. Also called **JDA.**

Joint Duty Assignment List — Positions designated as joint duty assignments are reflected in a list approved by the Secretary of Defense and maintained by the Joint Staff. The Joint Duty Assignment List is reflected in the Joint Duty Assignment Management Information System. Also called **JDAL.**

joint electromagnetic spectrum management operations — Those interrelated functions of frequency management, host nation coordination, and joint spectrum interference resolution that together enable the planning, management, and execution of operations within the electromagnetic operational environment during all phases of military operations. Also called **JEMSMO.** (JP 6-01)

joint electromagnetic spectrum operations — Those activities consisting of electronic warfare and joint electromagnetic spectrum management operations used to exploit, attack, protect, and manage the electromagnetic operational environment to achieve the commander's objectives. Also called **JEMSO.** (JP 6-01)

joint engagement zone — In air defense, that airspace of defined dimensions within which multiple air defense systems (surface-to-air missiles and aircraft) are simultaneously employed to engage air threats. Also called **JEZ.** (JP 3-01)

joint facilities utilization board — A joint board that evaluates and reconciles component requests for real estate, use of existing facilities, inter-Service support, and construction to ensure compliance with Joint Civil-Military Engineering Board priorities. Also called **JFUB.** (JP 3-34)

joint field office — A temporary multiagency coordination center established at the incident site to provide a central location for coordination of federal, state, local, tribal, nongovernmental, and private-sector organizations with primary responsibility for incident oversight, direction, or assistance to effectively coordinate protection, prevention, preparedness, response, and recovery actions. Also called **JFO.** (JP 3-28)

joint fires — Fires delivered during the employment of forces from two or more components in coordinated action to produce desired effects in support of a common objective. See also **fires.** (JP 3-0)

joint fires element — An optional staff element that provides recommendations to the operations directorate to accomplish fires planning and synchronization. Also called **JFE.** See also **fire support; joint fires.** (JP 3-60)

joint fires observer — A trained Service member who can request, adjust, and control surface-to-surface fires, provide targeting information in support of Type 2 and 3 close air support terminal attack control, and perform autonomous terminal guidance operations. Also called **JFO.** (JP 3-09.3)

joint fire support — Joint fires that assist air, land, maritime, and special operations forces to move, maneuver, and control territory, populations, airspace, and key waters. See also **fire support; joint fires.** (JP 3-0)

joint flow and analysis system for transportation — System that determines the transportation feasibility of a course of action or operation plan; provides daily lift assets needed to move forces and resupply; advises logistic planners of channel and port inefficiencies; and interprets shortfalls from various flow possibilities. Also called **JFAST.** See also **course of action; operation plan; system.** (JP 3-35)

joint force — A general term applied to a force composed of significant elements, assigned or attached, of two or more Military Departments operating under a single joint force commander. See also **joint force commander.** (JP 3-0)

joint force air component commander — The commander within a unified command, subordinate unified command, or joint task force responsible to the establishing commander for recommending the proper employment of assigned, attached, and/or made available for tasking air forces; planning and coordinating air operations; or accomplishing such operational missions as may be assigned. Also called **JFACC.** See also **joint force commander.** (JP 3-0)

joint force chaplain — The military chaplain designated by the joint force commander to serve as the senior chaplain for the joint force. Also called the **JFCH.** (JP 1-05)

joint force commander — A general term applied to a combatant commander, subunified commander, or joint task force commander authorized to exercise combatant command (command authority) or operational control over a joint force. Also called **JFC.** See also **joint force.** (JP 1)

joint force land component commander — The commander within a unified command, subordinate unified command, or joint task force responsible to the establishing commander for recommending the proper employment of assigned, attached, and/or made available for tasking land forces; planning and coordinating land operations; or accomplishing such operational missions as may be assigned. Also called **JFLCC**. See also **joint force commander.** (JP 3-0)

joint force maritime component commander — The commander within a unified command, subordinate unified command, or joint task force responsible to the establishing commander for recommending the proper employment of assigned, attached, and/or made available for tasking maritime forces and assets; planning and coordinating maritime operations; or accomplishing such operational missions as may be assigned. Also called **JFMCC**. See also **joint force commander.** (JP 3-0)

joint force special operations component commander — The commander within a unified command, subordinate unified command, or joint task force responsible to the establishing commander for recommending the proper employment of assigned, attached, and/or made available for tasking special operations forces and assets; planning and coordinating special operations; or accomplishing such operational missions as may be assigned. Also called **JFSOCC**. See also **joint force commander.** (JP 3-0)

joint force surgeon — A general term applied to a Department of Defense medical department officer appointed by the joint force commander to serve as the joint force special staff officer responsible for establishing, monitoring, or evaluating joint force health service support. Also called **JFS**. See also **health service support; joint force.** (JP 4-02)

joint functions — Related capabilities and activities placed into six basic groups of command and control, intelligence, fires, movement and maneuver, protection, and sustainment to help joint force commanders synchronize, integrate, and direct joint operations. (JP 3-0)

joint integrated prioritized target list — A prioritized list of targets approved and maintained by the joint force commander. Also called **JIPTL**. See also **target.** (JP 3-60)

joint intelligence — Intelligence produced by elements of more than one Service of the same nation. (JP 2-0)

joint intelligence architecture — A dynamic, flexible structure that consists of the Defense Joint Intelligence Operations Center, combatant command joint intelligence operations centers, and subordinate joint task force intelligence operations centers or joint intelligence support elements to provide national, theater, and tactical commanders with the full range of intelligence required for planning and conducting operations. See also **intelligence.** (JP 2-0)

joint intelligence operations center — An interdependent, operational intelligence organization at the Department of Defense, combatant command, or joint task force (if established) level, that is integrated with national intelligence centers, and capable of accessing all sources of intelligence impacting military operations planning, execution, and assessment. Also called **JIOC.** (JP 2-0)

joint intelligence preparation of the operational environment — The analytical process used by joint intelligence organizations to produce intelligence estimates and other intelligence products in support of the joint force commander's decision-making process. It is a continuous process that includes defining the operational environment; describing the impact of the operational environment; evaluating the adversary; and determining adversary courses of action. Also called **JIPOE.** (JP 2-01.3)

joint intelligence support element — A subordinate joint force element whose focus is on intelligence support for joint operations, providing the joint force commander, joint staff, and components with the complete air, space, ground, and maritime adversary situation. Also called **JISE.** See also **intelligence; joint force; joint operations.** (JP 2-01)

joint interagency coordination group — A staff group that establishes regular, timely, and collaborative working relationships between civilian and military operational planners. Also called **JIACG.** (JP 3-08)

joint interface control officer — The senior interface control officer for multi-tactical data link networks in the joint force who is responsible for development and validation of the architecture, joint interoperability and management of the multi-tactical data link networks, and overseeing operations of a joint interface control cell. Also called **JICO.** (JP 3-01)

joint interrogation and debriefing center — Physical location for the exploitation of intelligence information from detainees and other sources. Also called **JIDC.** See also **intelligence.** (JP 2-01.2)

joint interrogation operations — 1. Activities conducted by a joint or interagency organization to extract information for intelligence purposes from enemy prisoners of war, dislocated civilians, enemy combatants, or other uncategorized detainees. 2. Activities conducted in support of law enforcement efforts to adjudicate enemy combatants who are believed to have committed crimes against US persons or property. Also called **JIO.** See also **enemy combatant.** (JP 2-01)

joint land operations — Land operations performed across the range of military operations with land forces made available by Service components in support of the joint force commander's operation or campaign objectives, or in support of other components of the joint force. (JP 3-31)

joint land operations plan — A plan for a connected series of joint land operations to achieve the joint force commander's objectives within a given time and operational area. (JP 3-31)

joint logistics — The coordinated use, synchronization, and sharing of two or more Military Departments' logistic resources to support the joint force. See also **logistics.** (JP 4-0)

joint logistics enterprise — A multi-tiered matrix of key global logistics providers cooperatively engaged or structured to achieve a common purpose without jeopardizing the integrity of their own organizational missions and goals. Also called **JLEnt.** (JP 4-0)

Joint Logistics Operations Center — The Joint Logistics Operations Center is the current operations division within the Logistics Directorate of the Joint Staff, which monitors crises, exercises, and interagency actions and works acquisition and cross-servicing agreements as well as international logistics. Also called **JLOC.** See also **logistics.** (JP 4-01)

joint logistics over-the-shore commander — The commander selected by the joint force commander and tasked to organize the efforts of all elements participating in accomplishing the joint logistics over-the-shore mission. See also **joint logistics over-the-shore operations.** (JP 4-01.6)

joint logistics over-the-shore operations — Operations in which Navy and Army logistics over-the-shore forces conduct logistics over-the-shore operations together under a joint force commander. Also called **JLOTS operations.** See also **joint logistics; logistics over-the-shore operations.** (JP 4-01.6)

joint manpower program — The policy, processes, and systems used in determination and prioritization within and among joint Service manpower requirements. Also called **JMP.** (JP 1-0)

joint meteorological and oceanographic coordination cell — A subset of a joint meteorological and oceanographic coordination organization, which is delegated the responsibility of executing the coordination of meteorological and oceanographic support operations in the operational area. Also called **JMCC.** (JP 3-59)

joint meteorological and oceanographic coordination organization — A Service meteorological and oceanographic organization that is designated within the operations order as the lead organization responsible for coordinating meteorological and oceanographic operations support in the operational area. Also called **JMCO.** (JP 3-59)

joint meteorological and oceanographic officer — Officer designated to provide direct meteorological and oceanographic support to a joint force commander. Also called **JMO.** (JP 3-59)

joint military information support task force — A military information support operations task force composed of headquarters and operational assets that assists the joint force commander in developing strategic, operational, and tactical military information support operation plans for a theater campaign or other operations. Also called **JMISTF**. See also **joint special operations task force; military information support operations; special operations.** (JP 3-13.2)

joint mission-essential task — A mission task selected by a joint force commander deemed essential to mission accomplishment and defined using the common language of the Universal Joint Task List in terms of task, condition, and standard. Also called **JMET.** See also **condition, universal joint task list.** (JP 3-33)

joint mortuary affairs office — Plans and executes all mortuary affairs programs within a theater. Provides guidance to facilitate the conduct of all mortuary programs and to maintain data (as required) pertaining to recovery, identification, and disposition of all US dead and missing in the assigned theater. Serves as the central clearing point for all mortuary affairs and monitors the deceased and missing personal effects program. Also called **JMAO**. See also **mortuary affairs; personal effects.** (JP 4-06)

joint network operations control center — An element of the J-6 established to support a joint force commander. The joint network operations control center serves as the single control agency for the management and direction of the joint force communications systems. The joint network operations control center may include plans and operations, administration, system control, and frequency management sections. Also called **JNCC.** (JP 6-0)

joint operation planning — Planning activities associated with joint military operations by combatant commanders and their subordinate joint force commanders in response to contingencies and crises. See also **execution planning; Joint Operation Planning and Execution System; joint operation planning process.** (JP 5-0)

Joint Operation Planning and Execution System — An Adaptive Planning and Execution system technology. Also called **JOPES.** See also **joint operation planning; joint operations; level of detail.** (JP 5-0)

joint operation planning process — An orderly, analytical process that consists of a logical set of steps to analyze a mission, select the best course of action, and produce a joint operation plan or order. Also called **JOPP.** See also **joint operation planning; Joint Operation Planning and Execution System.** (JP 5-0)

joint operations — A general term to describe military actions conducted by joint forces and those Service forces employed in specified command relationships with each other, which of themselves, do not establish joint forces. (JP 3-0)

joint operations area — An area of land, sea, and airspace, defined by a geographic combatant commander or subordinate unified commander, in which a joint force commander (normally a joint task force commander) conducts military operations to accomplish a specific mission. Also called **JOA.** See also **area of responsibility; joint special operations area.** (JP 3-0)

joint operations area forecast — The official baseline meteorological and oceanographic forecast for operational planning and mission execution within the joint operations area. Also called **JOAF.** (JP 3-59)

joint operations center — A jointly manned facility of a joint force commander's headquarters established for planning, monitoring, and guiding the execution of the commander's decisions. Also called **JOC.** (JP 3-41)

joint patient movement requirements center — A joint activity established to coordinate the joint patient movement requirements function for a joint task force operating within a unified command area of responsibility. Also called **JPMRC.** See also **health service support; joint force surgeon; joint operations area; medical treatment facility.** (JP 4-02)

joint patient movement team — Teams comprised of personnel trained in medical regulating and movement procedures. Also called **JPMT.** (JP 4-02)

joint personnel accountability reconciliation and reporting — A data repository developed and implemented by the Defense Manpower Data Center that consumes and reconciles data from existing Service deployment systems. Also called **JPARR.** (JP 1-0)

joint personnel reception center — A center established in an operational area by the appropriate joint force commander with the responsibility for the in-processing and out-processing of personnel upon their arrival in and departure from the theater. Also called **JPRC.** (JP 1-0)

joint personnel recovery center — The primary joint force organization responsible for planning and coordinating personnel recovery for military operations within the assigned operational area. Also called **JPRC.** See also **combat search and rescue; search and rescue.** (JP 3-50)

joint personnel recovery support product — The basic reference document for personnel recovery-specific information on a particular country or region of interest. Also called **JPRSP.** (JP 3-50)

joint personnel training and tracking activity — The continental United States center established to facilitate the reception, accountability, processing, training, and onward movement of individual augmentees preparing for overseas movement to support a joint military operation. Also called **JPTTA.** (JP 1-0)

joint planning and execution community — Those headquarters, commands, and agencies involved in the training, preparation, mobilization, deployment, employment, support, sustainment, redeployment, and demobilization of military forces assigned or committed to a joint operation. Also called **JPEC.** (JP 5-0)

joint planning group — A planning organization consisting of designated representatives of the joint force headquarters principal and special staff sections, joint force components (Service and/or functional), and other supporting organizations or agencies as deemed necessary by the joint force commander. Also called **JPG.** See also **crisis action planning; joint operation planning.** (JP 5-0)

joint proponent — A Service, combatant command, or Joint Staff directorate assigned coordinating authority to lead the collaborative development and integration of joint capability with specific responsibilities designated by the Secretary of Defense. (SecDef Memo 03748-09)

Joint Public Affairs Support Element — A deployable unit assigned to assist a joint force commander in developing and training public affairs forces in joint, interagency, and multinational environments. Also called **JPASE.** (JP 3-61)

joint publication — A compilation of agreed to fundamental principles, considerations, and guidance on a particular topic, approved by the Chairman of the Joint Chiefs of Staff that guides the employment of a joint force toward a common objective. Also called **JP.** See also **Chairman of the Joint Chiefs of Staff instruction; Chairman of the Joint Chiefs of Staff manual; joint doctrine; joint test publication.** (CJCSI 5120.02)

joint reception coordination center — The organization, established by the Department of the Army as the designated Department of Defense executive agent for the repatriation of noncombatants, that ensures Department of Defense personnel and noncombatants receive adequate assistance and support for an orderly and expedient debarkation, movement to final destination in the United States, and appropriate follow-on assistance at the final destination. Also called **JRCC.** (JP 3-68)

joint reception, staging, onward movement, and integration — A phase of joint force projection occurring in the operational area during which arriving personnel, equipment, and materiel transition into forces capable of meeting operational requirements. Also called **JRSOI.** See also **integration; joint force; reception; staging.** (JP 3-35)

joint restricted frequency list — A time and geographically oriented listing of TABOO, PROTECTED, and GUARDED functions, nets, and frequencies and limited to the minimum number of frequencies necessary for friendly forces to accomplish objectives. Also called **JRFL.** See also **electronic warfare; guarded frequencies; protected frequencies; TABOO frequencies.** (JP 3-13.1)

joint security area — A specific surface area, designated by the joint force commander to facilitate protection of joint bases and their connecting lines of communications that support joint operations. Also called **JSA**. (JP 3-10)

joint security coordination center — A joint operations center tailored to assist the joint security coordinator in meeting the security requirements in the joint operational area. Also called **JSCC**. (JP 3-10)

joint security coordinator — The officer with responsibility for coordinating the overall security of the operational area in accordance with joint force commander directives and priorities. Also called **JSC**. (JP 3-10)

joint servicing — That function performed by a jointly staffed and financed activity in support of two or more Services. (JP 3-05)

joint special operations air component commander — The commander within a joint force special operations command responsible for planning and executing joint special operations air activities. Also called **JSOACC**. (JP 3-05)

joint special operations area — An area of land, sea, and airspace assigned by a joint force commander to the commander of a joint special operations force to conduct special operations activities. Also called **JSOA**. (JP 3-0)

joint special operations task force — A joint task force composed of special operations units from more than one Service, formed to carry out a specific special operation or prosecute special operations in support of a theater campaign or other operations. Also called **JSOTF**. (JP 3-05)

joint staff — 1. The staff of a commander of a unified or specified command, subordinate unified command, joint task force, or subordinate functional component (when a functional component command will employ forces from more than one Military Department), that includes members from the several Services comprising the force. 2. (capitalized as Joint Staff) The staff under the Chairman of the Joint Chiefs of Staff that assists the Chairman and the other members of the Joint Chiefs of Staff in carrying out their responsibilities. Also called **JS**. (JP 1)

Joint Staff doctrine sponsor — A Joint Staff directorate assigned to coordinate a specific joint doctrine project with the Joint Staff. Also called **JSDS**. See also **joint doctrine.** (CJCSM 5120.01)

Joint Strategic Capabilities Plan — A plan that provides guidance to the combatant commanders and the Joint Chiefs of Staff to accomplish tasks and missions based on current military capabilities. Also called **JSCP**. See also **combatant commander; joint.** (JP 5-0)

joint strategic exploitation center — Theater-level physical location for an exploitation facility that functions under the direction of the joint force commander and is used to hold detainees with potential long-term strategic intelligence value, deemed to be of interest to counterintelligence or criminal investigators, or who may be a significant threat to the United States, its citizens or interests, or US allies. Also called **JSEC.** (JP 2-01.2)

Joint Strategic Planning System — One of the primary means by which the Chairman of the Joint Chiefs of Staff, in consultation with the other members of the Joint Chiefs of Staff and the combatant commanders, carries out the statutory responsibilities to assist the President and Secretary of Defense in providing strategic direction to the Armed Forces. Also called **JSPS.** (JP 5-0)

joint table of distribution — A manpower document that identifies the positions and enumerates the spaces that have been approved for each organizational element of a joint activity for a specific fiscal year (authorization year), and those accepted for the four subsequent fiscal years (program years). Also called **JTD.** See also **joint manpower program.** (JP 1-0)

joint targeting coordination board — A group formed by the joint force commander to accomplish broad targeting oversight functions that may include but are not limited to coordinating targeting information, providing targeting guidance, synchronization, and priorities, and refining the joint integrated prioritized target list. Also called **JTCB.** See also **joint integrated prioritized target list; targeting.** (JP 3-60)

joint target list — A consolidated list of selected targets, upon which there are no restrictions placed, considered to have military significance in the joint force commander's operational area. Also called **JTL.** See also **joint; target.** (JP 3-60)

joint task force — A joint force that is constituted and so designated by the Secretary of Defense, a combatant commander, a subunified commander, or an existing joint task force commander. Also called **JTF.** (JP 1)

Joint Task Force-Civil Support — A standing joint task force established to plan and integrate Department of Defense support to the designated lead federal agency for domestic chemical, biological, radiological, nuclear, and high-yield explosives consequence management operations. Also called **JTF-CS.** (JP 3-41)

Joint Technical Coordinating Group for Munitions Effectiveness — A Joint Staff-level organization tasked to produce generic target vulnerability and weaponeering studies. The special operations working group is a subordinate organization specializing in studies for special operations. Also called **JTCG-ME.** (JP 3-05.1)

joint terminal attack controller — A qualified (certified) Service member who, from a forward position, directs the action of combat aircraft engaged in close air support and other offensive air operations. A qualified and current joint terminal attack controller will be recognized across the Department of Defense as capable and authorized to perform terminal attack control. Also called **JTAC.** See also **terminal attack control.** (JP 3-09.3)

joint test publication — A proposed publication produced for field-testing an emergent concept that has been validated through the Joint Experimentation Program or a similar joint process. Also called **JTP.** See also **Chairman of the Joint Chiefs of Staff instruction; joint doctrine; joint publication.** (CJCSM 5120.01)

Joint Transportation Board — Responsible to the Chairman of the Joint Chiefs of Staff, the Joint Transportation Board assures that common-user transportation resources assigned or available to the Department of Defense are allocated to achieve maximum benefit in meeting Department of Defense objectives. Also called **JTB.** See also **common-user transportation.** (JP 4-01)

joint urban operations — Joint operations planned and conducted on, or against objectives within a topographical complex and its adjacent natural terrain, where man-made construction or the density of population are the dominant features. Also called **JUOs.** See also **joint operations.** (JP 3-06)

Joint Worldwide Intelligence Communications System — The sensitive compartmented information portion of the Defense Information Systems Network, which incorporates advanced networking technologies that permit point-to-point or multipoint information exchange involving voice, text, graphics, data, and video teleconferencing. Also called **JWICS.** (JP 2-0)

judge advocate — An officer of the Judge Advocate General's Corps of the Army, Air Force, Marine Corps, Navy, and the United States Coast Guard who is designated as a judge advocate. Also called **JA.** (JP 1-04)

jumpmaster — The assigned airborne qualified individual who controls paratroops from the time they enter the aircraft until they exit. (JP 3-17)

K

key doctrine element — A foundational core concept, principle, or idea of joint operations as established in approved joint doctrine text; other information in joint doctrine expands on or supports these foundational doctrine elements. Also called **KDE.** (CJCSM 5120.01)

key position — A civilian position, public or private (designated by the employer and approved by the Secretary concerned), that cannot be vacated during war or national emergency. (JP 1-0)

keystone publications — Joint doctrine publications that establish the doctrinal foundation for a series of joint publications in the hierarchy of joint publications. See also **capstone publications; joint publication.** (CJCSM 5120.01)

key terrain — Any locality, or area, the seizure or retention of which affords a marked advantage to either combatant. (JP 2-01.3)

kill box — A three-dimensional area used to facilitate the integration of joint fires. (JP 3-09)

Intentionally Blank

L

land control operations — The employment of land forces, supported by maritime and air forces (as appropriate) to control vital areas of the land domain. Such operations are conducted to establish local military superiority in land operational areas. See also **sea control operations**. (JP 3-31)

land forces — Personnel, weapon systems, vehicles, and support elements operating on land to accomplish assigned missions and tasks. (JP 3-31)

landing aid — Any illuminating light, radio beacon, radar device, communicating device, or any system of such devices for aiding aircraft in an approach and landing. (JP 3-04)

landing area — 1. That part of the operational area within which are conducted the landing operations of an amphibious force. It includes the beach, the approaches to the beach, the transport areas, the fire support areas, the airspace above it, and the land included in the advance inland to the initial objective. 2. (Airborne) The general area used for landing troops and materiel either by airdrop or air landing. This area includes one or more drop zones or landing strips. 3. Any specially prepared or selected surface of land, water, or deck designated or used for takeoff and landing of aircraft. See also **airfield; amphibious force; landing beach; landing force.** (JP 3-02)

landing area diagram — A graphic means of showing, for amphibious operations, the beach designations, boat lanes, organization of the line of departure, scheduled waves, landing ship area, transport areas, and the fire support areas in the immediate vicinity of the boat lanes. (JP 3-02)

landing beach — That portion of a shoreline usually required for the landing of a battalion landing team. However, it may also be that portion of a shoreline constituting a tactical locality (such as the shore of a bay) over which a force larger or smaller than a battalion landing team may be landed. (JP 3-02)

landing craft — A craft employed in amphibious operations, specifically designed for carrying troops and their equipment and for beaching, unloading, and retracting. It is also used for resupply operations. (JP 3-02)

landing craft and amphibious vehicle assignment table — A table showing the assignment of personnel and materiel to each landing craft and amphibious vehicle and the assignment of the landing craft and amphibious vehicles to waves for the ship-to-shore movement. (JP 3-02)

landing craft availability table — A tabulation of the type and number of landing craft that will be available from each ship of the transport group. The table is the basis for the assignment of landing craft to the boat groups for the ship-to-shore movement. (JP 3-02)

landing diagram — A graphic means of illustrating the plan for the ship-to-shore movement. (JP 3-02)

landing force — A Marine Corps or Army task organization formed to conduct amphibious operations. The landing force, together with the amphibious task force and other forces, constitute the amphibious force. Also called **LF**. See also **amphibious force; amphibious operation; amphibious task force; task organization.** (JP 3-02)

landing force operational reserve material — Package of contingency supplies pre-positioned and maintained onboard selected amphibious ships to enhance reaction time and provide support for the embarked landing force in contingencies. Also called **LFORM.** (JP 3-02.1)

landing force support party — A temporary landing force organization composed of Navy and landing force elements, that facilitates the ship-to-shore movement and provides initial combat support and combat service support to the landing force. The landing force support party is brought into existence by a formal activation order issued by the commander, landing force. Also called **LFSP**. See also **combat service support; combat support; landing force; ship-to-shore movement.** (JP 3-02)

landing group — In amphibious operations, a subordinate task organization of the landing force capable of conducting landing operations, under a single tactical command, against a position or group of positions. (JP 3-02)

landing group commander — In amphibious operations, the officer designated by the commander, landing force as the single tactical commander of a subordinate task organization capable of conducting landing operations against a position or group of positions. See also **amphibious operation; commander, landing force.** (JP 3-02)

landing plan — In amphibious operations, a collective term referring to all individually prepared naval and landing force documents that, taken together, present in detail all instructions for execution of the ship-to-shore movement. (JP 3-02)

landing sequence table — A document that incorporates the detailed plans for ship-to-shore movement of nonscheduled units. (JP 3-02)

landing ship — An assault ship which is designed for long sea voyages and for rapid unloading over and on to a beach. (JP 3-02)

landing signalman enlisted — Enlisted man responsible for ensuring that helicopters/tiltrotor aircraft, on signal, are safely started, engaged, launched, recovered, and shut down. Also called **LSE.** (JP 3-04)

landing signals officer — Officer responsible for the visual control of aircraft in the terminal phase of the approach immediately prior to landing. Also called **LSO.** See also **terminal phase.** (JP 3-04)

landing site — 1. A site within a landing zone containing one or more landing points. See also airfield. 2. In amphibious operations, a continuous segment of coastline over which troops, equipment and supplies can be landed by surface means. (JP 3-02)

landing zone — Any specified zone used for the landing of aircraft. Also called **LZ.** See also **airfield.** (JP 3-17)

laser guided weapon — A weapon which uses a seeker to detect laser energy reflected from a laser marked/designated target and through signal processing provides guidance commands to a control system which guides the weapon to the point from which the laser energy is being reflected. Also called **LGW.** (JP 3-09)

laser rangefinder — A device which uses laser energy for determining the distance from the device to a place or object. (JP 3-09)

laser seeker — A device based on a direction sensitive receiver which detects the energy reflected from a laser designated target and defines the direction of the target relative to the receiver. See also **laser guided weapon.** (JP 3-09)

laser spot — The area on a surface illuminated by a laser. See also **spot.** (JP 3-09)

laser spot tracker — A device that locks on to the reflected energy from a laser-marked or designated target and defines the direction of the target relative to itself. Also called **LST.** (JP 3-09)

laser target designator — A device that emits a beam of laser energy which is used to mark a specific place or object. Also called **LTD.** See also **target.** (JP 3-09)

latest arrival date — A day, relative to C-Day, that is specified by the supported combatant commander as the latest date when a unit, a resupply shipment, or replacement personnel can arrive at the port of debarkation and support the concept of operations. Also called **LAD.** (JP 5-0)

law enforcement agency — Any of a number of agencies (outside the Department of Defense) chartered and empowered to enforce US laws in the United States, a state or territory (or political subdivision) of the United States, a federally recognized Native American tribe or Alaskan Native Village, or within the borders of a host nation. Also called **LEA.** (JP 3-28)

law of armed conflict — See **law of war.** (JP 1-04)

law of war — That part of international law that regulates the conduct of armed hostilities. Also called **the law of armed conflict.** See also **rules of engagement**. (JP 1-04)

lead — In intelligence usage, a person with potential for exploitation, warranting additional assessment, contact, and/or development. (JP 2-01.2)

lead agency — The US Government agency designated to coordinate the interagency oversight of the day-to-day conduct of an ongoing operation. (JP 3-08)

lead agent — 1. An individual Service, combatant command, or Joint Staff directorate assigned to develop and maintain a joint publication. (CJCSI 5120.02) 2. In medical materiel management, the designated unit or organization to coordinate or execute day-to-day conduct of an ongoing operation or function. Also called **LA.** (JP 4-02)

lead aircraft — 1. The airborne aircraft designated to exercise command of other aircraft within the flight. 2. An aircraft in the van of two or more aircraft.

lead federal agency — The federal agency that leads and coordinates the overall federal response to an emergency. Also called **LFA.** (JP 3-41)

lead nation — The nation with the will, capability, competence, and influence to provide the essential elements of political consultation and military leadership to coordinate the planning, mounting, and execution of a multinational operation. See also **logistic support; multinational force.** (JP 3-16)

lead Service or agency for common-user logistics — A Service component or Department of Defense agency that is responsible for execution of common-user item or service support in a specific combatant command or multinational operation as defined in the combatant or subordinate joint force commander's operation plan, operation order, and/or directives. See also **common-user logistics.** (JP 4-0)

letter of assist — A contractual document issued by the United Nations to a government authorizing it to provide goods or services to a peacekeeping operation. Also called **LOA.** See also **peacekeeping.** (JP 1-06)

letter of authorization — A document issued by the procuring contracting officer or designee that authorizes contractor personnel authorized to accompany the force to travel to, from, and within the operational area; and, outlines government furnished support authorizations within the operational area. Also called **LOA.** (JP 4-10)

letter of offer and acceptance — Standard Department of Defense form on which the United States Government documents its offer to transfer to a foreign government or international organization United States defense articles and services via foreign military sales pursuant to the Arms Export Control Act. Also called **LOA.** See also **foreign military sales.** (JP 4-08)

level of detail — Within the current joint planning and execution system, movement characteristics for both personnel and cargo are described at six distinct levels of detail. Levels I, V, and VI describe personnel and Levels I through IV and VI for cargo.

Levels I through IV are coded and visible in the Joint Operation Planning and Execution System automated data processing. Levels V and VI are used by Joint Operation Planning and Execution System automated data processing feeder systems. a. **level I** - personnel: expressed as total number of passengers by unit line number. Cargo: expressed in total short tons, total measurement tons, total square feet, and total thousands of barrels by unit line number. Petroleum, oils, and lubricants is expressed by thousands of barrels by unit line number. b. **level II** - cargo: expressed by short tons and measurement tons of bulk, oversize, outsize, and non-air transportable cargo by unit line number. Also square feet for vehicles and non self-deployable aircraft and boats by unit line number. c. **level III** - cargo: detail by cargo category code expressed as short tons and measurement tons as well as square feet associated to that cargo category code for an individual unit line number. d. **level IV** - cargo: detail for individual dimensional data expressed in length, width, and height in number of inches, and weight/volume in short tons/measurement tons, along with a cargo description. Each cargo item is associated with a cargo category code and a unit line number). e. **level V** - personnel: any general summarization/aggregation of level VI detail in distribution and deployment. f. **level VI** - personnel: detail expressed by name, Service, military occupational specialty and unique identification number. Cargo: detail expressed by association to a transportation control number or single tracking number or item of equipment to include federal stock number/national stock number and/or requisition number. Nested cargo, cargo that is contained within another equipment item, may similarly be identified. Also called **JOPES level of detail.** (CJCSM 3122.01A)

leverage — In the context of joint operation planning, a relative advantage in combat power and/or other circumstances against the adversary across one or more domains or the information environment sufficient to exploit that advantage. See also **operational art; operational design.** (JP 5-0)

L-hour — The specific hour on C-day at which a deployment operation commences or is to commence. (JP 5-0)

liaison — That contact or intercommunication maintained between elements of military forces or other agencies to ensure mutual understanding and unity of purpose and action. (JP 3-08)

life cycle — The total phases through which an item passes from the time it is initially developed until the time it is either consumed in use or disposed of as being excess to all known materiel requirements. (JP 4-02)

lighterage — The process in which small craft are used to transport cargo or personnel from ship-to-shore using amphibians, landing craft, discharge lighters, causeways, and barges. (JP 4-01.6)

limiting factor — A factor or condition that, either temporarily or permanently, impedes mission accomplishment. (JP 5-0)

line of communications — A route, either land, water, and/or air, that connects an operating military force with a base of operations and along which supplies and military forces move. Also called **LOC.** (JP 2-01.3)

line of demarcation — A line defining the boundary of a buffer zone used to establish the forward limits of disputing or belligerent forces after each phase of disengagement or withdrawal has been completed. See also **buffer zone; disengagement; peace operations.** (JP 3-07.3)

line of departure — 1. In land warfare, a line designated to coordinate the departure of attack elements. Also called **LD.** (JP 3-31) 2. In amphibious warfare, a suitably marked offshore coordinating line to assist assault craft to land on designated beaches at scheduled times the seaward end of a boat lane. Also called **LOD.** (JP 3-02)

line of effort — In the context of joint operation planning, using the purpose (cause and effect) to focus efforts toward establishing operational and strategic conditions by linking multiple tasks and missions. Also called **LOE.** (JP 5-0)

line of operation — A line that defines the interior or exterior orientation of the force in relation to the enemy or that connects actions on nodes and/or decisive points related in time and space to an objective(s). Also called **LOO.** (JP 5-0)

link — 1. A behavioral, physical, or functional relationship between nodes. 2. In communications, a general term used to indicate the existence of communications facilities between two points. 3. A maritime route, other than a coastal or transit route, which links any two or more routes. See also **node.** (JP 3-0)

listening watch — A continuous receiver watch established for the reception of traffic addressed to, or of interest to, the unit maintaining the watch, with complete log optional. (JP 3-50)

littoral — The littoral comprises two segments of operational environment: 1. Seaward: the area from the open ocean to the shore, which must be controlled to support operations ashore. 2. Landward: the area inland from the shore that can be supported and defended directly from the sea. (JP 2-01.3)

loading plan — All of the individually prepared documents which, taken together, present in detail all instructions for the arrangement of personnel, and the loading of equipment for one or more units or other special grouping of personnel or material moving by highway, water, rail, or air transportation. (JP 3-02.1)

load signal — In personnel recovery, a visual signal displayed in a covert manner to indicate the presence of an individual or object at a given location. See also **evasion; recovery operations.** (JP 3-50)

local procurement — The process of obtaining personnel, services, supplies, and equipment from local or indigenous sources.

local purchase — The function of acquiring a decentralized item of supply from sources outside the Department of Defense.

locate — In personnel recovery, the task where actions are taken to precisely find and authenticate the identity of isolated personnel. (JP 3-50)

lodgment — A designated area in a hostile or potentially hostile operational area that, when seized and held, makes the continuous landing of troops and materiel possible and provides maneuver space for subsequent operations. (JP 3-18)

logistics — Planning and executing the movement and support of forces. (JP 4-0)

logistics over-the-shore operation area — That geographic area required to conduct a logistics over-the-shore operation. Also called **LOA**. See also **logistics over-the-shore operations.** (JP 4-01.6)

logistics over-the-shore operations — The loading and unloading of ships without the benefit of deep draft-capable, fixed port facilities; or as a means of moving forces closer to tactical assembly areas dependent on threat force capabilities. Also called **LOTS operations**. See also **joint logistics over-the-shore operations.** (JP 4-01.6)

logistic support — Support that encompasses the logistic services, materiel, and transportation required to support the continental United States-based and worldwide deployed forces. (JP 4-0)

logistics supportability analysis — Combatant command internal assessment for the Joint Strategic Capabilities Plan on capabilities and shortfalls of key logistic capabilities required to execute and sustain the concept of support conducted on all level three plans with the time phased force deployment data. Also called **LSA**. (JP 4-0)

low-altitude missile engagement zone — In air defense, that airspace of defined dimensions within which the responsibility for engagement of air threats normally rests with low- to medium-altitude surface-to-air missiles. Also called **LOMEZ**. (JP 3-01)

low level flight — See **terrain flight.**

low-level transit route — A temporary corridor of defined dimensions established in the forward area to minimize the risk to friendly aircraft from friendly air defenses or surface forces. Also called **LLTR**. (JP 3-52)

low velocity drop — A drop procedure in which the drop velocity does not exceed 30 feet per second. (JP 3-17)

low visibility operations — Sensitive operations wherein the political-military restrictions inherent in covert and clandestine operations are either not necessary or not feasible; actions are taken as required to limit exposure of those involved and/or their activities. Execution of these operations is undertaken with the knowledge that the action and/or sponsorship of the operation may preclude plausible denial by the initiating power. (JP 3-05.1)

M

magnetic mine — A mine that responds to the magnetic field of a target. (JP 3-15)

mail embargo — A temporary shutdown or redirection of mail flow to or from a specific location. (JP 1-0)

main operating base — A facility outside the United States and US territories with permanently stationed operating forces and robust infrastructure. Main operating bases are characterized by command and control structures, enduring family support facilities, and strengthened force protection measures. Also called **MOB.** See also **cooperative security location; forward operating site.** (CJCS CM-0007-05)

main operations base — In special operations, a base established by a joint force special operations component commander or a subordinate special operations component commander in friendly territory to provide sustained command and control, administration, and logistic support to special operations activities in designated areas. Also called **MOB.** See also **advanced operations base; forward operations base.** (JP 3-05.1)

main supply route — The route or routes designated within an operational area upon which the bulk of traffic flows in support of military operations. Also called **MSR.** (JP 4-01.5)

maintenance — 1. All action, including inspection, testing, servicing, classification as to serviceability, repair, rebuilding, and reclamation, taken to retain materiel in a serviceable condition or to restore it to serviceability. 2. All supply and repair action taken to keep a force in condition to carry out its mission. 3. The routine recurring work required to keep a facility in such condition that it may be continuously used at its original or designed capacity and efficiency for its intended purpose. (JP 4-0)

major force — A military organization comprised of major combat elements and associated combat support, combat service support, and sustainment increments. (JP 5-0)

major operation — 1. A series of tactical actions (battles, engagements, strikes) conducted by combat forces of a single or several Services, coordinated in time and place, to achieve strategic or operational objectives in an operational area. 2. For noncombat operations, a reference to the relative size and scope of a military operation. See also **operation.** (JP 3-0)

maneuver — 1. A movement to place ships, aircraft, or land forces in a position of advantage over the enemy. 2. A tactical exercise carried out at sea, in the air, on the ground, or on a map in imitation of war. 3. The operation of a ship, aircraft, or vehicle, to cause it to perform desired movements. 4. Employment of forces in the operational area through movement in combination with fires to achieve a position of advantage in respect to the enemy. See also **mission; operation.** (JP 3-0)

manpower management — The means of manpower control to ensure the most efficient and economical use of available manpower. (JP 1-0)

manpower requirements — Human resources needed to accomplish specified work loads of organizations. (JP 1-0)

Marine air command and control system — A system that provides the aviation combat element commander with the means to command, coordinate, and control all air operations within an assigned sector and to coordinate air operations with other Services. It is composed of command and control agencies with communications-electronics equipment that incorporates a capability from manual through semiautomatic control. Also called **MACCS**. See also **direct air support center; tactical air operations center.** (JP 3-09.3)

Marine Corps special operations forces — Those Active Component Marine Corps forces designated by the Secretary of Defense that are specifically organized, trained, and equipped to conduct and support special operations. Also called **MARSOF.** (JP 3-05.1)

Maritime Administration — The Maritime Administration is the agency within the United States Department of Transportation dealing with waterborne transportation. Also called **MARAD.** (JP 4-01.2)

Maritime Administration Ready Reserve Force — The surge sealift assets owned and operated by the United States Department of Transportation/Maritime Administration and Military Sealift Command (in contingency), crewed by civilian mariners. Also called **MARAD RRF.** See also **National Defense Reserve Fleet.** (JP 4-01.6)

maritime domain — The oceans, seas, bays, estuaries, islands, coastal areas, and the airspace above these, including the littorals. (JP 3-32)

maritime domain awareness — The effective understanding of anything associated with the maritime domain that could impact the security, safety, economy, or environment of a nation. Also called **MDA**. (JP 3-32)

maritime forces — Forces that operate on, under, or above the sea to gain or exploit command of the sea, sea control, or sea denial and/or to project power from the sea. (JP 3-32)

maritime interception operations — Efforts to monitor, query, and board merchant vessels in international waters to enforce sanctions against other nations such as those in support of United Nations Security Council Resolutions and/or prevent the transport of restricted goods. Also called **MIO.** (JP 3-03)

maritime power projection — Power projection in and from the maritime environment, including a broad spectrum of offensive military operations to destroy enemy forces or logistic support or to prevent enemy forces from approaching within enemy weapons' range of friendly forces. (JP 3-32)

maritime pre-positioning force operation — A rapid deployment and assembly of a Marine expeditionary force in a secure area using a combination of intertheater airlift and forward-deployed maritime pre-positioning ships. Also called **MPF operation**. See also **maritime pre-positioning ships.** (JP 4-01.6)

maritime pre-positioning ships — Civilian-crewed, Military Sealift Command-chartered ships that are organized into three squadrons and are usually forward-deployed. These ships are loaded with pre-positioned equipment and 30 days of supplies to support three Marine expeditionary brigades. Also called **MPSs**. See also **Navy cargo handling battalion**. (JP 3-02.1)

maritime security operations — Those operations to protect maritime sovereignty and resources and to counter maritime-related terrorism, weapons proliferation, transnational crime, piracy, environmental destruction, and illegal seaborne immigration. Also called **MSO.** (JP 3-32)

Maritime Security Program — A program authorized in the Maritime Security Act of 2003 requiring the Secretary of Transportation, in consultation with the Secretary of Defense, to establish a fleet of active, commercially viable, militarily useful, privately-owned vessels to meet national defense and other security requirements. Also called **MSP.** (JP 4-01.2)

maritime superiority — That degree of dominance of one force over another that permits the conduct of maritime operations by the former and its related land, maritime, and air forces at a given time and place without prohibitive interference by the opposing force. (JP 3-32)

maritime terminal — A facility for berthing ships simultaneously at piers, quays, and/or working anchorages. Also known as a water terminal. (JP 4-01.5)

marking — To maintain contact on a target from such a position that the marking unit has an immediate offensive capability. (JP 3-09.3)

marshalling — 1. The process by which units participating in an amphibious or airborne operation group together or assemble when feasible or move to temporary camps in the vicinity of embarkation points, complete preparations for combat, or prepare for loading. 2. The process of assembling, holding, and organizing supplies and/or equipment, especially vehicles of transportation, for onward movement. See also **staging area.** (JP 3-17)

marshalling area — A location in the vicinity of a reception terminal or pre-positioned equipment storage site where arriving unit personnel, equipment, materiel, and accompanying supplies are reassembled, returned to the control of the unit commander, and prepared for onward movement. See also **marshalling**. (JP 3-35)

mass atrocity response operations — Military activities conducted to prevent or halt mass atrocities. Also called **MARO**. (JP 3-07.3)

mass casualty — Any large number of casualties produced in a relatively short period of time, usually as the result of a single incident such as a military aircraft accident, hurricane, flood, earthquake, or armed attack that exceeds local logistic support capabilities. Also called **MASCAL**. See also **casualty**. (JP4-02)

massed fire — 1. The fire of the batteries of two or more ships directed against a single target. 2. Fire from a number of weapons directed at a single point or small area. (JP 3-02)

master — The commanding officer of a United States naval ship, a commercial ship, or a government-owned general agency agreement ship operated for the Military Sealift Command by a civilian company to transport Department of Defense cargo. Also called **MA**. (JP 3-02.1)

master air attack plan — A plan that contains key information that forms the foundation of the joint air tasking order. Also called **MAAP**. See also **target**. (JP 3-60)

materiel — All items necessary to equip, operate, maintain, and support military activities without distinction as to its application for administrative or combat purposes. See also **equipment; personal property**. (JP 4-0)

materiel control — See **inventory control**.

materiel inventory objective — The quantity of an item required to be on hand and on order on M-day in order to equip, provide a materiel pipeline, and sustain the approved US force structure (active and reserve) and those Allied forces designated for US materiel support, through the period prescribed for war materiel planning purposes. It is the quantity by which the war materiel requirement exceeds the war materiel procurement capability and the war materiel requirement adjustment. It includes the M-day force materiel requirement and the war reserve materiel requirement. (JP 4-09)

materiel management — See **inventory control**.

materiel planning — A subset of logistic planning consisting of a four-step process. a. **requirements definition**. Requirements for significant items must be calculated at item level detail (i.e., National Stock Number) to support sustainability planning and analysis. Requirements include unit roundout, consumption and attrition replacement, safety stock, and the needs of allies. b. **apportionment**. Items are apportioned to the combatant commanders based on a global scenario to avoid sourcing of items to multiple theaters. The basis for apportionment is the capability provided by unit stocks, host-nation support, theater pre-positioned war reserve stocks and industrial base, and continental United States Department of Defense stockpiles and available production. Item apportionment cannot exceed total capabilities. c. **sourcing.** Sourcing is the matching of available capabilities on a given date against item requirements to support sustainability analysis and the identification of locations to support transportation planning. Sourcing of any item is done within the combatant commander's apportionment. d. **documentation.** Sourced item requirements and corresponding shortfalls are major inputs to the combatant commander's sustainability analysis. Sourced item requirements are translated into movement requirements and documented in the Joint Operation Planning and Execution System database for transportation feasibility analysis. Movement requirements for nonsignificant items are estimated in tonnage.

materiel readiness — The availability of materiel required by a military organization to support its wartime activities or contingencies, disaster relief (flood, earthquake, etc.), or other emergencies. (JP 4-03)

materiel release order — An order issued by an accountable supply system manager (usually an inventory control point or accountable depot or stock point) directing a non-accountable activity (usually a storage site or materiel drop point) within the same supply distribution complex to release and ship materiel. Also called **MRO.** (JP 4-09)

materiel requirements — Those quantities of items of equipment and supplies necessary to equip, provide a materiel pipeline, and sustain a Service, formation, organization, or unit in the fulfillment of its purposes or tasks during a specified period. (JP 4-09)

maximum ordinate — (*) In artillery and naval gunfire support, the height of the highest point in the trajectory of a projectile above the horizontal plane passing through its origin. Also called **vertex height.**

M-day — See **times.** (JP 4-06)

means of transport — See **mode of transport.**

measurement and signature intelligence — Information produced by quantitative and qualitative analysis of physical attributes of targets and events to characterize, locate, and identify targets and events, and derived from specialized, technically derived measurements of physical phenomenon intrinsic to an object or event. Also called **MASINT.** See also **intelligence; scientific and technical intelligence.** (JP 2-0)

Measurement and Signature Intelligence Requirements System — A system for the management of theater and national measurement and signature intelligence collection requirements, providing automated tools for users in support of submission, review, and validation of measurement and signature intelligence nominations of requirements to be tasked for national and Department of Defense measurement and signature intelligence collection, production, and exploitation resources. Also called **MRS.** See also **measurement and signature intelligence.** (JP 2-01)

measurement ton — The unit of volumetric measurement of equipment associated with surface-delivered cargo equal to the total cubic feet divided by 40. Also called **MTON.** (JP 4-01.5)

measure of effectiveness — A criterion used to assess changes in system behavior, capability, or operational environment that is tied to measuring the attainment of an end state, achievement of an objective, or creation of an effect. Also called **MOE.** See also **combat assessment; mission.** (JP 3-0)

measure of performance — A criterion used to assess friendly actions that is tied to measuring task accomplishment. Also called **MOP.** (JP 3-0)

mechanical sweep — In naval mine warfare, any sweep used with the object of physically contacting the mine or its appendages. (JP 3-15)

media operations center — A facility established by the joint force commander to serve as the focal point for the interface between the military and the media during the conduct of joint operations. Also called **MOC.** (JP 3-61)

media pool — A limited number of news media who represent a larger number of news media organizations for purposes of news gathering and sharing of material during a specified activity. Pooling is typically used when news media support resources cannot accommodate a large number of journalists. See also **news media representative; public affairs.** (JP 3-61)

medical civil-military operations — All military health-related activities in support of a joint force commander that establish, enhance, maintain or influence relations between the joint or multinational force and host nation, multinational governmental and nongovernmental civilian organizations and authorities, and the civilian populace in order to facilitate military operations, achieve United States operational objectives, and positively impact the health sector. Also called **MCMO.** (JP 4-02)

medical engagement protocols — Directives issued by competent military authority that delineate the circumstances and limitations under which United States medical forces will initiate medical care and support to those individuals that are not Department of Defense health care beneficiaries or designated eligible for care in a military medical treatment facility by the Secretary of Defense. (JP 4-02)

medical intelligence — That category of intelligence resulting from collection, evaluation, analysis, and interpretation of foreign medical, bio-scientific, and environmental information that is of interest to strategic planning and to military medical planning and operations for the conservation of the fighting strength of friendly forces and the formation of assessments of foreign medical capabilities in both military and civilian sectors. Also called **MEDINT.** See also **intelligence.** (JP 2-01)

medical intelligence preparation of the operational environment — A systematic continuing process that analyzes information on medical and disease threats, enemy capabilities, terrain, weather, local medical infrastructure, potential humanitarian and refugee situations, transportation issues, and political, religious and social issues for all types of operations. Also called **MIPOE.** (JP 4-02)

medical logistics support — A functional area of logistics support for the joint force surgeon's health service support mission and that includes supplying Class VIII medical supplies (medical material to include medical peculiar repair parts used to sustain the health service support system), optical fabrication, medical equipment maintenance, blood storage and distribution, and medical gases. Also called **MEDLOG support.** (JP 4-02)

medical regulating — The actions and coordination necessary to arrange for the movement of patients through the roles of care and to match patients with a medical treatment facility that has the necessary health service support capabilities and available bed space. See also **health service support; medical treatment facility.** (JP 4-02)

medical surveillance — The ongoing, systematic collection, analysis, and interpretation of data derived from instances of medical care or medical evaluation, and the reporting of population-based information for characterizing and countering threats to a population's health, well-being and performance. See also **surveillance.** (JP 4-02)

medical treatment facility — A facility established for the purpose of furnishing medical and/or dental care to eligible individuals. Also called **MTF.** (JP 4-02)

medium-range ballistic missile. A ballistic missile with a range capability from about 600 to 1,500 nautical miles. Also called **MRBM.** (JP 3-01)

mensuration — The process of measurement of a feature or location on the earth to determine an absolute latitude, longitude, and elevation. (JP 3-60)

message — 1. Any thought or idea expressed briefly in a plain or secret language and prepared in a form suitable for transmission by any means of communication. (JP 6-0) 2. A narrowly focused communication directed at a specific audience to support a specific theme. (JP 3-61)

meteorological and oceanographic — A term used to convey all environmental factors, from the sub-bottom of the Earth's oceans through maritime, land areas, airspace, ionosphere, and outward into space. Also called **METOC.** (JP 3-59)

meteorological and oceanographic data — Measurements or observations of meteorological and oceanographic variables. (JP 3-59)

meteorological and oceanographic environment — The surroundings that extend from the sub-bottom of the Earth's oceans, through maritime, land areas, airspace, ionosphere, and outward into space, which include conditions, resources, and natural phenomena, in and through which the joint force operates. (JP 3-59)

meteorological and oceanographic information — Actionable information to include meteorological, climatological, oceanographic, and space environment observations, analyses, prognostic data or products and meteorological and oceanographic effects. (JP 3-59)

meteorological and oceanographic operations support community — The collective of electronically connected, shore-based meteorological and oceanographic production facilities/centers, theater and/or regional meteorological and oceanographic production activities. Also called **MOSC.** See also **meteorological and oceanographic.** (JP 3-59)

meteorological watch — Monitoring the weather for a route, area, or terminal and advising concerned organizations when hazardous conditions that could affect their operations or pose a hazard to life or property are observed or forecast to occur. Also called **METWATCH.** (JP 3-59)

meteorology — The study dealing with the phenomena of the atmosphere including the physics, chemistry, and dynamics extending to the effects of the atmosphere on the Earth's surface and the oceans. (JP 3-59)

midcourse phase — That portion of the flight of a ballistic missile between the boost phase and the terminal phase. See also **boost phase; terminal phase.** (JP 3-01)

migrant — A person who (1) belongs to a normally migratory culture who may cross national boundaries, or (2) has fled his or her native country for economic reasons rather than fear of political or ethnic persecution. (JP 3-29)

Military Affiliate Radio System — A program conducted by the Departments of the Army, Navy, and Air Force in which amateur radio stations and operators participate in and contribute to the mission of providing auxiliary and emergency communications on a local, national, or international basis as an adjunct to normal military communications. Also called **MARS.**

military assistance advisory group — A joint Service group, normally under the military command of a commander of a unified command and representing the Secretary of Defense, which primarily administers the US military assistance planning and programming in the host country. Also called **MAAG.** (JP 3-22)

Military Assistance Program — That portion of the US security assistance authorized by the Foreign Assistance Act of 1961, as amended, which provides defense articles and services to recipients on a nonreimbursable (grant) basis. Also called **MAP.** (JP 3-22)

military civic action — Programs and projects managed by United States forces but executed primarily by indigenous military or security forces that contribute to the economic and social development of a host nation civil society thereby enhancing the legitimacy and social standing of the host nation government and its military forces. Also called **MCA.** (JP 3-57)

military construction — Any construction, alteration, development, conversion, or extension of any kind carried out with respect to a military installation. Also called **MILCON.** (JP 3-34)

military deception — Actions executed to deliberately mislead adversary military, paramilitary, or violent extremist organization decision makers, thereby causing the adversary to take specific actions (or inactions) that will contribute to the accomplishment of the friendly mission. Also called **MILDEC.** (JP 3-13.4)

Military Department — One of the departments within the Department of Defense created by the National Security Act of 1947, which are the Department of the Army, the Department of the Navy, and the Department of the Air Force. Also called **MILDEP.** See also **Department of the Air Force; Department of the Army; Department of the Navy.** (JP 1)

military engagement — Routine contact and interaction between individuals or elements of the Armed Forces of the United States and those of another nation's armed forces, or foreign and domestic civilian authorities or agencies to build trust and confidence, share information, coordinate mutual activities, and maintain influence. (JP 3-0)

military geography — The specialized field of geography dealing with natural and manmade physical features that may affect the planning and conduct of military operations.

military government — The supreme authority the military exercises by force or agreement over the lands, property, and indigenous populations and institutions of domestic, allied, or enemy territory therefore substituting sovereign authority under rule of law for the previously established government. (JP 3-57)

military health system — A health system that supports the military mission by fostering, protecting, sustaining, and restoring health and providing the direction, resources, health care providers, and other means necessary for promoting the health of the beneficiary population. (JP 4-02)

military information support operations — Planned operations to convey selected information and indicators to foreign audiences to influence their emotions, motives, objective reasoning, and ultimately the behavior of foreign governments, organizations, groups, and individuals in a manner favorable to the originator's objectives. Also called **MISO.** (JP 3-13.2)

military information support operations impact indicators — An observable event or a discernible subjectively determined behavioral change that represents an effect of a military information support operations activity on the intended foreign target audience at a particular point in time. (JP 3-13.2)

Military Intelligence Board — A decision-making forum which formulates Department of Defense intelligence policy and programming priorities. Also called **MIB.** See also **intelligence.** (JP 2-0)

military intervention — The deliberate act of a nation or a group of nations to introduce its military forces into the course of an existing controversy. (JP 3-0)

military journalist — A US Service member or Department of Defense civilian employee providing photographic, print, radio, or television command information for military internal audiences. See also **command information.** (JP 3-61)

military occupation — A condition in which territory is under the effective control of a foreign armed force. See also **occupied territory.** (JP 3-0)

Military Postal Service — The command, organization, personnel, and facilities established to provide a means for the transmission of mail to and from the Department of Defense, members of the US Armed Forces, and other authorized agencies and individuals. Also called **MPS.** (JP 1-0)

Military Postal Service Agency — The single manager operating agency established to manage the Military Postal Service. Also called **MPSA.** (JP 1-0)

military post office — A branch of a designated US-based post office established by US Postal Service authority and operated by one of the Services. Also called **MPO.** (JP 1-0)

military resources — Military and civilian personnel, facilities, equipment, and supplies under the control of a Department of Defense component. (JP 4-01.5)

Military Sealift Command — A major command of the United States Navy reporting to Commander Fleet Forces Command, and the United States Transportation Command's component command responsible for designated common-user sealift transportation services to deploy, employ, sustain, and redeploy United States forces on a global basis. Also called **MSC.** See also **transportation component command.** (JP 4-01.2)

Military Sealift Command force — Common-user sealift consisting of three subsets: the Naval Fleet Auxiliary Force, common-user ocean transportation, and the special mission support force. See also **common-user sealift; Military Sealift Command.** (JP 4-01.2)

military source operations — The collection, from, by and/or via humans, of foreign and military and military-related intelligence. (JP 2-01.2)

military specification container — A container that meets specific written standards. Also called **MILSPEC container.** (JP 4-09)

military standard requisitioning and issue procedure — A uniform procedure established by the Department of Defense for use within the Department of Defense to govern requisition and issue of materiel within standardized priorities. Also called **MILSTRIP.** (JP 4-01)

military standard transportation and movement procedures — Uniform and standard transportation data, documentation, and control procedures applicable to all cargo movements in the Department of Defense transportation system. Also called **MILSTAMP.** (JP 4-01.5)

military technician — A Federal civilian employee providing full-time support to a National Guard, Reserve, or Active Component organization for administration, training, and maintenance of the Selected Reserve. Also called **MILTECH.** (CJCSM 3150.13)

military van (container) — Military-owned, demountable container, conforming to United States and international standards, operated in a centrally controlled fleet for movement of military cargo. Also called **MILVAN.** (JP 4-02)

MILSPEC container — See **military specification container.** (JP 4-09)

mine — 1. In land mine warfare, an explosive or other material, normally encased, designed to destroy or damage ground vehicles, boats, or aircraft, or designed to wound, kill, or otherwise incapacitate personnel and designed to be detonated by the action of its victim, by the passage of time, or by controlled means. 2. In naval mine warfare, an explosive device laid in the water with the intention of damaging or sinking ships or of deterring shipping from entering an area. See also **mine warfare.** (JP 3-15)

mine countermeasures — All methods for preventing or reducing damage or danger from mines. Also called **MCM.** (JP 3-15)

minefield — 1. In land warfare, an area of ground containing mines emplaced with or without a pattern. 2. In naval warfare, an area of water containing mines emplaced with or without a pattern. See also **mine; mine warfare.** (JP 3-15)

minefield record — A complete written record of all pertinent information concerning a minefield, submitted on a standard form by the officer in charge of the emplacement operations. (JP 3-15)

minefield report — An oral, electronic, or written communication concerning mining activities (friendly or enemy) submitted in a standard format by the fastest secure means available. (JP 3-15)

minehunting — Employment of sensor and neutralization systems, whether air, surface, or subsurface, to locate and dispose of individual mines in a known field, or to verify the presence or absence of mines in a given area. See also **minesweeping.** (JP 3-15)

minesweeping — The technique of clearing mines using either mechanical sweeping to remove, disturb, or otherwise neutralize the mine; explosive sweeping to cause sympathetic detonations, damage, or displace the mine; or influence sweeping to produce either the acoustic or magnetic influence required to detonate the mine. See also **minehunting.** (JP 3-15)

mine warfare — The strategic, operational, and tactical use of mines and mine countermeasures either by emplacing mines to degrade the enemy's capabilities to wage land, air, and maritime warfare or by countering of enemy-emplaced mines to permit friendly maneuver or use of selected land or sea areas. Also called **MW.** (JP 3-15)

minimize — A condition wherein normal message and telephone traffic is drastically reduced in order that messages connected with an actual or simulated emergency shall not be delayed. (JP 6-0)

minimum force — Those minimum actions, including the use of armed force, sufficient to bring a situation under control or to defend against hostile act or hostile intent, where the firing of weapons is to be considered as a means of last resort. (JP 3-07.3)

minimum-risk route — A temporary corridor of defined dimensions recommended for use by high-speed, fixed-wing aircraft that presents the minimum known hazards to low-flying aircraft transiting the combat zone. Also called **MRR.** (JP 3-52)

missile defense — Defensive measures designed to destroy attacking enemy missiles, or to nullify or reduce the effectiveness of such attack. (JP 3-01)

missile engagement zone — In air defense, that airspace of defined dimensions within which the responsibility for engagement of air threats normally rests with surface-to-air missile systems. Also called **MEZ.** (JP 3-01)

mission — 1. The task, together with the purpose, that clearly indicates the action to be taken and the reason therefore. (JP 3-0) 2. In common usage, especially when applied to lower military units, a duty assigned to an individual or unit; a task. (JP 3-0) 3. The dispatching of one or more aircraft to accomplish one particular task. (JP 3-30)

mission assignment — The vehicle used by the Department of Homeland Security/Emergency Preparedness and Response/Federal Emergency Management Agency to support federal operations in a Stafford Act major disaster or emergency declaration that orders immediate, short-term emergency response assistance when an applicable state or local government is overwhelmed by the event and lacks the capability to perform, or contract for, the necessary work. (JP 3-28)

mission command — The conduct of military operations through decentralized execution based upon mission-type orders. (JP 3-31)

mission needs statement — A formatted non-system-specific statement containing operational capability needs and written in broad operational terms. Also called **MNS.** (CJCSI 3180.01)

mission-oriented protective posture — A flexible system of protection against chemical, biological, radiological, and nuclear contamination in which personnel are required to wear only that protective clothing and equipment appropriate to the threat level, work rate imposed by the mission, temperature, and humidity. Also called **MOPP.** See also **mission-oriented protective posture gear.** (JP 3-11)

mission-oriented protective posture gear — Military term for individual protective equipment including suit, boots, gloves, mask with hood, first aid treatments, and decontamination kits issued to military members. Also called **MOPP gear.** See also **decontamination; mission-oriented protective posture.** (JP 3-11)

mission statement — A short sentence or paragraph that describes the organization's essential task(s), purpose, and action containing the elements of who, what, when, where, and why. See also **mission.** (JP 5-0)

mission type order — 1. An order issued to a lower unit that includes the accomplishment of the total mission assigned to the higher headquarters. 2. An order to a unit to perform a mission without specifying how it is to be accomplished. (JP 3-50)

mobile mine — A mine propelled to its laying position by propulsion equipment, such as a torpedo that sinks at the end of its run placing the mine. See also **mine.** (JP 3-34)

mobile security force — A dedicated security force designed to defeat Level I and II threats on a base and/or base cluster. Also called **MSF.** (JP 3-10)

mobile training team — A team consisting of one or more US military or civilian personnel sent on temporary duty, often to a foreign nation, to give instruction. The mission of the team is to train indigenous personnel to operate, maintain, and employ weapons and support systems, or to develop a self-training capability in a particular skill. The Secretary of Defense may direct a team to train either military or civilian indigenous personnel, depending upon host-nation requests. Also called **MTT.**

mobility — A quality or capability of military forces which permits them to move from place to place while retaining the ability to fulfill their primary mission. (JP 3-17)

mobility air forces — Air components and Service components that are assigned and/or routinely exercise command authority over mobility operations. Also called **MAF.** (JP 3-17)

mobility corridor — Areas where a force will be canalized due to terrain restrictions. They allow military forces to capitalize on the principles of mass and speed and are therefore relatively free of obstacles. (JP 2-01.3)

mobilization — 1. The act of assembling and organizing national resources to support national objectives in time of war or other emergencies. See also **industrial mobilization.** 2. The process by which the Armed Forces or part of them are brought to a state of readiness for war or other national emergency. This includes activating all or part of the Reserve Component as well as assembling and organizing personnel, supplies, and materiel. Mobilization of the Armed Forces includes but is not limited to the following categories: a. **selective mobilization** — Expansion of the active Armed Forces resulting from action by Congress and/or the President to mobilize Reserve Component units, Individual Ready Reservists, and the resources needed for their support to meet the requirements of a domestic emergency that is not the result of an enemy attack. b. **partial mobilization** — Expansion of the active Armed Forces resulting from action by Congress (up to full mobilization) or by the President (not more than 1,000,000 for not more than 24 consecutive months) to mobilize Ready Reserve Component units, individual reservists, and the resources needed for their support to meet the requirements of a war or other national emergency involving an external threat to the national security. c. **full mobilization** — Expansion of the active Armed Forces resulting from action by Congress and the President to mobilize all Reserve Component units and individuals in the existing approved force structure, as well as all retired military personnel, and the resources needed for their support to meet the requirements of a war or other national emergency involving an external threat to the national security. Reserve personnel can be placed on active duty for the duration of the emergency plus six months. d. **total mobilization** — Expansion of the active Armed Forces resulting from action by Congress and the President to organize and/or generate additional units or personnel beyond the existing force structure, and the resources needed for their support, to meet the total requirements of a war or other national emergency involving an external threat to the national security. Also called **MOB.** (JP 4-05)

mobilization base — The total of all resources available, or that can be made available, to meet foreseeable wartime needs. Such resources include the manpower and materiel resources and services required for the support of essential military, civilian, and survival activities, as well as the elements affecting their state of readiness, such as (but not limited to) the following: manning levels, state of training, modernization of equipment, mobilization materiel reserves and facilities, continuity of government, civil defense plans and preparedness measures, psychological preparedness of the people, international agreements, planning with industry, dispersion, and standby legislation and controls. (JP 4-05)

mobilization exercise — An exercise involving, either completely or in part, the implementation of mobilization plans. (JP 4-05)

mobilization site — The designated location where a Reserve Component unit or individual mobilizes or moves after mobilization for further processing, training, and employment. This differs from a mobilization station in that it is not necessarily a military installation. See also **mobilization; mobilization station; Reserve Component.** (JP 4-05)

mobilization staff officer — The action officer assigned the principle responsibility or additional duties related to Reserve Component mobilization actions. See also **mobilization; Reserve Component.** (JP 4-05)

mobilization station — The designated military installation to which a Reserve Component unit or individual is moved for further processing, organizing, equipping, training, and employment and from which the unit or individual may move to an aerial port of embarkation or seaport of embarkation. See also **mobilization; mobilization site; Reserve Component.** (JP 4-05)

mode (identification, friend or foe) — The number or letter referring to the specific pulse spacing of the signals transmitted by an interrogator or transponder used for radar identification of aircraft. (JP 3-01)

mode of transport — One of the various modes used for a movement. For each mode, there are several means of transport. They are: a. inland surface transportation (rail, road, and inland waterway); b. sea transport (coastal and ocean); c. air transportation; and d. pipelines. (JP 4-09)

Modernized Integrated Database — The national level repository for the general military intelligence available to the entire Department of Defense Intelligence Information System community and, through Global Command and Control System integrated imagery and intelligence, to tactical units. Also called **MIDB.** (JP 2-01)

modified combined obstacle overlay — A joint intelligence preparation of the operational environment product used to portray the militarily significant aspects of the operational environment, such as obstacles restricting military movement, key geography, and military objectives. Also called **MCOO.** See also **joint intelligence preparation of the operational environment.** (JP 2-01.3)

moored mine — A contact or influence-operated mine of positive buoyancy held below the surface by a mooring attached to a sinker or anchor on the bottom. See also **mine.** (JP 3-15)

morale, welfare, and recreation — The merging of multiple unconnected disciplines into programs which improve unit readiness, promote fitness, build unit morale and cohesion, enhance quality of life, and provide recreational, social, and other support services. Also called **MWR.** (JP 1-0)

mortuary affairs — Provides for the search for, recovery, identification, preparation, and disposition of human remains of persons for whom the Services are responsible by status and executive order. Also called **MA.** See also **joint mortuary affairs office.** (JP 4-06)

mounting —1. All preparations made in areas designated for the purpose, in anticipation of an operation. It includes the assembly in the mounting area, preparation and maintenance within the mounting area, movement to loading points, and subsequent embarkation into ships, craft, or aircraft if applicable. 2. A carriage or stand upon which a weapon is placed. (JP 3-02.1)

mounting area — A general locality where assigned forces of an amphibious or airborne operation, with their equipment, are assembled, prepared, and loaded in shipping and/or aircraft preparatory to an assault. See also **embarkation area.** (JP 3-02.1)

movement control — The planning, routing, scheduling, and control of personnel and cargo movements over lines of communications; includes maintaining in-transit visibility of forces and material through the deployment and/or redeployment process. See also **line of communications; movement control teams; non-unit cargo; non-unit-related personnel.** (JP 4-01.5)

movement control team — An Army team used to decentralize the execution of movement responsibilities on an area basis or at key transportation nodes. Also called **MCT.** (JP 4-09)

movement data — Those essential elements of information to schedule lift, obtain transportation assets, manage movement of forces, and report in-transit visibility of movements and associated forces (people, equipment, and supplies). (JP 4-09)

movement group — Those ships and embarked units that load out and proceed to rendezvous in the objective area. (JP 3-02)

movement phase — In amphibious operations, the period during which various elements of the amphibious force move from points of embarkation to the operational area. This move may be via rehearsal, staging, or rendezvous areas. The movement phase is completed when the various elements of the amphibious force arrive at their assigned positions in the operational area. See also **amphibious force; amphibious operation.** (JP 3-02)

movement plan — In amphibious operations, the naval plan providing for the movement of the amphibious task force to the objective area. It includes information and instructions concerning departure of ships from embarkation points, the passage at sea, and the approach to and arrival in assigned positions in the objective area. See also **amphibious operation; amphibious task force.** (JP 3-02)

movement requirement — A stated movement mode and time-phased need for the transport of units, personnel, and/or materiel from a specified origin to a specified destination. (JP 4-09)

movement schedule — A schedule developed to monitor or track a separate entity, whether it is a force requirement, cargo or personnel increment, or lift asset. The schedule reflects the assignment of specific lift resources (such as an aircraft or ship) that will be used to move the personnel and cargo included in a specific movement increment. Arrival and departure times at ports of embarkation, etc., are detailed to show a flow and workload at each location. Movement schedules are detailed enough to support plan implementation. (JP 4-09)

movement table — A table giving detailed instructions or data for a move. When necessary it will be qualified by the words road, rail, sea, air, etc., to signify the type of movement. Normally issued as an annex to a movement order or instruction. (JP 4-09)

movement to contact — A form of the offense designed to develop the situation and to establish or regain contact. (JP 3-50)

multinational — Between two or more forces or agencies of two or more nations or coalition partners. See also **alliance; coalition.** (JP 5-0)

multinational doctrine — The agreed upon fundamental principles that guide the employment of forces of two or more nations in coordinated action toward a common objective. See also **doctrine; joint doctrine.** (JP 3-16)

multinational force — A force composed of military elements of nations who have formed an alliance or coalition for some specific purpose. Also called **MNF.** See also **multinational force commander; multinational operations.** (JP 1)

multinational force commander — A general term applied to a commander who exercises command authority over a military force composed of elements from two or more nations. Also called **MNFC.** See also **multinational force.** (JP 3-16)

multinational integrated logistic unit — An organization resulting when two or more nations agree to provide logistics assets to a multinational logistic force under the operational control of a multinational commander for the logistic support of a multinational force. Also called **MILU**. See also **logistic support; multinational.** (JP 4-08)

multinational logistics — Any coordinated logistic activity involving two or more nations supporting a multinational force conducting military operations under the auspices of an alliance or coalition, including those conducted under United Nations mandate. Also called **MNL**. See also **logistics; multinational.** (JP 4-08)

multinational operations — A collective term to describe military actions conducted by forces of two or more nations, usually undertaken within the structure of a coalition or alliance. See also **alliance; coalition.** (JP 3-16)

multinational staff — A staff composed of personnel of two or more nations within the structure of a coalition or alliance. See also **integrated staff; joint staff.** (JP 3-16)

multipoint refueling system — KC-135 aircraft equipped with external wing-mounted pods to conduct drogue air refueling, while still maintaining boom air refueling capability on the same mission. Also called **MPRS**. See also **air refueling.** (JP 3-17)

multi-Service publication — A publication containing principles, terms, tactics, techniques, and procedures used and approved by the forces of two or more Services to perform a common military function consistent with approved joint doctrine. (CJCSM 5120.01)

multispectral imagery — The image of an object obtained simultaneously in a number of discrete spectral bands. Also called **MSI**. (JP 3-14)

multispot ship — Those ships certified to have two or more adjacent landing areas. See also **spot.** (JP 3-04)

munitions effectiveness assessment — Conducted concurrently and interactively with battle damage assessment, the assessment of the military force applied in terms of the weapon system and munitions effectiveness to determine and recommend any required changes to the methodology, tactics, weapon system, munitions, fusing, and/or weapon delivery parameters to increase force effectiveness. Munitions effectiveness assessment is primarily the responsibility of operations with required inputs and coordination from the intelligence community. Also called **MEA**. See also **assessment; battle damage assessment.** (JP 2-01)

mutual support — That support which units render each other against an enemy, because of their assigned tasks, their position relative to each other and to the enemy, and their inherent capabilities. See also **close support; direct support; support.** (JP 3-31)

N

named area of interest — The geospatial area or systems node or link against which information that will satisfy a specific information requirement can be collected. Named areas of interest are usually selected to capture indications of adversary courses of action, but also may be related to conditions of the operational environment. Also called **NAI.** See also **area of interest.** (JP 2-01.3)

nap-of-the-earth flight — See **terrain flight.**

narcoterrorism — Terrorism that is linked to illicit drug trafficking. (JP 3-07.4)

National Capital Region — A geographic area encompassing the District of Columbia and eleven local jurisdictions in the State of Maryland and the Commonwealth of Virginia. Also called **NCR.** (JP 3-28)

National Communications System — The telecommunications system that results from the technical and operational integration of the separate telecommunications systems of the several executive branch departments and agencies having a significant telecommunications capability. Also called **NCS.** (JP 6-0)

National Defense Reserve Fleet — 1. Including the Maritime Administration Ready Reserve Force, a fleet composed of ships acquired and maintained by the Maritime Administration for use in mobilization or emergency. 2. Less the Maritime Administration Ready Reserve Force, a fleet composed of the older dry cargo ships, tankers, troop transports, and other assets in Maritime Administration's custody that are maintained at a relatively low level of readiness. Also called **NDRF.** See also **Maritime Administration Ready Reserve Force.** (JP 4-01.2)

national defense strategy — A document approved by the Secretary of Defense for applying the Armed Forces of the United States in coordination with Department of Defense agencies and other instruments of national power to achieve national security strategy objectives. Also called **NDS.** (JP 1)

national detainee reporting center — National-level center that obtains and stores information concerning enemy prisoners of war, civilian internees, and retained personnel and their confiscated personal property. May be established upon the outbreak of an armed conflict or when persons are captured or detained by U.S. military forces in the course of the full range of military operations. Accounts for all persons who pass through the care, custody, and control of the U.S. Department of Defense. Also called **NDRC.** (JP 3-63)

National Disaster Medical System — A coordinated partnership between Departments of Homeland Security, Health and Human Services, Defense, and Veterans Affairs established for the purpose of responding to the needs of casualties of a public health emergency. Also called **NDMS.** (JP 3-41)

national emergency — A condition declared by the President or the Congress by virtue of powers previously vested in them that authorize certain emergency actions to be undertaken in the national interest. See also **mobilization.** (JP 3-28)

National Incident Management System — A national crisis response system that provides a consistent, nationwide approach for federal, state, local, and tribal governments; the private sector; and nongovernmental organizations to work effectively and efficiently together to prepare for, respond to, and recover from domestic incidents, regardless of cause, size, or complexity. Also called **NIMS.** (JP 3-41)

national intelligence — All intelligence, regardless of the source from which derived, and including that which is gathered within or outside of the United States, that pertains to more than one agency, and involves (1) threats to the United States, its people, property, or interests, (2) the development, proliferation, or use of weapons of mass destruction, or (3) any other matter bearing on US national or homeland security. (JP 2-01)

national intelligence estimate — A strategic estimate of the capabilities, vulnerabilities, and probable courses of action of foreign nations produced at the national level as a composite of the views of the intelligence community. Also called **NIE.** (JP 2-01)

National Military Command System — The priority component of the Global Command and Control System designed to support the President, Secretary of Defense and Joint Chiefs of Staff in the exercise of their responsibilities. Also called **NMCS.** (JP 6-0)

national military strategy — A document approved by the Chairman of the Joint Chiefs of Staff for distributing and applying military power to attain national security strategy and national defense strategy objectives. Also called **NMS.** See also **national security strategy; strategy; theater strategy.** (JP 1)

national operations center — The primary national hub for domestic incident management operational coordination and situational awareness. Also called **NOC.** (JP 3-28)

national policy — A broad course of action or statements of guidance adopted by the government at the national level in pursuit of national objectives. (JP 1)

national security — A collective term encompassing both national defense and foreign relations of the United States with the purpose of gaining: a. A military or defense advantage over any foreign nation or group of nations; b. A favorable foreign relations position; or c. A defense posture capable of successfully resisting hostile or destructive action from within or without, overt or covert. See also **security.** (JP 1)

National Security Agency/Central Security Service Representative — The senior theater or military command representative of the Director, National Security Agency/Chief, Central Security Service in a specific country or military command headquarters who provides the Director, National Security Agency, with information on command plans requiring cryptologic support. The National Security Agency/Central Security Service representative serves as a special advisor to the combatant commander for cryptologic matters, to include signals intelligence, communications security, and computer security. Also called **NCR.** See also **counterintelligence.** (JP 2-01.2)

National Security Council — A governmental body specifically designed to assist the President in integrating all spheres of national security policy. Also called **NSC.** (JP 1)

national security interests — The foundation for the development of valid national objectives that define United States goals or purposes. (JP 1)

national security strategy — A document approved by the President of the United States for developing, applying, and coordinating the instruments of national power to achieve objectives that contribute to national security. Also called **NSS.** See also **national military strategy; strategy; theater strategy.** (JP 1)

national shipping authority — The organization within each Allied government responsible in time of war for the direction of its own merchant shipping. Also called **NSA.** (JP 4-01.2)

national special security event — A designated event that, by virtue of its political, economic, social, or religious significance, may be the target of terrorism or other criminal activity. Also called **NSSE.** (JP 3-28)

national stock number — The 13-digit stock number replacing the 11-digit federal stock number. It consists of the 4-digit federal supply classification code and the 9-digit national item identification number. The national item identification number consists of a 2-digit National Codification Bureau number designating the central cataloging office (whether North Atlantic Treaty Organization or other friendly country) that assigned the number and a 7-digit (xxx-xxxx) nonsignificant number. The number shall be arranged as follows: 9999-00-999-9999. Also called **NSN.** (JP 4-09)

national support element — Any national organization or activity that supports national forces that are a part of a multinational force. See also **multinational force; support.** (JP 1)

National System for Geospatial Intelligence — The combination of technology, policies, capabilities, doctrine, activities, people, data, and organizations necessary to produce geospatial intelligence in an integrated, multi-intelligence environment. Also called **NSG.** (JP 2-03)

nation assistance — Assistance rendered to a nation by foreign forces within that nation's territory based on agreements mutually concluded between nations. (JP 3-0)

natural disaster — An emergency situation posing significant danger to life and property that results from a natural cause. See also **domestic emergencies.** (JP 3-29)

naval advanced logistic support site — An overseas location used as the primary transshipment point in the theater of operations for logistic support. Also called **NALSS.** See also **logistic support; naval forward logistic site; support; theater of operations.** (JP 3-35)

Naval Air Training and Operating Procedures Standardization manual — Series of general and specific aircraft procedural manuals that govern the operations of naval aircraft. Also called **NATOPS manual.** (JP 3-04)

naval beach group — A permanently organized naval command within an amphibious force composed of a commander and staff, a beachmaster unit, an amphibious construction battalion, and assault craft units, designed to provide an administrative group from which required naval tactical components may be made available to the attack force commander and to the amphibious landing force commander. Also called **NBG.** See also **shore party.** (JP 3-02)

naval construction force — The combined construction units of the Navy that are part of the operating forces and represent the Navy's capability for advanced base construction. Also called **NCF.** (JP 3-34)

naval forward logistic site — An overseas location, with port and airfield facilities nearby, which provides logistic support to naval forces within the theater of operations during major contingency and wartime periods. Also called **NFLS.** See also **logistic support; naval advanced logistic support site; staging.** (JP 3-35)

naval gunfire support — Fire provided by Navy surface gun systems in support of a unit or units tasked with achieving the commander's objectives. A subset of naval surface fire support. Also called **NGFS.** See also **naval surface fire support.** (JP 3-09)

naval operation — 1. A naval action (or the performance of a naval mission) that may be strategic, operational, tactical, logistic, or training. 2. The process of carrying on or training for naval combat in order to gain the objectives of any battle or campaign. (JP 3-32)

naval special warfare — A naval warfare specialty that conducts special operations with an emphasis on maritime, coastal, and riverine environments using small, flexible, mobile units operating under, on, and from the sea. Also called **NSW.** (JP 3-05)

naval special warfare forces — Those Active and Reserve Component Navy forces designated by the Secretary of Defense that are specifically organized, trained, and equipped to conduct and support special operations. Also called **NSW forces** or **NAVSOF.** (JP 3-05.1)

naval special warfare group — A permanent Navy echelon III major command to which most naval special warfare forces are assigned for some operational and all administrative purposes. It consists of a group headquarters with command and control, communications, and support staff; sea-air-land teams; and sea-air-land team delivery vehicle teams. Also called **NSWG.** (JP 3-05.1)

naval special warfare task element — A provisional subordinate element of a naval special warfare task unit, employed to extend the command and control and support capabilities of its parent task unit. Also called **NSWTE.** See also **naval special warfare task unit.** (JP 3-05.1)

naval special warfare task group — A provisional naval special warfare organization that plans, conducts, and supports special operations in support of fleet commanders and joint force special operations component commanders. Also called **NSWTG.** (JP 3-05.1)

naval special warfare task unit — A provisional subordinate unit of a naval special warfare task group. Also called **NSWTU.** See also **naval special warfare task group.** (JP 3-05.1)

naval special warfare unit — A permanent Navy organization forward based to control and support attached naval special warfare forces. Also called **NSWU.** (JP 3-05.1)

naval surface fire support — Fire provided by Navy surface gun and missile systems in support of a unit or units. Also called **NSFS.** See also **fire support.** (JP 3-09.3)

navigation warfare — Deliberate defensive and offensive action to assure and prevent positioning, navigation, and timing information through coordinated employment of space, cyberspace, and electronic warfare operations. Also called **NAVWAR.** (JP-3-14)

Navy cargo-handling battalion — A mobile logistic support unit capable of worldwide deployment in its entirety or in specialized detachments. It is organized, trained, and equipped to: a. load and offload Navy and Marine Corps cargo carried in maritime pre-positioning ships and merchant breakbulk or container ships in all environments; b. operate an associated temporary ocean cargo terminal; c. load and offload Navy and Marine Corps cargo carried in military-controlled aircraft; and d. operate an associated expeditionary air cargo terminal. Also called **NCHB or Navy CHB.** See also **maritime pre-positioning ships.** (JP 3-02.1)

Navy expeditionary logistics support group — The quick response cargo-handling units of the Navy specialize in open ocean cargo handling. Also called **NAVELSG.** (JP 4-01.6)

Navy special operations component — The Navy component of a joint force special operations component. Also called **NAVSOC.** See also **Air Force special operations component; Army special operations component.** (JP 3-05.1)

Navy support element — The maritime pre-positioning force element that is composed of naval beach group staff and subordinate unit personnel, a detachment of Navy cargo handling force personnel, and other Navy components, as required. It is tasked with conducting the off-load and ship-to-shore movement of maritime pre-positioned equipment and/or supplies. Also called **NSE.** (JP 3-02.1)

Navy tactical air control center — See **tactical air control center.** (JP 3-09.3)

Navy-unique fleet essential aircraft — Combatant commander-controlled airlift assets deemed essential for providing air transportation in support of naval operations' transportation requirements. Also called **NUFEA.** (JP 3-17)

near miss (aircraft) — Any circumstance in flight where the degree of separation between two aircraft is considered by either pilot to have constituted a hazardous situation involving potential risk of collision.

need to know — A criterion used in security procedures that requires the custodians of classified information to establish, prior to disclosure, that the intended recipient must have access to the information to perform his or her official duties. (JP 2-01.2)

negation — In space operations, active and offensive measures to deceive, disrupt, degrade, deny or destroy space capabilities being used to interfere with or attack United States/allied systems. See also **space control.** (JP 3-14)

nerve agent — A potentially lethal chemical agent that interferes with the transmission of nerve impulses. (JP 3-11)

net explosive weight — The actual weight in pounds of explosive mixtures or compounds, including the trinitrotoluene equivalent of energetic material, that is used in determination of explosive limits and explosive quantity data arcs. Also called **NEW.** (JP 4-09)

networked munitions. Remotely controlled, interconnected, weapons system designed to provide rapidly emplaced ground-based countermobility and protection capability through scalable application of lethal and nonlethal means. (JP 3-15)

network operations — Activities conducted to operate and defend the Global Information Grid. Also called **NETOPS.** (JP 6-0)

neutral — In combat and combat support operations, an identity applied to a track whose characteristics, behavior, origin, or nationality indicate that it is neither supporting nor opposing friendly forces. See also **suspect; unknown.** (JP 3-0)

neutrality — In international law, the attitude of impartiality during periods of war adopted by third states toward a belligerent and subsequently recognized by the belligerent, which creates rights and duties between the impartial states and the belligerent. (JP 3-0)

neutralize — 1. As pertains to military operations, to render ineffective or unusable. 2. To render enemy personnel or materiel incapable of interfering with a particular operation. 3. To render safe mines, bombs, missiles, and booby traps. 4. To make harmless anything contaminated with a chemical agent. (JP 3-0)

neutral state — In international law, a state that pursues a policy of neutrality during war. See also **neutrality.** (JP 3-50)

news media representative — An individual employed by a civilian radio or television station, newspaper, newsmagazine, periodical, or news agency to gather and report on a newsworthy event. Also called **NMR.** See also **public affairs.** (JP 3-61)

night vision device — Any electro-optical device that is used to detect visible and infrared energy and provide a visible image. Night vision goggles, forward-looking infrared, thermal sights, and low-light level television are night vision devices. Also called **NVD.** See also **forward-looking infrared; night vision goggles(s).** (JP 3-09.3)

night vision goggle(s) — An electro-optical image intensifying device that detects visible and near-infrared energy, intensifies the energy, and provides a visible image for night viewing. Night vision goggles can be either hand-held or helmet-mounted. Also called **NVG.** See also **night vision device.** (JP 3-09.3)

node — 1. A location in a mobility system where a movement requirement is originated, processed for onward movement, or terminated. (JP 3-17) 2. In communications and computer systems, the physical location that provides terminating, switching, and gateway access services to support information exchange. (JP 6-0) 3. An element of a system that represents a person, place, or physical thing. (JP 3-0)

no-fire area — An area designated by the appropriate commander into which fires or their effects are prohibited. Also called **NFA.** See also **fires.** (JP 3-09.3)

nonappropriated funds — Funds generated by Department of Defense personnel and their dependents used to augment funds appropriated by the Congress to provide a comprehensive, morale-building welfare, religious, educational, and recreational programs. Also called **NAF.** (JP 1-0)

nonbattle injury — A person who becomes a casualty due to circumstances not directly attributable to hostile action or terrorist activity. Also called **NBI.** (JP 4-02)

noncombatant evacuation operations — Operations directed by the Department of State or other appropriate authority, in conjunction with the Department of Defense, whereby noncombatants are evacuated from foreign countries when their lives are endangered by war, civil unrest, or natural disaster to safe havens as designated by the Department of State. Also called **NEOs.** See also **evacuation; noncombatant evacuees; operation; safe haven.** (JP 3-68)

noncombatant evacuation operations tracking system — An automated data processing hardware and software package that has the capability to provide evacuee in-transit visibility to combatant commanders and senior leadership during the conduct of a noncombatant evacuation operation. Also called **NTS.** (JP 3-68)

noncombatant evacuees — 1. US citizens who may be ordered to evacuate by competent authority include: a. civilian employees of all agencies of the US Government and their dependents, except as noted in 2a below; b. military personnel of the Armed Forces of the United States specifically designated for evacuation as noncombatants; and c. dependents of members of the Armed Forces of the United States. 2. US (and non-US) citizens who may be authorized or assisted (but not necessarily ordered to evacuate) by competent authority include: a. civilian employees of US Government agencies and their dependents, who are residents in the country concerned on their own volition, but express the willingness to be evacuated; b. private US citizens and their dependents; c. military personnel and dependents of members of the Armed Forces of the United States outlined in 1c above, short of an ordered evacuation; and d. designated personnel, including dependents of persons listed in 1a through 1c above, as prescribed by the Department of State. See also **noncombatant evacuation operations.** (JP 3-68)

nonconventional assisted recovery —˙ Personnel recovery conducted by indigenous/surrogate personnel that are trained, supported, and led by special operations forces, unconventional warfare ground and maritime forces, or other government agencies' personnel that have been specifically trained and directed to establish and operate indigenous or surrogate infrastructures. Also called **NAR.** (JP 3-50)

nondestructive electronic warfare — Those electronic warfare actions, not including employment of wartime reserve modes, that deny, disrupt, or deceive rather than damage or destroy. See also **electronic warfare.** (JP 3-13.1)

nongovernmental organization — A private, self-governing, not-for-profit organization dedicated to alleviating human suffering; and/or promoting education, health care, economic development, environmental protection, human rights, and conflict resolution; and/or encouraging the establishment of democratic institutions and civil society. Also called **NGO.** (JP 3-08)

nonlethal weapon — A weapon that is explicitly designed and primarily employed so as to incapacitate personnel or materiel, while minimizing fatalities, permanent injury to personnel, and undesired damage to property and the environment. Also called **NLW.** (JP 3-28)

nonpersistent agent — A chemical agent that when released dissipates and/or loses its ability to cause casualties after 10 to 15 minutes. (JP 3-11)

nonproliferation — Actions to prevent the proliferation of weapons of mass destruction by dissuading or impeding access to, or distribution of, sensitive technologies, material, and expertise. Also called **NP.** See also **counterproliferation.** (JP 3-40)

nonscheduled units — Units of the landing force held in readiness for landing during the initial unloading period, but not included in either scheduled or on-call waves. This category usually includes certain of the combat support units and most of the combat service support units with higher echelon (division and above) reserve units of the landing force. Their landing is directed when the need ashore can be predicted with a reasonable degree of accuracy. (JP 3-02)

non-unit cargo — All equipment and supplies requiring transportation to an operational area, other than those identified as the equipment or accompanying supplies of a specific unit. (JP 4-01.5).

non-unit-related personnel — All personnel requiring transportation to or from an operational area, other than those assigned to a specific unit. Also called **NRP** or **NUP.** (JP 1-0)

no-strike list — A list of objects or entities characterized as protected from the effects of military operations under international law and/or rules of engagement. Also called **NSL.** See also **law of armed conflict.** (JP 3-60)

not mission capable, maintenance — Material condition indicating that systems and equipment are not capable of performing any of their assigned missions because of maintenance requirements. Also called **NMCM.** See also **not mission capable, supply.**

not mission capable, supply — Material condition indicating that systems and equipment are not capable of performing any of their assigned missions because of maintenance work stoppage due to a supply shortage. Also called **NMCS.** See also **not mission capable, maintenance.**

nuclear incident — An unexpected incident involving a nuclear weapon, facility, or component, but not constituting a nuclear weapon(s) accident, resulting in any of the following: a. an increase in the possibility of explosion or radioactive contamination; b. errors committed in the assembly, testing, loading, or transportation of equipment, and/or the malfunctioning of equipment and materiel which could lead to an unintentional operation of all or part of the weapon arming and/or firing sequence, or which could lead to a substantial change in yield, or increased dud probability; and c. any act of God, unfavorable environment, or condition resulting in damage to the weapon, facility, or component. (JP 3-41)

nuclear reactor — A facility in which fissile material is used in a self-supporting chain reaction (nuclear fission) to produce heat and/or radiation for both practical application and research and development. (JP 3-40)

nuclear weapon(s) accident — An unexpected incident involving nuclear weapons or radiological nuclear weapon components that results in any of the following; a. accidental or unauthorized launching, firing, or use by United States forces or United States supported allied forces of a nuclear-capable weapon system that could create the risk of an outbreak of war; b. nuclear detonation; c. nonnuclear detonation or burning of a nuclear weapon or radiological nuclear weapon component; d. radioactive contamination; e. seizure, theft, loss, or destruction of a nuclear weapon or radiological nuclear weapon component, including jettisoning; and f. public hazard, actual or implied. (JP 3-41)

nuisance minefield — A minefield laid to delay and disorganize the enemy and to hinder the use of an area or route. See also **minefield**. (JP 3-15)

numbered beach — In amphibious operations, a subdivision of a colored beach, designated for the assault landing of a battalion landing team or similarly sized unit, when landed as part of a larger force. (JP 3-02)

numbered fleet — A major tactical unit of the Navy immediately subordinate to a major fleet command and comprising various task forces, elements, groups, and units for the purpose of prosecuting specific naval operations. See also **fleet**. (JP 3-32)

numbered wave — See **wave.**

O

objective — 1. The clearly defined, decisive, and attainable goal toward which every operation is directed. 2. The specific target of the action taken which is essential to the commander's plan. See also **target.** (JP 5-0)

objective area — A geographical area, defined by competent authority, within which is located an objective to be captured or reached by the military forces. Also called **OA.** (JP 3-06)

obstacle — Any natural or man-made obstruction designed or employed to disrupt, fix, turn, or block the movement of an opposing force, and to impose additional losses in personnel, time, and equipment on the opposing force. (JP 3-15)

obstacle belt — A brigade-level command and control measure, normally given graphically, to show where within an obstacle zone the ground tactical commander plans to limit friendly obstacle employment and focus the defense. See also **obstacle.** (JP 3-15)

obstacle clearing — The total elimination or neutralization of obstacles. (JP 3-15)

obstacle intelligence — Those collection efforts to detect the presence of enemy and natural obstacles, determine their types and dimensions, and provide the necessary information to plan appropriate combined arms breaching, clearance, or bypass operations to negate the impact on the friendly scheme of maneuver. Also called **OBSTINT.** (JP 3-15)

obstacle restricted areas — A command and control measure used to limit the type or number of obstacles within an area. See also **obstacle.** (JP 3-15)

obstacle zone — A division-level command and control measure, normally done graphically, to designate specific land areas where lower echelons are allowed to employ tactical obstacles. See also **obstacle.** (JP 3-15)

occupational and environmental health surveillance — The regular or repeated collection, analysis, archiving, interpretation, and dissemination of occupational and environmental health-related data for monitoring the health of, or potential health hazard impact on, a population and individual personnel, and for intervening in a timely manner to prevent, treat, or control the occurrence of disease or injury when determined necessary. (JP 4-02)

occupational and environmental health threats — Threats to the health of military personnel and to military readiness created by exposure to hazardous agents, environmental contamination, or toxic industrial materials. See also **health threat.** (JP 4-02)

occupied territory — Territory under the authority and effective control of a belligerent armed force and not being administered pursuant to peace terms, treaty, or other agreement, express or implied, with the civil authority of the territory. (JP 4-02)

oceanography — The study of the sea, embracing and integrating all knowledge pertaining to the sea and its physical boundaries, the chemistry and physics of seawater, and marine biology. (JP 3-59)

offensive counterair — Offensive operations to destroy, disrupt, or neutralize enemy aircraft, missiles, launch platforms, and their supporting structures and systems both before and after launch, and as close to their source as possible. Also called **OCA.** See also **counterair; defensive counterair; operation.** (JP 3-01)

offensive counterair attack operations — Offensive action by any part of the joint force in support of the offensive counterair mission against surface targets which contribute to the enemy's air and missile capabilities. Also called **OCA attack operations.** See also **counterair; offensive counterair.** (JP 3-01)

offensive counterintelligence operation — A clandestine counterintelligence activity conducted for military, strategic, Department of Defense, or national counterintelligence and security purposes against a target having suspected or known affiliation with foreign intelligence entities, international terrorism, or other foreign persons or organizations, to counter terrorism, espionage, or other clandestine intelligence activities that threaten the security of the Department or the United States. The two types of offensive counterintelligence operations are double agent operation and controlled source operation. Also called **OFCO.** (JP 2-01.2)

offensive cyberspace operations — Cyberspace operations intended to project power by the application of force in or through cyberspace. Also called **OCO.** (JP 3-12)

offensive space control — Those operations to prevent an adversary's hostile use of United States/third-party space capabilities and services or negate (deceive, disrupt, degrade, deny, or destroy) an adversary's efforts to interfere with or attack United States/allied space systems. Also called **OSC.** (JP 3-14)

office — An enduring organization that is formed around a specific function within a joint force commander's headquarters to coordinate and manage support requirements. (JP 3-33)

officer in tactical command — In maritime usage, the senior officer present eligible to assume command, or the officer to whom the senior officer has delegated tactical command. Also called **OTC.** (JP 3-32)

officer of the deck — 1. When underway, the officer designated by the commanding officer to be in charge of the ship, including its safe and proper operation. 2. When in port or at anchor, the officer of the deck is designated by the command duty officer, has similar responsibilities, and may be enlisted. Also called **OOD.** (JP 3-04)

official information — Information that is owned by, produced for or by, or is subject to the control of the United States Government. (JP 3-61)

offset costs — Costs for which funds have been appropriated that may not be incurred as a result of a contingency operation. See also **contingency operation.** (JP 1-06)

offshore bulk fuel system — The system used for transferring fuel from points offshore to reception facilities on the beach. Also called **OBFS.** See also **amphibious bulk liquid transfer system; offshore petroleum discharge system.** (JP 4-01.6)

offshore petroleum discharge system — Provides bulk transfer of petroleum directly from an offshore tanker to a beach termination unit located immediately inland from the high watermark. Bulk petroleum then is either transported inland or stored in the beach support area. Also called **OPDS.** See also **facility; petroleum, oils, and lubricants; single-anchor leg mooring.** (JP 4-03)

off-the-shelf item — An item that has been developed and produced to military or commercial standards and specifications, is readily available for delivery from an industrial source, and may be procured without change to satisfy a military requirement.

on-call — 1. A term used to signify that a prearranged concentration, air strike, or final protective fire may be called for. 2. Preplanned, identified force or materiel requirements without designated time-phase and destination information. (JP 3-01)

on-call resupply — A resupply mission planned before insertion of a special operations team into the operations area but not executed until requested by the operating team. See also **emergency resupply.** (JP 3-05.1)

on-call target — Planned target upon which fires or other actions are determined using deliberate targeting and triggered, when detected or located, using dynamic targeting. See also **dynamic targeting; on-call; operational area; planned target; target.** (JP 3-60)

on-call wave — See **wave.**

on hand — The quantity of an item that is physically available in a storage location and contained in the accountable property book records of an issuing activity. (JP 4-09)

on-scene commander — 1. An individual in the immediate vicinity of an isolating event who temporarily assumes command of the incident. 2. The federal officer designated to direct federal crisis and consequence management efforts at the scene of a terrorist or weapons of mass destruction incident. Also called **OSC.** (JP 3-50)

on-station time — The time an aircraft can remain on station, that may be determined by endurance or orders. (JP 3-50)

open ocean — Ocean limit defined as greater than 12 nautical miles from shore, as compared with high seas that are over 200 nautical miles from shore. See also **contiguous zone.** (JP 3-32)

open-source information — Information that any member of the public could lawfully obtain by request or observation as well as other unclassified information that has limited public distribution or access. (JP 2-0)

open-source intelligence — Relevant information derived from the systematic collection, processing, and analysis of publicly available information in response to known or anticipated intelligence requirements. Also called **OSINT.** See also **intelligence.** (JP 2-0)

operation — 1. A sequence of tactical actions with a common purpose or unifying theme. (JP 1) 2. A military action or the carrying out of a strategic, operational, tactical, service, training, or administrative military mission. (JP 3-0).

operational approach — A description of the broad actions the force must take to transform current conditions into those desired at end state. (JP 5-0)

operational area — An overarching term encompassing more descriptive terms (such as area of responsibility and joint operations area) for geographic areas in which military operations are conducted. Also called **OA.** See also **amphibious objective area; area of operations; area of responsibility; joint operations area; joint special operations area; theater of operations; theater of war.** (JP 3-0)

operational art — The cognitive approach by commanders and staffs — supported by their skill, knowledge, experience, creativity, and judgment — to develop strategies, campaigns, and operations to organize and employ military forces by integrating ends, ways, and means. (JP 3-0)

operational characteristics — Those military characteristics that pertain primarily to the functions to be performed by equipment, either alone or in conjunction with other equipment; e.g., for electronic equipment, operational characteristics include such items as frequency coverage, channeling, type of modulation, and character of emission. (JP 5-0)

operational contract support — The process of planning for and obtaining supplies, services, and construction from commercial sources in support of joint operations along with the associated contractor management functions. (JP 4-10)

operational control — The authority to perform those functions of command over subordinate forces involving organizing and employing commands and forces, assigning tasks, designating objectives, and giving authoritative direction necessary to accomplish the mission. Also called **OPCON.** See also **combatant command; combatant command (command authority); tactical control.** (JP 1)

operational control authority — The naval commander responsible within a specified geographical area for the naval control of all merchant shipping under Allied naval control. Also called **OCA.** (JP 3-15)

operational decontamination — Decontamination carried out by an individual and/or a unit, restricted to specific parts of operationally essential equipment, materiel and/or working areas, in order to minimize contact and transfer hazards and to sustain operations. See also **decontamination; immediate decontamination; thorough decontamination.** (JP 3-11)

operational design — The conception and construction of the framework that underpins a campaign or major operation plan and its subsequent execution. See also **campaign; major operation.** (JP 5-0)

operational design element — A key consideration used in operational design. (JP 5-0)

operational energy — The energy required for training, moving, and sustaining military forces and weapons platforms for military operations. (JP 4-0)

operational environment — A composite of the conditions, circumstances, and influences that affect the employment of capabilities and bear on the decisions of the commander. Also called **OE.** (JP 3-0)

operational exposure guidance — The maximum amount of nuclear/external ionizing radiation that the commander considers a unit may be permitted to receive while performing a particular mission or missions. Also called **OEG.** See also **radiation exposure status.** (JP 3-11)

operational intelligence — Intelligence that is required for planning and conducting campaigns and major operations to accomplish strategic objectives within theaters or operational areas. See also **intelligence; strategic intelligence; tactical intelligence.** (JP 2-0)

operational level of war — The level of war at which campaigns and major operations are planned, conducted, and sustained to achieve strategic objectives within theaters or other operational areas. See also **strategic level of war; tactical level of war.** (JP 3-0)

operational limitation — An action required or prohibited by higher authority, such as a constraint or a restraint, and other restrictions that limit the commander's freedom of action, such as diplomatic agreements, rules of engagement, political and economic conditions in affected countries, and host nation issues. See also **constraint; restraint.** (JP 5-0)

operationally ready — 1. A unit, ship, or weapon system capable of performing the missions or functions for which organized or designed. 2. Personnel available and qualified to perform assigned missions or functions. (JP 4-01.5)

operational necessity — A mission associated with war or peacetime operations in which the consequences of an action justify the risk of loss of aircraft and crew. See also **mission.** (JP 3-04)

operational pause — A temporary halt in operations. (JP 5-0)

operational preparation of the environment — The conduct of activities in likely or potential areas of operations to prepare and shape the operational environment. Also called **OPE.** (CJCS interim)

operational reach — The distance and duration across which a joint force can successfully employ military capabilities. (JP 3-0)

operational readiness — The capability of a unit/formation, ship, weapon system, or equipment to perform the missions or functions for which it is organized or designed. Also called **OR.** See also **combat readiness.** (JP 1-0)

operational reserve — An emergency reserve of men and/or materiel established for the support of a specific operation. (JP 5-0)

operational support airlift — Airlift movements of high-priority passengers and cargo with time, place, or mission-sensitive requirements. Also called **OSA.** (JP 3-17)

operational testing — A continuing process of evaluation that may be applied to either operational personnel or situations to determine their validity or reliability. (JP 4-02)

operation and maintenance — Maintenance and repair of real property, operation of utilities, and provision of other services such as refuse collection and disposal, entomology, snow removal, and ice alleviation. Also called **O&M.** (JP 3-34)

operation order — A directive issued by a commander to subordinate commanders for the purpose of effecting the coordinated execution of an operation. Also called **OPORD.** (JP 5-0)

operation plan — 1. Any plan for the conduct of military operations prepared in response to actual and potential contingencies. 2. A complete and detailed joint plan containing a full description of the concept of operations, all annexes applicable to the plan, and a time-phased force and deployment data. Also called **OPLAN.** See also **operation order.** (JP 5-0)

operations center — The facility or location on an installation, base, or facility used by the commander to command, control, and coordinate all operational activities. Also called **OC.** See also **base defense operations center.** (JP 3-07.2)

operations research — The analytical study of military problems undertaken to provide responsible commanders and staff agencies with a scientific basis for decision on action to improve military operations. Also called **operational research; operations analysis.**

operations security — A process of identifying critical information and subsequently analyzing friendly actions attendant to military operations and other activities. Also called **OPSEC.** See also **operations security indicators; operations security measures; operations security planning guidance; operations security vulnerability.** (JP 3-13.3)

operations security assessment — An evaluative process, usually exercise, or support function to determine the likelihood that critical information can be protected from the adversary's intelligence. (JP 3-13.3)

operations security countermeasures — Methods and means to gain and maintain essential secrecy about critical information. (JP 3-13.3)

operations security indicators — Friendly detectable actions and open-source information that can be interpreted or pieced together by an adversary to derive critical information. (JP 3-13.3)

operations security planning guidance — Guidance that defines the critical information requiring protection from the adversary and outlines provisional measures to ensure secrecy. (JP 3-13.3)

operations security survey — A collection effort by a team of subject matter experts to reproduce the intelligence image projected by a specific operation or function simulating hostile intelligence processes. (JP 3-13.3)

operations security vulnerability — A condition in which friendly actions provide operations security indicators that may be obtained and accurately evaluated by an adversary in time to provide a basis for effective adversary decision making. (JP 3-13.3)

operations support element — An element that is responsible for all administrative, operations support and services support functions within the counterintelligence and human intelligence staff element of a joint force intelligence directorate. Also called **OSE.** (JP 2-01.2)

operations to restore order — Operations intended to halt violence and support, reinstate, or establish civil authorities so that indigenous police forces can effectively enforce the law and restore civil authority. See also **operation; peace operations.** (JP 3-07.3)

opportune lift — That portion of lift capability available for use after planned requirements have been met. (JP 4-02)

ordered departure — A procedure by which the number of US Government personnel, their dependents, or both are reduced at a foreign service post. Departure is directed by the Department of State (initiated by the chief of mission or the Secretary of State) to designated safe havens with implementation of the combatant commander noncombatant evacuation operations plan. (JP 3-68)

order of battle — The identification, strength, command structure, and disposition of the personnel, units, and equipment of any military force. Also called **OB; OOB.** (JP 2-01.3)

ordnance — Explosives, chemicals, pyrotechnics, and similar stores, e.g., bombs, guns and ammunition, flares, smoke, or napalm. (JP 3-15)

ordnance handling — Applies to those individuals who engage in the breakout, lifting, or repositioning of ordnance or explosive devices in order to facilitate storage or stowage, assembly or disassembly, loading or downloading, or transporting. (JP 3-04)

organic — Assigned to and forming an essential part of a military organization as listed in its table of organization for the Army, Air Force, and Marine Corps, and are assigned to the operating forces for the Navy. (JP 1)

organizational maintenance — That maintenance that is the responsibility of and performed by a using organization on its assigned equipment. (JP 4-01.5)

organization for combat — In amphibious operations, task organization of landing force units for combat, involving combinations of command, ground and aviation combat, combat support, and combat service support units for accomplishment of missions ashore. See also **amphibious operation; task organization.** (JP 3-02)

organization for embarkation — In amphibious operations, the organization for embarkation consisting of temporary landing force task organizations established by the commander, landing force and a temporary organization of Navy forces established by the commander, amphibious task force for the purpose of simplifying planning and facilitating the execution of embarkation. See also **amphibious operation; embarkation; landing force; task organization.** (JP 3-02)

organization for landing — In amphibious operations, the specific tactical grouping of the landing force for the assault. (JP 3-02)

Organized Crime and Drug Enforcement Task Force — The network of regional task forces that coordinates federal law enforcement efforts to combat the national and international organizations that cultivate, process, and distribute illicit drugs. Also called **OCDETF.** (JP 3-07.4)

originating medical treatment facility — A medical facility that initially transfers a patient to another medical facility. (JP 4-02)

originator — The command by whose authority a message is sent, which includes the responsibility for the functions of the drafter and the releasing officer. (JP 2-01)

oscillating mine — A mine, hydrostatically controlled, which maintains a pre-set depth below the surface of the water independently of the rise and fall of the tide. See also **mine.** (JP 3-15)

other detainee — Person in the custody of the US Armed Forces who has not been classified as an enemy prisoner of war (article 4, Geneva Convention of 1949 Relative to the Treatment of Prisoners of War (GPW)), retained personnel (article 33, GPW), or civilian internee (article 78, Geneva Convention). Also called **OD.** See also **civilian internee; custody; detainee; prisoner of war; retained personnel.** (JP 1-0)

outer transport area — In amphibious operations, an area inside the antisubmarine screen to which assault transports proceed initially after arrival in the objective area. See also **inner transport area; transport area.** (JP 3-02)

outsized cargo — A single item that exceeds 1,000 inches long by 117 inches wide by 105 inches high in any one dimension. See also **oversized cargo.** (JP 4-01.6)

overhead persistent infrared — Those systems originally developed to detect and track foreign intercontinental ballistic missile systems. Also called **OPIR.** (JP 3-14)

overpressure — The pressure resulting from the blast wave of an explosion referred to as "positive" when it exceeds atmospheric pressure and "negative" during the passage of the wave when resulting pressures are less than atmospheric pressure. (JP 3-11)

Overseas Environmental Baseline Guidance Document — A set of objective criteria and management practices developed by the Department of Defense to protect human health and the environment. Also called **OEBGD.** (JP 3-34)

oversized cargo — 1. Large items of specific equipment such as a barge, side loadable warping tug, causeway section, powered, or causeway section, nonpowered that require transport by sea. 2. Air cargo exceeding the usable dimension of a 463L pallet loaded to the design height of 96 inches, but equal to or less than 1,000 inches in length, 117 inches in width, and 105 inches in height. See also **outsized cargo.** (JP 3-17)

over-the-horizon amphibious operations — An operational initiative launched from beyond visual and radar range of the shoreline. (JP 3-02)

overt — Activities that are openly acknowledged by or are readily attributable to the United States Government, including those designated to acquire information through legal and open means without concealment. Overt information may be collected by observation, elicitation, or from knowledgeable human sources. (JP 2-01.2)

overt operation — An operation conducted openly, without concealment. See also **clandestine operation; covert operation.** (JP 2-01.2)

P

packaged petroleum product — A petroleum product (generally a lubricant, oil, grease, or specialty item) normally packaged by a manufacturer and procured, stored, transported, and issued in containers having a fill capacity of 55 United States gallons (or 45 Imperial gallons, or 205 liters) or less.

packup kit — Service-provided maintenance gear sufficient for a short-term deployment, including spare parts and consumables most commonly needed by the deployed helicopter detachment. Supplies are sufficient for a short-term deployment but do not include all material needed for every maintenance task. Also called **PUK.** (JP 3-04)

parallel chains of command — In amphibious operations, a parallel system of command, responding to the interrelationship of Navy, landing force, Air Force, and other major forces assigned, wherein corresponding commanders are established at each subordinate level of all components to facilitate coordinated planning for, and execution of, the amphibious operation. (JP 3-02)

paramilitary forces — Forces or groups distinct from the regular armed forces of any country, but resembling them in organization, equipment, training, or mission. (JP 3-24)

partial mobilization — See **mobilization, Part 2.**

partisan warfare — Not to be used. See **guerrilla warfare.**

partner nation — A nation that the United States works with in a specific situation or operation. Also called **PN.** (JP 1)

passage of lines — An operation in which a force moves forward or rearward through another force's combat positions with the intention of moving into or out of contact with the enemy. (JP 3-18)

passive air defense — All measures, other than active air defense, taken to minimize the effectiveness of hostile air and missile threats against friendly forces and assets. See also **air defense.** (JP 3-01)

passive defense — Measures taken to reduce the probability of and to minimize the effects of damage caused by hostile action without the intention of taking the initiative. See also **active defense.** (JP 3-60)

passive mine — 1. A mine whose anticountermining device has been operated preventing the firing mechanism from being actuated. 2. A mine which does not emit a signal to detect the presence of a target. (JP 3-15)

pathfinders — 1. Experienced aircraft crews who lead a formation to the drop zone, release point, or target. 2. Teams dropped or air landed at an objective to establish and operate navigational aids for the purpose of guiding aircraft to drop and landing zones. 3. A radar device used for navigating or homing to an objective when visibility precludes accurate visual navigation. 4. Teams air delivered into enemy territory for the purpose of determining the best approach and withdrawal lanes, landing zones, and sites for helicopterborne forces. (JP 3-13.3)

patient movement — The act or process of moving a sick, injured, wounded, or other person to obtain medical and/or dental care or treatment. Functions include medical regulating, patient evacuation, and en route medical care. See also **patient movement items; patient movement requirements center.** (JP 4-02)

patient movement items — The medical equipment and supplies required to support patients during aeromedical evacuation, which is part of a standardized list of approved safe-to-fly equipment. Also called **PMIs.** (JP 4-02)

patient movement policy — Command decision establishing the maximum number of days that patients may be held within the command for treatment. See also **evacuation.** (JP 4-02)

patient movement requirements center — 1. A joint activity that coordinates patient movement by functionally merging of joint medical regulating processes, Services' medical regulating processes, and patient movement evacuation requirements planning (transport to bed plan). 2. Term used to represent any theater, joint or the Global Patient Movement Requirements Center function. Also called **PMRC.** (JP 4-02)

Patriot — A point and limited area defense surface-to-air missile system capable of intercepting aircraft and theater missiles, including short-, medium-, and intermediate-range ballistic missiles in the terminal phase. (JP 3-01)

peace building — Stability actions, predominately diplomatic and economic, that strengthen and rebuild governmental infrastructure and institutions in order to avoid a relapse into conflict. Also called **PB.** See also **peace enforcement; peacekeeping; peacemaking; peace operations.** (JP 3-07.3)

peace enforcement — Application of military force, or the threat of its use, normally pursuant to international authorization, to compel compliance with resolutions or sanctions designed to maintain or restore peace and order. See also **peace building; peacekeeping; peacemaking; peace operations.** (JP 3-07.3)

peacekeeping — Military operations undertaken with the consent of all major parties to a dispute, designed to monitor and facilitate implementation of an agreement (cease fire, truce, or other such agreement) and support diplomatic efforts to reach a long-term political settlement. See also **peace building; peace enforcement; peacemaking; peace operations.** (JP 3-07.3)

peacemaking — The process of diplomacy, mediation, negotiation, or other forms of peaceful settlements that arranges an end to a dispute and resolves issues that led to it. See also **peace building; peace enforcement; peacekeeping; peace operations.** (JP 3-07.3)

peace operations — A broad term that encompasses multiagency and multinational crisis response and limited contingency operations involving all instruments of national power with military missions to contain conflict, redress the peace, and shape the environment to support reconciliation and rebuilding and facilitate the transition to legitimate governance. Also called **PO.** See also **peace building; peace enforcement; peacekeeping; and peacemaking.** (JP 3-07.3)

peacetime operating stocks — Logistic resources on hand or on order necessary to support day-to-day operational requirements, and which, in part, can also be used to offset sustaining requirements. Also called **POS.** (JP 4-03)

penetration (air traffic control) — That portion of a published high altitude instrument approach procedure that prescribes a descent path from the fix on which the procedure is based to a fix or altitude from which an approach to the airport is made.

performance work statement — A statement of work for performance based acquisitions that describe the results in clear, specific, and objective terms with measurable outcomes. Also called **PWS.** (JP 4-10)

period — The time it takes for a satellite to complete one orbit around the earth. (JP 3-14)

perishable cargo — Cargo requiring refrigeration, such as meat, fruit, fresh vegetables, and medical department biologicals. (JP 4-01.5)

permissive environment — Operational environment in which host country military and law enforcement agencies have control as well as the intent and capability to assist operations that a unit intends to conduct. (JP 3-0)

persistent agent — A chemical agent that, when released, remains able to cause casualties for more than 24 hours to several days or weeks. (JP 3-11)

personal effects — All privately owned moveable, personal property of an individual. Also called **PE.** See also **mortuary affairs; personal property.** (JP 4-06)

personal locator beacon — An emergency device carried by individuals, to assist locating during personnel recovery. Also called **PLB.** See also **emergency locator beacon.** (JP 3-50)

personal locator system — A system that provides rough range and bearing to isolated personnel by integrating the survival radio (if equipped with a transponder) with an airborne locating system, based on an encrypted communications homing system. Also called **PLS.** (JP 3-50)

personal property — Property of any kind or any interest therein, except real property, records of the United States Government, and naval vessels of the following categories: surface combatants, support ships, and submarines. (JP 4-06)

personal protective equipment — The protective clothing and equipment provided to shield or isolate a person from the chemical, physical, and thermal hazards that can be encountered at a hazardous materials incident. Also called **PPE.** See also **individual protective equipment.** (JP 3-11)

personal staff — Aides and staff officers handling special matters over which the commander wishes to exercise close personal control. (JP 3-33)

person authorized to direct disposition of human remains — A person, usually primary next of kin, who is authorized to direct disposition of human remains. Also called **PADD.** See also **mortuary affairs.** (JP 4-06)

person eligible to receive effects — The person authorized by law to receive the personal effects of a deceased military member. Receipt of personal effects does not constitute ownership. Also called **PERE.** See also **mortuary affairs; personal effects.** (JP 4-06)

personnel — Those individuals required in either a military or civilian capacity to accomplish the assigned mission. (JP 1-0)

personnel accountability — The process of identifying, capturing, and recording the personal identification information of an individual usually through the use of a database. (JP 1-0)

personnel effects inventory officer — An officer appointed to establish clear chain of custody for all personal effects of an individual from the time they establish control of the effects until they release the effect to mortuary affairs personnel. Also called **PEIO.** (JP 4-06)

personnel increment number — A seven-character, alphanumeric field that uniquely describes a non-unit-related personnel entry (line) in a Joint Operation Planning and Execution System time-phased force and deployment data. Also called **PIN.** (JP 5-0)

personnel recovery — The sum of military, diplomatic, and civil efforts to prepare for and execute the recovery and reintegration of isolated personnel. Also called **PR.** See also **combat search and rescue; evasion; personnel; recovery; search and rescue.** (JP 3-50)

personnel recovery coordination cell — The primary joint force component organization responsible for coordinating and controlling component personnel recovery missions. Also called **PRCC.** (JP 3-50)

personnel recovery task force — A force comprised of US or multinational military forces and/or other US agencies formed to execute a specific personnel recovery mission to locate, support, and recover isolated personnel. Also called **PRTF.** (JP 3-50)

personnel security investigation — An inquiry into the activities of an individual, designed to develop pertinent information pertaining to trustworthiness and suitability for a position of trust as related to loyalty, character, emotional stability, and reliability. Also called **PSI.** (JP 2-01)

personnel services support — Service-provided sustainment activities that support a Service member during both joint exercises and joint operations. Also called **PSS.** (JP 1-0)

petroleum, oils, and lubricants — A broad term that includes all petroleum and associated products used by the Armed Forces. Also called **POL.** (JP 4-01.6)

phase — In joint operation planning, a definitive stage of an operation or campaign during which a large portion of the forces and capabilities are involved in similar or mutually supporting activities for a common purpose. (JP 5-0)

phase line — A line utilized for control and coordination of military operations, usually an easily identified feature in the operational area. Also called **PL.** (JP 3-09)

phony minefield — An area free of live mines used to simulate a minefield, or section of a minefield, with the object of deceiving the enemy. See also **minefield.** (JP 3-15)

physical characteristics — Those military characteristics of equipment that are primarily physical in nature. (JP 3-60)

physical damage assessment — The estimate of the quantitative extent of physical damage to a target resulting from the application of military force. **See also battle damage assessment.** (JP 3-60)

physical security —1. That part of security concerned with physical measures designed to safeguard personnel; to prevent unauthorized access to equipment, installations, material, and documents; and to safeguard them against espionage, sabotage, damage, and theft. (JP 3-0) 2. In communications security, the component that results from all physical measures necessary to safeguard classified equipment, material, and documents from access thereto or observation thereof by unauthorized persons. See also **communications security; security.** (JP 6-0)

placement — An individual's proximity to information of intelligence interest. (JP 2-01.2)

plan for landing — In amphibious operations, a collective term referring to all individually prepared naval and landing force documents which, taken together, present in detail all instructions for execution of the ship-to-shore movement. (JP 3-02)

plan identification number — 1. A command-unique four-digit number followed by a suffix indicating the Joint Strategic Capabilities Plan year for which the plan is written. 2. A five-digit number representing the command-unique four-digit identifier, followed by a one-character, alphabetic suffix indicating the operation plan option, or a one-digit number numeric value indicating the Joint Strategic Capabilities Plan year for which the plan is written. Also called **PID**. (JP 5-0)

planned target — Target that is known to exist in the operational environment, upon which actions are planned using deliberate targeting, creating effects which support commander's objectives. There are two subcategories of planned targets: scheduled and on-call. See also **on-call target; operational area; scheduled target; target.** (JP 3-60)

planning and direction — In intelligence usage, the determination of intelligence requirements, development of appropriate intelligence architecture, preparation of a collection plan, and issuance of orders and requests to information collection agencies. See also **intelligence process.** (JP 2-01)

planning directive — In amphibious operations, the plan issued by the designated commander, following receipt of the initiating directive, to ensure that the planning process and interdependent plans developed by the amphibious force will be coordinated, completed in the time allowed, and important aspects not overlooked. See also **amphibious force; amphibious operation.** (JP 3-02)

planning factor — A multiplier used in planning to estimate the amount and type of effort involved in a contemplated operation. (JP 5-0)

planning factors database — Databases created and maintained by the Services for the purpose of identifying all geospatial information and services requirements for emerging and existing forces and systems. Also called **PFDB.** See also **geospatial information and services.** (JP 2-03)

planning order — A planning directive that provides essential planning guidance and directs the initiation of execution planning before the directing authority approves a military course of action. Also called **PLANORD.** See also **execution planning.** (JP 5-0)

planning phase — In amphibious operations, the phase normally denoted by the period extending from the issuance of the initiating directive up to the embarkation phase. The planning phase may occur during movement or at any other time upon receipt of a new mission or change in the operational situation. See also **amphibious operation.** (JP 3-02)

planning team — A functional element within a joint force commander's headquarters established to solve problems related to a specific task or requirement, and which dissolves upon completion of the assigned task. (JP 3-33)

point defense — The defense or protection of special vital elements and installations; e.g., command and control facilities or air bases. (JP 3-52)

pointee-talkee — A language aid containing selected phrases in English opposite a translation in a foreign language used by pointing to appropriate phrases. See also **evasion aid.** (JP 3-50)

point of employment — In distribution operations, a physical location designated by the commander at the tactical level where force employment, emplacement, or commodity consumption occurs. (JP 4-09)

point of need — In distribution operations, a physical location within a desired operational area designated by the geographic combatant commander or subordinate commander as a receiving point for forces or materiel, for subsequent use or consumption. (JP 4-09)

point of origin — In distribution operations, the beginning point of a deployment, redeployment, or movement where forces or materiel are located. (JP 4-09)

polar orbit — A satellite orbit that passes over the North and South Poles on each orbit, has an angle of inclination relative to the equator of 90 degrees, and eventually passes over all points on the earth. (JP 3-14)

population at risk — The strength in personnel of a given force structure in terms of which casualty rates are stated. Also called **PAR.** (JP 4-02)

port complex — One or more port areas whose activities are geographically linked either because these areas are dependent on a common inland transport system or because they constitute a common initial destination for convoys. (JP 4-01.5)

port of debarkation — The geographic point at which cargo or personnel are discharged. Also called **POD.** See also **port of embarkation.** (JP 4-0)

port of embarkation — The geographic point in a routing scheme from which cargo or personnel depart. Also called **POE.** See also **port of debarkation.** (JP 4-01.2)

port operations group — A task-organized unit, located at the seaport of embarkation and/or debarkation that assists and provides support in the loading and/or unloading and staging of personnel, supplies, and equipment from shipping. Also called **POG.** See also **landing force support party; task organization.** (JP 3-35)

port security — The safeguarding of vessels, harbors, ports, waterfront facilities, and cargo from internal threats such as destruction, loss, or injury from sabotage or other subversive acts; accidents; thefts; or other causes of similar nature. See also **physical security; security.** (JP 3-10)

port support activity — A tailorable support organization composed of mobilization station assets that ensures the equipment of the deploying units is ready to load. Also called **PSA.** See also **support.** (JP 3-35)

positive control — A method of airspace control that relies on positive identification, tracking, and direction of aircraft within an airspace, conducted with electronic means by an agency having the authority and responsibility therein. (JP 3-52)

positive identification — An identification derived from observation and analysis of target characteristics including visual recognition, electronic support systems, non-cooperative target recognition techniques, identification friend or foe systems, or other physics-based identification techniques. (JP 3-01)

prearranged fire — Fire that is formally planned and executed against targets or target areas of known location. Such fire is usually planned well in advance and is executed at a predetermined time or during a predetermined period of time. See also **on-call.** (JP 3-09)

preassault operations — Operations conducted by the amphibious force upon its arrival in the operational area and prior to H-hour and/or L-hour. See also **amphibious force; times.** (JP 3-02)

precipitation static — Charged precipitation particles that strike antennas and gradually charge the antenna, which ultimately discharges across the insulator, causing a burst of static. Also called **P-STATIC.** (JP 3-13.1)

precise time and time interval — A reference value of time and time interval (frequency). Also called **PTTI.** (JP 3-59)

precision-guided munition — A guided weapon intended to destroy a point target and minimize collateral damage. Also called **PGM, smart weapon, smart munition.** (JP 3-03)

preparation of the environment — An umbrella term for operations and activities conducted by selectively trained special operations forces to develop an environment for potential future special operations. Also called **PE.** (JP 3-05)

prepare to deploy order — An order issued by competent authority to move forces or prepare forces for movement (e.g., increase deployability posture of units). Also called **PTDO.** (JP 5-0)

preplanned air support — Air support in accordance with a program, planned in advance of operations. (JP 3-09.3)

preplanned mission request — A request for an air strike on a target that can be anticipated sufficiently in advance to permit detailed mission coordination and planning. (JP 3-09.3)

pre-position — To place military units, equipment, or supplies at or near the point of planned use or at a designated location to reduce reaction time, and to ensure timely support of a specific force during initial phases of an operation. (JP 4-0)

pre-positioned war reserve materiel requirement, balance — That portion of the pre-positioned war reserve materiel requirement that has not been acquired or funded. This level consists of the pre-positioned war reserve materiel requirement, less the pre-positioned war reserve requirement, protectable.

pre-positioned war reserve materiel requirement, protectable — That portion of the pre-positioned war reserve materiel requirement that is protected for purposes of procurement, funding, and inventory management.

pre-positioned war reserve stock — The assets that are designated to satisfy the pre-positioned war reserve materiel requirement. Also called **PWRS.**

presail — The time prior to a ship getting under way used to prepare for at-sea events. (JP 3-04)

Presidential Call-up — Procedures by which the President brings all or a part of the Army National Guard or the Air National Guard to active federal service under Title 10, United States Code, Section 12406 and Chapter 15. See also **active duty; federal service; Presidential Reserve Call-up.** (JP 4-05)

Presidential Reserve Call-up — Provision of a public law (Title 10, United States Code, Section 12304) that provides the President a means to activate, without a declaration of national emergency, not more than 200,000 members of the Selected Reserve and the Individual Ready Reserve (of whom not more than 30,000 may be members of the Individual Ready Reserve), for not more than 365 days to meet the requirements of any operational mission. Members called under this provision may not be used for disaster relief or to suppress insurrection. This authority has particular utility when used in circumstances in which the escalatory national or international signals of partial or full mobilization would be undesirable. Forces available under this authority can provide a tailored, limited-scope, deterrent, or operational response, or may be used as a precursor to any subsequent mobilization. Also called **PRC.** See also **Individual Ready Reserve; mobilization; Presidential Call-up; Selected Reserve.** (JP 4-05)

pressure mine — 1. In land mine warfare, a mine whose fuse responds to the direct pressure of a target. 2. In naval mine warfare, a mine whose circuit responds to the hydrodynamic pressure field of a target. See also **mine.** (JP 3-15)

prevention —In space usage, measures to preclude an adversary's hostile use of United States or third-party space systems and services. See also **space control.** (JP 3-14)

prevention of mutual interference — In submarine operations, procedures established to prevent submerged collisions between friendly submarines, between submarines and friendly surface ship towed bodies and arrays, and between submarines and any other hazards to submerged navigation. Also called **PMI.** (JP 3-32)

preventive maintenance — The care and servicing by personnel for the purpose of maintaining equipment and facilities in satisfactory operating condition by providing for systematic inspection, detection, and correction of incipient failures either before they occur or before they develop into major defects. (JP 4-02)

preventive medicine — The anticipation, communication, prediction, identification, prevention, education, risk assessment, and control of communicable diseases, illnesses and exposure to endemic, occupational, and environmental threats. Also called **PVNTMED.** (JP 4-02)

primary agency — The federal department or agency assigned primary responsibility for managing and coordinating a specific emergency support function in the National Response Framework. (JP 3-28)

primary control officer — In amphibious operations, the officer embarked in a primary control ship assigned to control the movement of landing craft, amphibious vehicles, and landing ships to and from a colored beach. Also called **PCO.** (JP 3-02)

primary control ship — In amphibious operations, a ship of the task force designated to provide support for the primary control officer and a combat information center control team for a colored beach. Also called **PCS.** (JP 3-02)

primary flight control — The controlling agency on air-capable ships that is responsible for air traffic control of aircraft within 5 nautical miles of the ship. On most Coast Guard cutters, primary flight control duties are performed by a combat information center, and the term "PRIFLY" is not used. Also called **PRIFLY.** (JP 3-04)

primary review authority — The organization, within the lead agent's chain of command, that is assigned by the lead agent to perform the actions and coordination necessary to develop and maintain the assigned joint publication under the cognizance of the lead agent. Also called **PRA.** See also **joint publication; lead agent.** (CJCSM 5120.01)

prime contract — A contract or contractual action entered into by the United States Government for the purpose of obtaining supplies, materials, equipment, or services of any kind. (JP 4-10)

prime vendor — A contracting process that provides commercial products to regionally grouped military and federal customers from commercial distributors using electronic commerce. Customers typically receive materiel delivery through the vendor's commercial distribution system. Also called **PV.** See also **distribution system.** (JP 4-09)

principal federal official — The federal official designated by the Secretary of Homeland Security to act as his/her representative locally to oversee, coordinate, and execute the Secretary's incident management responsibilities under Homeland Security Presidential Directive 5. Also called **PFO.** (JP 3-41)

principal officer — The officer in charge of a diplomatic mission, consular office, or other Foreign Service post, such as a United States liaison office. (JP 3-08)

priority designator — A two-digit issue and priority code placed in military standard requisitioning and issue procedure requisitions to provide a means of assigning relative rankings to competing demands placed on the Department of Defense supply system. (JP 4-01)

priority intelligence requirement — An intelligence requirement, stated as a priority for intelligence support, that the commander and staff need to understand the adversary or other aspects of the operational environment. Also called **PIR.** See also **information requirements; intelligence; intelligence process; intelligence requirement.** (JP 2-01)

prisoner of war — A detained person (as defined in Articles 4 and 5 of the Geneva Convention Relative to the Treatment of Prisoners of War of August 12, 1949) who, while engaged in combat under orders of his or her government, is captured by the armed forces of the enemy. Also called **POW or PW.** (JP 3-50)

prisoner of war camp — An installation established for the internment and administration of prisoners of war.

private sector — An umbrella term that may be applied to any or all of the nonpublic or commercial individuals and businesses, specified nonprofit organizations, most of academia and other scholastic institutions, and selected nongovernmental organizations. (JP 3-57)

privity of contract — The legal relationship that exists between two contracting parties, for example, between the government and the prime contractor (JP 4-10)

probability of damage — The probability that damage will occur to a target expressed as a percentage or as a decimal. Also called **PD.** (JP 3-60)

procedural control — A method of airspace control which relies on a combination of previously agreed and promulgated orders and procedures. (JP 3-52)

procedural identification — An identification based on observation and analysis of target behaviors including location and trajectory, as well as compliance with airspace control measures. (JP 3-01)

procedures — Standard, detailed steps that prescribe how to perform specific tasks. See also **tactics; techniques.** (CJCSM 5120.01)

procedure word — A word or phrase limited to radio telephone procedure used to facilitate communication by conveying information in a condensed standard form. Also called **proword.** (JP 3-09.3)

processing — A system of operations designed to convert raw data into useful information. (JP 2-0)

processing and exploitation — In intelligence usage, the conversion of collected information into forms suitable to the production of intelligence. See also **intelligence process.** (JP 2-01)

process owner — The head of a Department of Defense component assigned a responsibility by the Secretary of Defense when process improvement involves more than one Service or Department of Defense component. (JP 4-0)

procurement lead time — The interval in time between the initiation of procurement action and receipt into the supply system of the production model (excludes prototypes) purchased as the result of such actions. It is composed of two elements, production lead time and administrative lead time. See also **production lead time; receipt into the supply system.** (JP 4-10)

procuring contracting officer — A contracting officer who initiates and signs the contract. Also called **PCO.** See also **administrative contracting officer; contracting officer.** (JP 4-10)

production base — The total national industrial production capacity available for the manufacture of items to meet materiel requirements. (JP 4-05)

production lead time — The time interval between the placement of a contract and receipt into the supply system of materiel purchased. Two entries are provided: a. **initial** — The time interval if the item is not under production as of the date of contract placement; and b. **reorder** — The time interval if the item is under production as of the date of contract placement. See also **procurement lead time.** (JP 4-10)

production logistics — That part of logistics concerning research, design, development, manufacture, and acceptance of materiel. In consequence, production logistics includes: standardization and interoperability, contracting, quality assurance, initial provisioning, transportability, reliability and defect analysis, safety standards, specifications and production processes, trials and testing (including provision of necessary facilities), equipment documentation, configuration control, and modifications.

production requirement — An intelligence requirement that cannot be met by current analytical products resulting in tasking to produce a new product that can meet this intelligence requirement. Also called **PR**. (JP 2-0)

prohibited area — A specified area within the land areas of a state or its internal waters, archipelagic waters, or territorial sea adjacent thereto over which the flight of aircraft is prohibited. May also refer to land or sea areas to which access is prohibited. See also **restricted area.** (JP 3-52)

proliferation — The transfer of weapons of mass destruction, related materials, technology, and expertise from suppliers to hostile state or non-state actors. (JP 3-40)

proof — To verify that a breached lane is free of live mines by passing a mine roller or other mine-resistant vehicle through as the lead vehicle. (JP 3-15)

propaganda — Any form of adversary communication, especially of a biased or misleading nature, designed to influence the opinions, emotions, attitudes, or behavior of any group in order to benefit the sponsor, either directly or indirectly. (JP 3-13.2)

proper clearance — A clearance for entry of units into specified defense areas by civil or military authorities having responsibility for granting such clearance.

property — 1. Anything that may be owned. 2. As used in the military establishment, this term is usually confined to tangible property, including real estate and materiel. For special purposes and as used in certain statutes, this term may exclude such items as the public domain, certain lands, certain categories of naval vessels, and records of the Federal Government.

protected emblems — The red cross, red crescent, and other symbols that designate that persons, places, or equipment so marked have a protected status under the law of war. (JP 3-60)

protected frequencies — Friendly, generally time-oriented, frequencies used for a particular operation, identified and protected to prevent them from being inadvertently jammed by friendly forces while active electronic warfare operations are directed against hostile forces. See also **electronic warfare.** (JP 3-13.1)

protected persons/places — Persons (such as enemy prisoners of war) and places (such as hospitals) that enjoy special protections under the law of war. They may or may not be marked with protected emblems. (JP 1-04)

protection — 1. Preservation of the effectiveness and survivability of mission-related military and nonmilitary personnel, equipment, facilities, information, and infrastructure deployed or located within or outside the boundaries of a given operational area. (JP 3-0) 2. In space usage, active and passive defensive measures to ensure that United States and friendly space systems perform as designed by seeking to overcome an adversary's attempts to negate them and to minimize damage if negation is attempted. See also **mission-oriented protective posture; space control.** (JP 3-14)

protection of shipping — The use of proportionate force, when necessary for the protection of US flag vessels and aircraft, US citizens (whether embarked in US or foreign vessels), and their property against unlawful violence. (JP 3-0)

protective clothing — Clothing especially designed, fabricated, or treated to protect personnel against hazards. (JP 3-11)

protective minefield — 1. In land mine warfare, a minefield employed to assist a unit in its local, close-in protection. 2. In naval mine warfare, a minefield emplaced in friendly territorial waters to protect ports, harbors, anchorages, coasts, and coastal routes. See also **minefield.** (JP 3-15)

provincial reconstruction team — A civil-military team designed to improve stability in a given area by helping build the legitimacy and effectiveness of a host nation local or provincial government in providing security to its citizens and delivering essential government services. Also called **PRT.** (JP 3-57)

proword — See **procedure word.**

public affairs — Those public information, command information, and community engagement activities directed toward both the external and internal publics with interest in the Department of Defense. Also called **PA.** See also **command information; public information.** (JP 3-61)

public affairs assessment — An analysis of the news media and public environments to evaluate the degree of understanding about strategic and operational objectives and military activities and to identify levels of public support. It includes judgments about the public affairs impact of pending decisions and recommendations about the structure of public affairs support for the assigned mission. See also **assessment; public affairs.** (JP 3-61)

public affairs guidance — Constraints and restraints established by proper authority regarding public information, command information, and community relations activities. It may also address the method(s), timing, location, and other details governing the release of information to the public. Also called **PAG.** See also **public affairs.** (JP 3-61)

public diplomacy — 1. Those overt international public information activities of the United States Government designed to promote United States foreign policy objectives by seeking to understand, inform, and influence foreign audiences and opinion makers, and by broadening the dialogue between American citizens and institutions and their counterparts abroad. 2. In peace building, civilian agency efforts to promote an understanding of the reconstruction efforts, rule of law, and civic responsibility through public affairs and international public diplomacy operations. (JP 3-07.3)

public information — Within public affairs, that information of a military nature, the dissemination of which is consistent with security and approved for release. (JP 3-61)

public key infrastructure — An enterprise-wide service that supports digital signatures and other public key-based security mechanisms for Department of Defense functional enterprise programs, including generation, production, distribution, control, and accounting of public key certificates. Also called **PKI.** (JP 2-03)

purchasing office — Any installation or activity, or any division, office, branch, section, unit, or other organizational element of an installation or activity charged with the functions of procuring supplies or services.

Intentionally Blank

Q

Q-route — A system of preplanned shipping lanes in mined or potentially mined waters used to minimize the area the mine countermeasures commander has to keep clear of mines in order to provide safe passage for friendly shipping. (JP 3-15)

quadruple container — A quadruple container box 57.5inches x 96inches x 96inches with a metal frame, pallet base, and International Organization for Standardization corner fittings. Four of these boxes can be lashed together to form a 20-foot American National Standards Institute or International Organization for Standardization intermodal container. Also called **QUADCON**. (JP 4-09)

qualifying entity — A non-governmental organization to which the Department of Defense may provide assistance for special events by virtue of statute, regulation, policy, or other approval by the Secretary of Defense or his or her authorized designee. (DODD 3025.18)

quay — A structure of solid construction along a shore or bank that provides berthing and generally provides cargo-handling facilities. See also **wharf**. (JP 4-01.5)

Intentionally Blank

R

radar advisory — The term used to indicate that the provision of advice and information is based on radar observation. (JP 3-52)

radial — A magnetic bearing extending from a very high frequency omni-range and/or tactical air navigation station.

radiation dose — The total amount of ionizing radiation absorbed by material or tissues. See also **exposure dose.** (JP 3-11)

radiation dose rate — Measurement of radiation dose per unit of time. (JP 3-11)

radiation exposure status — Criteria to assist the commander in measuring unit exposure to radiation based on total past cumulative dose, normally expressed in centigray. Also called **RES.** (JP 3-11)

radio frequency countermeasures — Any device or technique employing radio frequency materials or technology that is intended to impair the effectiveness of enemy activity, particularly with respect to precision guided weapons and sensor systems. Also called **RF CM.** (JP 3-13.1)

radiological dispersal device — An improvised assembly or process, other than a nuclear explosive device, designed to disseminate radioactive material in order to cause destruction, damage, or injury. Also called **RDD.** (JP 3-11)

radiological exposure device — A radioactive source placed to cause injury or death. Also called **RED.** (JP 3-11)

raid — An operation to temporarily seize an area in order to secure information, confuse an adversary, capture personnel or equipment, or to destroy a capability culminating with a planned withdrawal. (JP 3-0)

railhead — A point on a railway where loads are transferred between trains and other means of transport. (JP 4-09)

Rangers — Rapidly deployable airborne light infantry organized and trained to conduct highly complex joint direct action operations in coordination with or in support of other special operations units of all Services. (JP 3-05)

Rapid Engineer Deployable Heavy Operational Repair Squadron Engineer. Air Force units wartime-structured to provide a heavy engineer capability that are mobile, rapidly deployable, and largely self-sufficient for limited periods of time. Also called **RED HORSE.** (JP 3-34)

rapid global mobility — The timely movement, positioning, and sustainment of military forces and capabilities across the range of military operations. See also **mobility.** (JP 3-17)

ratification — 1. The declaration by which a nation formally accepts, with our without reservation, the content of a standardization agreement. 2. The process of approving an unauthorized commitment by an official who has the authority to do so. See also **implementation; unauthorized commitment.** (JP 4-10)

rationalization — Any action that increases the effectiveness of allied forces through more efficient or effective use of defense resources committed to the alliance. (JP 3-16)

reachback — The process of obtaining products, services, and applications, or forces, or equipment, or material from organizations that are not forward deployed. (JP 3-30)

readiness — The ability of military forces to fight and meet the demands of assigned missions. See also **national military strategy.** (JP 1)

Ready Reserve — The Selected Reserve, Individual Ready Reserve, and Inactive National Guard liable for active duty as prescribed by law (Title 10, United States Code, Sections 10142, 12301, and 12302). See also **active duty; Inactive National Guard; Individual Ready Reserve; Selected Reserve.** (JP 4-05)

ready-to-load date — The date when a unit will be ready to move from the origin, i.e., mobilization station. Also called **RLD.** (JP 5-0)

Realistic Military Training — Department of Defense training conducted off federal property utilizing private or non-federal public property and infrastructure. (DODI 1322.28)

real property — Lands, buildings, structures, utilities systems, improvements, and appurtenances, thereto that includes equipment attached to and made part of buildings and structures, but not movable equipment. JP 3-34

rear area operations center/rear tactical operations center — A command and control facility that serves as an area and/or subarea commander's planning, coordinating, monitoring, advising, and directing agency for area security operations. (JP 3-10)

reattack recommendation — An assessment, derived from the results of battle damage assessment and munitions effectiveness assessment, providing the commander systematic advice on reattack of a target. Also called **RR.** See also **assessment; battle damage assessment; munitions effectiveness assessment; target.** (JP 3-60)

receipt into the supply system — That point in time when the first item or first quantity of the item of the contract has been received at or is en route to point of first delivery after inspection and acceptance. See also **procurement lead time.** (JP 4-10)

receiving ship — The ship in a replenishment unit that receives the rig(s). (JP 4-03)

reception — 1. All ground arrangements connected with the delivery and disposition of air or sea drops. 2. Arrangements to welcome and provide secure quarters or transportation for defectors, escapees, evaders, or incoming agents. 3. The process of receiving, off-loading, marshalling, accounting for, and transporting of personnel, equipment, and materiel from the strategic and/or intratheater deployment phase to a sea, air, or surface transportation point of debarkation to the marshalling area. (JP 3-35)

recognition — 1. The determination by any means of the individuality of persons, or of objects such as aircraft, ships, or tanks, or of phenomena such as communications-electronics patterns. 2. In ground combat operations, the determination that an object is similar within a category of something already known. (JP 3-01)

recognition signal — Any prearranged signal by which individuals or units may identify each other. (JP 3-50)

reconnaissance — A mission undertaken to obtain, by visual observation or other detection methods, information about the activities and resources of an enemy or adversary, or to secure data concerning the meteorological, hydrographic, or geographic characteristics of a particular area. Also called **RECON.** (JP 2-0)

recovery — 1. In air (aviation) operations, that phase of a mission that involves the return of an aircraft to a land base or platform afloat. (JP 3-52) 2. The retrieval of a mine from the location where emplaced. (JP 3-15) 3. In personnel recovery, actions taken to physically gain custody of isolated personnel and return them to friendly control. (JP 3-50) 4. Actions taken to extricate damaged or disabled equipment for return to friendly control or repair at another location. See also **evader; evasion; recovery force.** (JP 3-34)

recovery and reconstitution — 1. Those actions taken by one nation prior to, during, and following an attack by an enemy nation to minimize the effects of the attack, rehabilitate the national economy, provide for the welfare of the populace, and maximize the combat potential of remaining forces and supporting activities. 2. Those actions taken by a military force during or after operational employment to restore its combat capability to full operational readiness. See also **recovery.** (JP 3-35)

recovery force — In personnel recovery, an organization consisting of personnel and equipment with a mission of locating, supporting, and recovering isolated personnel, and returning them to friendly control. See also **evader; evasion; recovery operations.** (JP 3-50)

recovery mechanism — Designated indigenous or surrogate infrastructure that is specifically developed, trained, and directed by US forces to contact, authenticate, support, move, and exfiltrate designated isolated personnel from uncertain or hostile areas back to friendly control. Also called **RM**. (JP 3-50)

recovery operations — Operations conducted to search for, locate, identify, recover, and return isolated personnel, human remains, sensitive equipment, or items critical to national security. (JP 3-50)

recovery site — In personnel recovery, an area from which isolated personnel can be recovered. See also **escapee; evader; evasion**. (JP 3-50)

recovery team — In personnel recovery, designated US or US-directed forces, who are specifically trained to operate in conjunction with indigenous or surrogate forces, and are tasked to contact, authenticate, support, move, and exfiltrate isolated personnel. Also called **RT**. (JP 3-50)

recovery vehicle — In personnel recovery, the vehicle on which isolated personnel are boarded and transported from the recovery site. (JP 3-50)

recovery zone — A designated geographic area from which special operations forces can be extracted by air, boat, or other means. Also called **RZ**. (JP 3-05.1)

redeployment — The transfer or rotation of forces and materiel to support another joint force commander's operational requirements, or to return personnel, equipment, and materiel to the home and/or demobilization stations for reintegration and/or out-processing. See also **deployment**. (JP 3-35)

red team — An organizational element comprised of trained and educated members that provide an independent capability to fully explore alternatives in plans and operations in the context of the operational environment and from the perspective of adversaries and others. (JP 2-0)

reduced charge — 1. The smaller of the two propelling charges available for naval guns. 2. Charge employing a reduced amount of propellant to fire a gun at short ranges as compared to a normal charge.

reduced operating status — Military Sealift Command ships withdrawn from full operating status because of decreased operational requirements. Also called **ROS**. See also **Military Sealift Command**. (JP 4-01.2)

reduction — The creation of lanes through a minefield or obstacle to allow passage of the attacking ground force. (JP 3-15)

refraction — The process by which the direction of a wave is changed when moving into shallow water at an angle to the bathymetric contours. (JP 4-01.6)

refugee — A person who owing to a well-founded fear of being persecuted for reasons of race, religion, nationality, membership of a particular social group or political opinion, is outside the country of his or her nationality and is unable or, owing to such fear, is unwilling to avail himself or herself of the protection of that country. See also **dislocated civilian; displaced person; evacuee; stateless person.** (JP 3-29)

regimental landing team — A task organization for landing composed of an infantry regiment reinforced by those elements that are required for initiation of its combat function ashore. (JP 3-02)

regional air defense commander — Commander, subordinate to the area air defense commander, who is responsible for air and missile defenses in the assigned regionand exercises authorities as delegated by the area air defense commander. Also called **RADC.** (JP 3-01)

regional liaison group — A combined Department of State-Department of Defense element collocated with a combatant command for the purpose of coordinating post emergency evacuation plans. Also called **RLG.** (JP 3-68)

regional response coordination center — A standing facility that is activated to coordinate regional response efforts, until a joint field office is established and/or the principal federal official, federal or coordinating officer can assume their National Response Framework coordination responsibilities. Also called **RRCC.** (JP 3-28)

regional satellite communications support center — United States Strategic Command operational element responsible for providing the operational communications planners with a point of contact for accessing and managing satellite communications resources. Also called **RSSC.** (JP 3-14)

regional security officer — A security officer responsible to the chief of mission (ambassador), for security functions of all US embassies and consulates in a given country or group of adjacent countries. Also called **RSO.** (JP 3-10)

regulating point — An anchorage, port, or ocean area to which assault and assault follow-on echelons and follow-up shipping proceed on a schedule, and at which they are normally controlled by the commander, amphibious task force, until needed in the transport area for unloading. See also **assault; commander, amphibious task force.** (JP 3-02)

regulating station — A command agency established to control all movements of personnel and supplies into or out of a given area.

rehabilitative care — Therapy that provides evaluations and treatment programs using exercises, massage, or electrical therapeutic treatment to restore, reinforce, or enhance motor performance and restores patients to functional health allowing for their return to duty or discharge from the Service. Also called **restorative care.** See also **patient movement policy; theater.** (JP 4-02)

rehearsal phase — In amphibious operations, the period during which the prospective operation is practiced for the purpose of: (1) testing adequacy of plans, the timing of detailed operations, and the combat readiness of participating forces; (2) ensuring that all echelons are familiar with plans; and (3) testing communications-information systems. See also **amphibious operation.** (JP 3-02)

reinforcing obstacles — Those obstacles specifically constructed, emplaced, or detonated through military effort and designed to strengthen existing terrain to disrupt, fix, turn, or block enemy movement. See also **obstacle.** (JP 3-15)

reintegrate — In personnel recovery, the task of conducting appropriate debriefings and reintegrating recovered isolated personnel back to duty and their family. (JP 3-50)

release altitude — Altitude of an aircraft above the ground at the time of release of bombs, rockets, missiles, tow targets, etc.

relief in place — An operation in which, by direction of higher authority, all or part of a unit is replaced in an area by the incoming unit and the responsibilities of the replaced elements for the mission and the assigned zone of operations are transferred to the incoming unit. (JP 3-07.3)

religious advisement — The practice of informing the commander on the impact of religion on joint operations to include, but not limited to: worship, rituals, customs, and practices of US military personnel, international forces, and the indigenous population; as well as the impact of military operations on the religious and humanitarian dynamics in the operational area. (JP 1-05)

religious affairs — The combination of religious support and religious advisement. (JP 1-05)

religious support — Chaplain-facilitated free exercise of religion through worship, religious and pastoral counseling services, ceremonial honors for the fallen, crisis intervention, and advice to the commander on ethical and moral issues, and morale as affected by religion. Also called **RS.** See also **combatant command chaplain; command chaplain; religious support team.** (JP 1-05)

religious support team — A team, comprised of at least one chaplain and one enlisted support person, that works together in designing, implementing, and executing the command religious program. Also called **RST.** See also **combatant command chaplain; command chaplain; religious support.** (JP 1-05)

remain-behind equipment — Unit equipment left by deploying forces at their bases when they deploy. (JP 4-05)

render safe procedures — The portion of the explosive ordnance disposal procedures involving the application of special explosive ordnance disposal methods and tools to provide for the interruption of functions or separation of essential components of unexploded explosive ordnance to prevent an unacceptable detonation. (JP 3-15.1)

rendezvous area — In an amphibious operation, the area in which the landing craft and amphibious vehicles rendezvous to form waves after being loaded, and prior to movement to the line of departure. (JP 3-02)

reorder point — 1. That point at which time a stock replenishment requisition would be submitted to maintain the predetermined or calculated stockage objective. 2. The sum of the safety level of supply plus the level for order and shipping time equals the reorder point.

repairable item — An item that can be reconditioned or economically repaired for reuse when it becomes unserviceable. (JP 4-09)

repair cycle — The stages through which a repairable item passes from the time of its removal or replacement until it is reinstalled or placed in stock in a serviceable condition. (JP 4-09)

repatriate — A person who returns to his or her country or citizenship, having left said native country either against his or her will, or as one of a group who left for reason of politics, religion, or other pertinent reasons.

repatriation — 1. The procedure whereby American citizens and their families are officially processed back into the United States subsequent to an evacuation. See also **evacuation.** (JP 3-68) 2. The release and return of enemy prisoners of war to their own country in accordance with the 1949 Geneva Convention Relative to the Treatment of Prisoners of War. (JP 1-0)

reportable incident — Any suspected or alleged violation of Department of Defense policy or of other related orders, policies, procedures or applicable law, for which there is credible information. (JP 3-63)

request for assistance — A request based on mission requirements and expressed in terms of desired outcome, formally asking the Department of Defense to provide assistance to a local, state, tribal, or other federal agency. Also called **RFA.** (JP 3-28)

request for information — 1. Any specific time-sensitive ad hoc requirement for intelligence information or products to support an ongoing crisis or operation not necessarily related to standing requirements or scheduled intelligence production. 2. A term used by the National Security Agency/Central Security Service to state ad hoc signals intelligence requirements. Also called **RFI.** See also **intelligence.** (JP 2-0)

required delivery date — The date that a force must arrive at the destination and complete unloading. Also called **RDD.** (JP 5-0)

requirements management system — A system for the management of theater and national imagery collection requirements that provides automated tools for users in support of submission, review, and validation of imagery nominations as requirements to be tasked on national or Department of Defense imagery collection, production, and exploitation resources. Also called **RMS.** See also **imagery.** (JP 2-01)

requiring activity — A military or other designated supported organization that identifies and receives contracted support during military operations. See also **supported unit.** (JP 4-10)

requisitioning objective — The maximum quantities of materiel to be maintained on hand and on order to sustain current operations. It will consist of the sum of stocks represented by the operating level, safety level, and the order and shipping time or procurement lead time, as appropriate.

rescue combat air patrol — An aircraft patrol provided over that portion of an objective area in which recovery operations are being conducted for the purpose of intercepting and destroying hostile aircraft. Also called **RESCAP.** See also **combat air patrol.** (JP 3-50)

rescue coordination center — A unit, recognized by International Civil Aviation Organization, International Maritime Organization, or other cognizant international body, responsible for promoting efficient organization of search and rescue services and coordinating the conduct of search and rescue operations within a search and rescue region. Also called **RCC.** (JP 3-50)

reserve — 1. Portion of a body of troops that is kept to the rear, or withheld from action at the beginning of an engagement, in order to be available for a decisive movement. 2. Members of the Military Services who are not in active service but who are subject to call to active duty. 3. Portion of an appropriation or contract authorization held or set aside for future operations or contingencies and, in respect to which, administrative authorization to incur commitments or obligations has been withheld. See also **operational reserve.**

Reserve Component — The Armed Forces of the United States Reserve Component consists of: a. the Army National Guard of the United States; b. the Army Reserve; c. the Navy Reserve; d. the Marine Corps Reserve; e. the Air National Guard of the United States; f. the Air Force Reserve; and g. the Coast Guard Reserve. Also called **RC.** See also **component; reserve.** (JP 4-05)

Reserve Component category — The category that identifies an individual's status in the Reserve Component. The three Reserve Component categories are Ready Reserve,

Standby Reserve, and Retired Reserve. Each reservist is identified by a specific Reserve Component category designation. (JP 4-05)

reserved obstacles — Those demolition obstacles that are deemed critical to the plan for which the authority to detonate is reserved by the designating commander. See also **obstacle.** (JP 3-15)

reset — A set of actions to restore equipment to a desired level of combat capability commensurate with a unit's future mission. (JP 4-0)

resettled person — A refugee or an internally displaced person wishing to return somewhere other than his or her previous home or land within the country or area of original displacement. (JP 3-29)

residual forces — Unexpended portions of the remaining United States forces that have an immediate combat potential for continued military operations, and that have been deliberately withheld from utilization.

residual radiation — Nuclear radiation caused by fallout, artificial dispersion of radioactive material, or irradiation that results from a nuclear explosion and persists longer than one minute after burst. See also **contamination; initial radiation.** (JP 3-11)

resistance movement — An organized effort by some portion of the civil population of a country to resist the legally established government or an occupying power and to disrupt civil order and stability. (JP 3-05)

resource management — A financial management function that provides advice and guidance to the commander to develop command resource requirements. Also called **RM.** See also **financial management.** (JP 1-06)

resources — The forces, materiel, and other assets or capabilities apportioned or allocated to the commander of a unified or specified command. (JP 1)

response force — A mobile force with appropriate fire support designated, usually by the area commander, to deal with Level II threats in the operational area. Also called **RF.** (JP 3-10)

rest and recuperation — The withdrawal of individuals from combat or duty in a combat area for short periods of rest and recuperation. Also called **R&R.** (JP 1-0)

restraint — In the context of joint operation planning, a requirement placed on the command by a higher command that prohibits an action, thus restricting freedom of action. See also **constraint; operational limitation.** (JP 5-0)

restricted area — 1. An area (land, sea, or air) in which there are special restrictive measures employed to prevent or minimize interference between friendly forces. 2. An area under military jurisdiction in which special security measures are employed to prevent unauthorized entry. See also **controlled firing area; restricted areas (air).**

restricted areas (air) — Designated areas established by appropriate authority over which flight of aircraft is restricted. They are shown on aeronautical charts, published in notices to airmen, and provided in publications of aids to air navigation. See also **restricted area**. (JP 3-52)

restricted items list — A document listing those logistic goods and services for which nations must coordinate any contracting activity with a commander's centralized contracting organization. (JP 4-08)

restricted operations area — Airspace of defined dimensions, designated by the airspace control authority, in response to specific operational situations/requirements within which the operation of one or more airspace users is restricted. Also called **ROA**. (JP 3-52)

restricted reporting — Reporting option that allows sexual assault victims to confidentially disclose the assault to specified individual and receive medical treatment without triggering an official investigation. (JP 1-0)

restricted target — A valid target that has specific restrictions placed on the actions authorized against it due to operational considerations. See also **target**. (JP 3-60)

restricted target list — A list of restricted targets nominated by elements of the joint force and approved by the joint force commander or directed by higher authorities. Also called **RTL**. See also **restricted target; target**. (JP 3-60)

restrictive fire area — An area in which specific restrictions are imposed and into which fires that exceed those restrictions will not be delivered without coordination with the establishing headquarters. Also called **RFA**. See also **fires**. (JP 3-09)

restrictive fire line — A line established between converging friendly surface forces that prohibits fires or their effects across that line. Also called **RFL**. See also **fires**. (JP 3-09)

resupply — The act of replenishing stocks in order to maintain required levels of supply. (JP 4-09)

resuscitative care — Advanced emergency medical treatment required to prevent immediate loss of life or limb and to attain stabilization to ensure the patient could tolerate evacuation. (JP 4-02)

retained personnel — Enemy medical personnel and medical staff administrators who are engaged in either the search for, collection, transport, or treatment of the wounded or sick, or the prevention of disease; chaplains attached to enemy armed forces; and, staff of National Red Cross Societies and that of other volunteer aid societies, duly recognized and authorized by their governments to assist medical service personnel of their own armed forces, provided they are exclusively engaged in the search for, or the collection, transport or treatment of wounded or sick, or in the prevention of disease, and provided that the staff of such societies are subject to military laws and regulations. Also called **RP.** See also **personnel.** (JP 3-63)

Retired Reserve — All Reserve members who receive retirement pay on the basis of their active duty and/or Reserve service; those members who are otherwise eligible for retirement pay but have not reached age 60 and who have not elected discharge and are not voluntary members of the Ready or Standby Reserve. See also **active duty; Ready Reserve; Standby Reserve.** (JP 4-05)

retrograde — The process for the movement of non-unit equipment and materiel from a forward location to a reset (replenishment, repair, or recapitalization) program or to another directed area of operations to replenish unit stocks, or to satisfy stock requirements. (JP 4-09)

returned to military control — The status of a person whose casualty status of "duty status - whereabouts unknown" or "missing" has been changed due to the person's return or recovery by US military authority. Also called **RMC.**

returnee — A displaced person who has returned voluntarily to his or her former place of residence. (JP 3-29)

return to base — An order to proceed to the point indicated by the displayed information or by verbal communication. Also called **RTB.** (JP 3-01)

revolving fund account — An account authorized by specific provisions of law to finance a continuing cycle of business-type operations, and which are authorized to incur obligations and expenditures that generate receipts. (JP 1-06)

riot control agent — Any chemical, not listed in a schedule of the Convention on the Prohibition of the Development, Production, Stockpiling and Use of Chemical Weapons and on their Destruction that can produce rapidly in humans sensory irritation or disabling physical effects that disappear within a short time following termination of exposure. Also called **RCA.** See also **chemical warfare.** (JP 3-11)

rising mine — In naval mine warfare, a mine having positive buoyancy which is released from a sinker by a ship influence or by a timing device. (JP 3-15)

risk — Probability and severity of loss linked to hazards. See also **hazard; risk management.** (JP 5-0)

risk assessment — The identification and assessment of hazards (first two steps of risk management process). Also called **RA.** (JP 3-07.2)

risk management — The process of identifying, assessing, and controlling risks arising from operational factors and making decisions that balance risk cost with mission benefits. Also called **RM.** See also **risk.** (JP 3-0)

riverine operations — Operations conducted by forces organized to cope with and exploit the unique characteristics of a riverine area, to locate and destroy hostile forces, and/or to achieve or maintain control of the riverine area. (JP 3-32)

road net — The system of roads available within a particular locality or area. (JP 4-01.5)

role specialist nation — A nation that has agreed to assume responsibility for providing a particular class of supply or service for all or part of the multinational force. Also called **RSN.** See also **lead nation; multinational force.** (JP 4-08)

roll-on/roll-off discharge facility — A platform made up of causeway sections that provide a means of embarking and disembarking vehicles from a roll-on and roll-off ship at sea to lighterage. Also called **RRDF.** See also **facility; lighterage.** (JP 4-01.6)

rough terrain container handler — A piece of materials handling equipment used to pick up and move containers. Also called **RTCH.** (JP 4-01.6)

route classification — Classification assigned to a route using factors of minimum width, worst route type, least bridge, raft, or culvert military load classification, and obstructions to traffic flow. (JP 3-34)

rules of engagement — Directives issued by competent military authority that delineate the circumstances and limitations under which United States forces will initiate and/or continue combat engagement with other forces encountered. Also called **ROE.** See also **law of war.** (JP 1-04)

ruse — In military deception, a trick of war designed to deceive the adversary, usually involving the deliberate exposure of false information to the adversary's intelligence collection system. (JP 3-13.4)

S

sabotage — An act or acts with intent to injure, interfere with, or obstruct the national defense of a country by willfully injuring or destroying, or attempting to injure or destroy, any national defense or war materiel, premises, or utilities, to include human and natural resources. (JP 2-01.2)

safe area — A designated area in hostile territory that offers the evader or escapee a reasonable chance of avoiding capture and of surviving until he or she can be evacuated. (JP 3-50)

safe haven — 1. Designated area(s) to which noncombatants of the United States Government's responsibility and commercial vehicles and materiel may be evacuated during a domestic or other valid emergency. (JP 3-68) 2. A protected body of water or the well deck of an amphibious ship used by small craft operating offshore for refuge from storms or heavy seas. (JP 4-01.6)

safe house — An innocent-appearing house or premises established by an organization for the purpose of conducting clandestine or covert activity in relative security. (JP 3-07.2)

safing — As applied to weapons and ammunition, the changing from a state of readiness for initiation to a safe condition. Also called **de-arming.**

salvage — 1. Property that has some value in excess of its basic material content but is in such condition that it has no reasonable prospect of use for any purpose as a unit and its repair or rehabilitation for use as a unit is clearly impractical. 2. The saving or rescuing of condemned, discarded, or abandoned property, and of materials contained therein for reuse, refabrication, or scrapping. (JP 4-0)

salvo — 1. In naval gunfire support, a method of fire in which a number of weapons are fired at the same target simultaneously. 2. In close air support or air interdiction operations, a method of delivery in which the release mechanisms are operated to release or fire all ordnance of a specific type simultaneously.

sanction enforcement — Operations that employ coercive measures to control the movement of certain types of designated items into or out of a nation or specified area. (JP 3-0)

scheduled target — Planned target upon which fires or other actions are scheduled for prosecution at a specified time. See also **planned target; target.** (JP 3-60)

scheduled wave — See **wave.**

schedule of fire — Groups of fires or series of fires fired in a definite sequence according to a definite program. The time of starting the schedule may be on call. For identification purposes, schedules may be referred to by a code name or other designation. (JP 3-09)

scheme of maneuver — The central expression of the commander's concept for operations that governs the design of supporting plans or annexes of how arrayed forces will accomplish the mission. (JP 5-0)

scientific and technical intelligence — The product resulting from the collection, evaluation, analysis, and interpretation of foreign scientific and technical information that covers: a. foreign developments in basic and applied research and in applied engineering techniques; and b. scientific and technical characteristics, capabilities, and limitations of all foreign military systems, weapons, weapon systems, and materiel; the research and development related thereto; and the production methods employed for their manufacture. Also called **S&TI**. See also **intelligence; technical intelligence.** *(JP 2-01)

screening — In intelligence, the evaluation of an individual or a group of individuals to determine their potential to answer collection requirements or to identify individuals who match a predetermined source profile coupled with the process of identifying and assessing the areas of knowledge, cooperation, and possible approach techniques for an individual who has information of intelligence value. (JP 2-01.2)

S-Day — See **times.**

sea areas — Areas in the amphibious objective area designated for the stationing of amphibious task force ships. Sea areas include inner transport area, sea echelon area, fire support area, etc. See also **amphibious objective area; fire support area; inner transport area; sea echelon area.** (JP 3-02)

sea barge — A type of barge-ship that can carry up to 38 loaded barges and also carry tugs, stacked causeway sections, various watercraft, or heavy lift equipment to better support joint logistics over-the-shore operations. Also called **SEABEE**. (JP 4-01.2)

seabasing — The deployment, assembly, command projection, reconstitution, and re-employment of joint power from the sea without reliance on land bases within the operational area. See also **amphibious operation.** (JP 3-02)

sea control operations — The employment of forces to destroy enemy naval forces, suppress enemy sea commerce, protect vital sea lanes, and establish local military superiority in vital sea areas. See also **land control operations.** (JP 3-32)

sea echelon — A portion of the assault shipping which withdraws from or remains out of the transport area during an amphibious landing and operates in designated areas to seaward in an on-call or unscheduled status. (JP 3-02)

sea echelon area — In amphibious operations, an area to seaward of a transport area from which assault shipping is phased into the transport area, and to which assault shipping withdraws from the transport area. (JP 3-02)

sea echelon plan — In amphibious operations, the distribution plan for amphibious shipping in the transport area to minimize losses due to enemy attack by weapons of mass destruction and to reduce the area to be swept of mines. See also **amphibious operation.** (JP 3-02)

SEAL delivery vehicle team — United States Navy forces organized, trained, and equipped to conduct special operations with SEAL delivery vehicles, dry deck shelters, and other submersible platforms. (JP 3-05)

sealift enhancement features — Special equipment and modifications that adapt merchant-type dry cargo ships and tankers to specific military missions. Also called **SEFs.** See also **Military Sealift Command; Ready Reserve.** (JP 4-01.2)

SEAL team — United States Navy forces organized, trained, and equipped to conduct special operations with an emphasis on maritime, coastal, and riverine environments. (JP 3-05)

seaport — A land facility designated for reception of personnel or materiel moved by sea, and that serves as an authorized port of entrance into or departure from the country in which located. See also **port of debarkation; port of embarkation.** (JP 4-01.2)

search — A systematic reconnaissance of a defined area, so that all parts of the area have passed within visibility. (JP 3-50)

search and rescue — The use of aircraft, surface craft, submarines, and specialized rescue teams and equipment to search for and rescue distressed persons on land or at sea in a permissive environment. Also called **SAR.** See also **combat search and rescue; isolated personnel; joint personnel recovery center; personnel recovery coordination cell.** (JP 3-50)

search and rescue numerical encryption grid — A predesignated ten-letter word without repeated letters used exclusively by recovery forces or isolated personnel to encrypt numerical data such as position, time, headings, etc., in a covert manner. Also called **SARNEG.** (JP 3-50)

search and rescue point — A predesignated specific location, relative to which isolated personnel provide their position to recovery forces. Also called **SARDOT.** (JP 3-50)

search and rescue region — An area of defined dimensions, recognized by the International Civil Aviation Organization, International Maritime Organization, or other cognizant international body, and associated with a rescue coordination center within which search and rescue services are provided. See also **inland search and rescue region.** (JP 3-50)

sea state — A scale that categorizes the force of progressively higher seas by wave height. (JP 4-01.6)

seaward launch point — A designated point off the coast from which special operations forces will launch to proceed to the beach to conduct operations. Also called **SLP.** See also **seaward recovery point.** (JP 3-05.1)

seaward recovery point — A designated point off the coast to which special operations forces will proceed for recovery by submarine or other means of recovery. Also called **SRP.** See also **seaward launch point.** (JP 3-05.1)

secondary loads — Unit equipment, supplies, and major end items that are transported in the beds of organic vehicles.

secret — Security classification that shall be applied to information, the unauthorized disclosure of which reasonably could be expected to cause serious damage to the national security that the original classification authority is able to identify or describe. (EO 13526)

SECRET Internet Protocol Router Network — The worldwide SECRET-level packet switch network that uses high-speed internet protocol routers and high-capacity Defense Information Systems Network circuitry. Also called **SIPRNET.** See also **Defense Information Systems Network.** (JP 6-0)

section — 1. As applied to ships or naval aircraft, a tactical subdivision of a division. 2. A subdivision of an office, installation, territory, works, or organization; especially a major subdivision of a staff. 3. A tactical unit of the Army and Marine Corps smaller than a platoon and larger than a squad. 4. An area in a warehouse extending from one wall to the next; usually the largest subdivision of one floor. (JP 3-33)

sector air defense commander — Commander subordinate to an area/regional air defense commander, who is responsible for air and missile defenses in the assigned sector and exercises authorities delegated by the area/regional air defense commander. Also called **SADC.** (JP 3-01)

security — 1. Measures taken by a military unit, activity, or installation to protect itself against all acts designed to, or which may, impair its effectiveness. (JP 3-10) 2. A condition that results from the establishment and maintenance of protective measures that ensure a state of inviolability from hostile acts or influences. (JP 3-10) 3. With respect to classified matter, the condition that prevents unauthorized persons from having access to official information that is safeguarded in the interests of national security. See also **national security.** (JP 2-0)

security assistance — Group of programs authorized by the Foreign Assistance Act of 1961, as amended, and the Arms Export Control Act of 1976, as amended, or other related statutes by which the United States provides defense articles, military training, and other defense-related services by grant, loan, credit, or cash sales in furtherance of national policies and objectives. Security assistance is an element of security cooperation funded and authorized by Department of State to be administered by Department of Defense/Defense Security Cooperation Agency. Also called **SA.** See also **security cooperation.** (JP 3-22)

security classification — A category to which national security information and material is assigned to denote the degree of damage that unauthorized disclosure would cause to national defense or foreign relations of the United States and to denote the degree of protection required. There are three such categories: **top secret**, **secret**, and **confidential**. See also **classification; security.** (EO 13526)

security clearance — An administrative determination by competent authority that an individual is eligible for access to classified information. (JP 1-0)

security cooperation — All Department of Defense interactions with foreign defense establishments to build defense relationships that promote specific US security interests, develop allied and friendly military capabilities for self-defense and multinational operations, and provide US forces with peacetime and contingency access to a host nation. Also called **SC.** See also **security assistance.** (JP 3-22)

security cooperation organization — All Department of Defense elements located in a foreign country with assigned responsibilities for carrying out security assistance/cooperation management functions. It includes military assistance advisory groups, military missions and groups, offices of defense and military cooperation, liaison groups, and defense attaché personnel designated to perform security assistance/cooperation functions. Also called **SCO.** (JP 3-22)

security countermeasures — Those protective activities required to prevent espionage, sabotage, theft, or unauthorized use of classified or controlled information, systems, or material of the Department of Defense. See also **counterintelligence.** (JP 2-01.2)

security force assistance — The Department of Defense activities that contribute to unified action by the US Government to support the development of the capacity and capability of foreign security forces and their supporting institutions. Also called **SFA.** (JP 3-22)

security forces — Duly constituted military, paramilitary, police, and constabulary forces of a state. (JP 3-22)

security review — The process of reviewing information and products prior to public release to ensure the material will not jeopardize ongoing or future operations. See also **security.** (JP 3-61)

security sector reform — A comprehensive set of programs and activities undertaken to improve the way a host nation provides safety, security, and justice. Also called **SSR.** (JP 3-07)

security service — Entity or component of a foreign government charged with responsibility for counterespionage or internal security functions. (JP 2-01.2)

sedition — Willfully advocating or teaching the duty or necessity of overthrowing the US government or any political subdivision by force or violence. See also **counterintelligence.** (JP 2-01.2)

segregation — In detainee operations, the removal of a detainee from other detainees and their environment for legitimate purposes unrelated to interrogation, such as when necessary for the movement, health, safety, and/or security of the detainee, the detention facility, or its personnel. (JP 3-63)

seize — To employ combat forces to occupy physically and to control a designated area. (JP 3-18)

seizures — In counterdrug operations, includes drugs and conveyances seized by law enforcement authorities and drug-related assets confiscated based on evidence that they have been derived from or used in illegal narcotics activities. See also **counterdrug operations; law enforcement agency.** (JP 3-07.4)

Selected Reserve — Those units and individuals within the Ready Reserve designated by their respective Services and approved by the Joint Chiefs of Staff as so essential to initial wartime missions that they have priority over all other Reserves. Selected Reservists actively participate in a Reserve Component training program. The Selected Reserve also includes persons performing initial active duty for training. See also **Ready Reserve.** (JP 4-05)

selective identification feature — A capability that, when added to the basic identification friend or foe system, provides the means to transmit, receive, and display selected coded replies. (JP 3-52)

selective loading — The arrangement and stowage of equipment and supplies aboard ship in a manner designed to facilitate issues to units. (JP 3-02.1)

selective mobilization — See **mobilization, Part 2.**

selective off-loading — The capability to access and off-load vehicles, supplies, and equipment without having to conduct a major reconfiguration or total off-load; influenced by the number and types of ships allocated, and the space made available for the embarkation of the landing force. (JP 3-02.1)

selective unloading — In an amphibious operation, the controlled unloading from assault shipping, and movement ashore, of specific items of cargo at the request of the landing force commander. Normally, selective unloading parallels the landing of nonscheduled units during the initial unloading period of the ship-to-shore movement. (JP 3-02)

senior airfield authority — An individual designated by the joint force commander to be responsible for the control, operation, and maintenance of an airfield to include the runways, associated taxiways, parking ramps, land, and facilities whose proximity directly affects airfield operations. Also called SAA. (JP 3-17)

senior contracting official — The lead Service or joint command designated contracting official who has direct managerial responsibility over theater support contracting. Also called SCO. (JP 4-10)

senior meteorological and oceanographic officer — Meteorological and oceanographic officer responsible for assisting the combatant commander and staff in developing and executing operational meteorological and oceanographic service concepts in support of a designated joint force. Also called SMO. See also meteorological and oceanographic. (JP 3-59)

sensitive — An agency, installation, person, position, document, material, or activity requiring special protection from disclosure that could cause embarrassment, compromise, or threat to the security of the sponsoring power. (JP 2-01)

sensitive compartmented information — All information and materials bearing special community controls indicating restricted handling within present and future community intelligence collection programs and their end products for which community systems of compartmentation have been or will be formally established. (These controls are over and above the provisions of DOD 5200.1-R, Information Security Program Regulation.) Also called SCI. (JP 2-01)

sensitive compartmented information facility — An accredited area, room, group of rooms, or installation where sensitive compartmented information may be stored, used, discussed, and/or electronically processed, where procedural and physical measures prevent the free access of persons unless they have been formally indoctrinated for the particular sensitive compartmented information authorized for use or storage within the sensitive compartmented information facility. Also called SCIF. See also sensitive compartmented information. (JP 2-01)

sensitive site — A geographically limited area that contains, but is not limited to, adversary information systems, war crimes sites, critical government facilities, and areas suspected of containing high value targets. (JP 3-31)

sequel — The subsequent major operation or phase based on the possible outcomes (success, stalemate, or defeat) of the current major operation or phase. See also branch. (JP 5-0)

serial — 1. An element or a group of elements within a series which is given a numerical or alphabetical designation for convenience in planning, scheduling, and control. 2. A serial can be a group of people, vehicles, equipment, or supplies and is used in airborne, air assault, amphibious operations, and convoys. (JP 3-02)

serial assignment table — A table that is used in amphibious operations and shows the serial number, the title of the unit, the approximate number of personnel; the material, vehicles, or equipment in the serial; the number and type of landing craft and/or amphibious vehicles required to boat the serial; and the ship on which the serial is embarked. (JP 3-02)

Service — A branch of the Armed Forces of the United States, established by act of Congress, which are: the Army, Marine Corps, Navy, Air Force, and Coast Guard. (JP 1)

Service-common — Equipment, material, supplies, and services adopted by a Military Service for use by its own forces and activities. These include standard military items, base operating support, and the supplies and services provided by a Military Service to support and sustain its own forces, including those assigned to the combatant commands. Items and services defined as Service-common by one Military Service are not necessarily Service-common for all other Military Services. See also **special operations-peculiar.** (JP 3-05)

Service component command — A command consisting of the Service component commander and all those Service forces, such as individuals, units, detachments, organizations, and installations under that command, including the support forces that have been assigned to a combatant command or further assigned to a subordinate unified command or joint task force. See also **component; functional component command.** (JP 1)

Service-organic transportation assets — Transportation assets that are assigned to a Military Department for functions of the Secretaries of the Military Departments set forth in Title 10, United States Code, Sections 3013(b), 5013(b), and 8013(b). (JP 4-01)

service troops — Those units designed to render supply, maintenance, transportation, evacuation, hospitalization, and other services required by air and ground combat units to carry out effectively their mission in combat. See also **combat service support elements.**

Service-unique container — Any 20- or 40-foot International Organization for Standardization container procured or leased by a Service to meet Service-unique requirements. Also called **component-owned container.** See also **common-use container; component-owned container.** (JP 4-09)

sexual assault forensic examination kit — The medical and forensic examination kit used to ensure controlled procedures and safekeeping of any bodily specimens in a sexual assault case. Also called **SAFE kit.** (JP 1-0)

sexual assault prevention and response program — A Department of Defense program for the Military Departments and Department of Defense components that establishes sexual assault prevention and response policies to be implemented worldwide. Also called **SAPR program.** (JP 1-0)

sexual assault response coordinator — The single point of contact at an installation or within a geographic area who overseas sexual assault awareness, prevention, and response. Also called **SARC.** (JP 1-0)

shelter — An International Organization for Standardization container outfitted with live- or work-in capability. See also **International Organization for Standardization.** (JP 4-09)

shielding — 1. Material of suitable thickness and physical characteristics used to protect personnel from radiation during the manufacture, handling, and transportation of fissionable and radioactive materials. 2. Obstructions that tend to protect personnel or materials from the effects of a nuclear explosion. (JP 3-11)

shifting fire — Fire delivered at constant range at varying deflections; used to cover the width of a target that is too great to be covered by an open sheaf.

ship-to-shore movement — That portion of the action phase of an amphibious operation which includes the deployment of the landing force from the assault shipping to designated landing areas. (JP 3-02)

shore fire control party — A specially trained unit for control of naval gunfire in support of troops ashore. It consists of a spotting team to adjust fire and a naval gunfire liaison team to perform liaison functions for the supported battalion commander. Also called **SFCP.**

shore party — A task organization of the landing force, formed for the purpose of facilitating the landing and movement off the beaches of troops, equipment, and supplies; for the evacuation from the beaches of casualties and enemy prisoners of war; and for facilitating the beaching, retraction, and salvaging of landing ships and craft. It comprises elements of both the naval and landing forces. Also called **beach group.** See also **beachmaster unit; beach party; naval beach group.** (JP 3-02)

shortfall — The lack of forces, equipment, personnel, materiel, or capability, reflected as the difference between the resources identified as a plan requirement and those apportioned to a combatant commander for planning, that would adversely affect the command's ability to accomplish its mission. (JP 5-0)

short-range air defense engagement zone — In air defense, that airspace of defined dimensions within which the responsibility for engagement of air threats normally rests with short-range air defense weapons and may be established within a low- or high-altitude missile engagement zone. Also called **SHORADEZ.** (JP 3-01)

short-range ballistic missile — A land-based ballistic missile with a range capability up to about 600 nautical miles. Also called **SRBM.** (JP 3-01)

short takeoff and landing — The ability of an aircraft to clear a 50-foot (15 meters) obstacle within 1,500 feet (450 meters) of commencing takeoff or in landing, to stop within 1,500 feet (450 meters) after passing over a 50-foot (15 meters) obstacle. Also called **STOL.** (JP 3-04)

short title — A short, identifying combination of letters, and/or numbers assigned to a document or device for purposes of brevity and/or security. (JP 2-01)

show of force — An operation designed to demonstrate US resolve that involves increased visibility of US deployed forces in an attempt to defuse a specific situation that, if allowed to continue, may be detrimental to US interests or national objectives. (JP 3-0)

signal operating instructions — A series of orders issued for technical control and coordination of the signal communication activities of a command. In Marine Corps usage, these instructions are designated communication operation instructions. (JP 6-0)

signal security — A generic term that includes both communications security and electronics security. See also **security.** (JP 3-13.3)

signals intelligence — 1. A category of intelligence comprising either individually or in combination all communications intelligence, electronic intelligence, and foreign instrumentation signals intelligence, however transmitted. 2. Intelligence derived from communications, electronic, and foreign instrumentation signals. Also called **SIGINT.** See also **communications intelligence; electronic intelligence; foreign instrumentation signals intelligence; intelligence.** (JP 2-0)

signals intelligence operational control — The authoritative direction of signals intelligence activities, including tasking and allocation of effort, and the authoritative prescription of those uniform techniques and standards by which signals intelligence information is collected, processed, and reported. (JP 2-01)

signals intelligence operational tasking authority — A military commander's authority to operationally direct and levy signals intelligence requirements on designated signals intelligence resources; includes authority to deploy and redeploy all or part of the signals intelligence resources for which signals intelligence operational tasking authority has been delegated. Also called **SOTA.** (JP 2-01)

significant wave height — The average height of the third of waves observed during a given period of time. See also **surf zone.** (JP 4-01.6)

simultaneous engagement — The concurrent engagement of hostile targets by combination of interceptor aircraft and surface-to-air missiles. (JP 3-01)

single-anchor leg mooring — A mooring facility dedicated to the offshore petroleum discharge system, which permits a tanker to remain on station and pump in much higher sea states than is possible with a spread moor. Also called **SALM**. See also **offshore petroleum discharge system.** (JP 4-01.6)

single manager— A Military Department or agency designated by the Secretary of Defense to be responsible for management of specified commodities or common service activities on a Department of Defense-wide basis. (JP 4-01)

single manager for transportation — The United States Transportation Command is the Department of Defense single manager for transportation, other than Service-organic or theater-assigned transportation assets. See also **Service-organic transportation assets; theater-assigned transportation assets** (JP 4-01)

single port manager — The transportation component, designated by the Department of Defense through the US Transportation Command, responsible for management of all common-user aerial and seaports worldwide. Also called **SPM**. See also **Surface Deployment and Distribution Command; transportation component command.** (JP 4-01.5)

single-service manager — A Service component commander who is assigned the responsibility and delegated the authority to coordinate and/or perform specified personnel support or personnel service support functions in the theater of operations. See also **component.** (JP 1-0)

site exploitation — A series of activities to recognize, collect, process, preserve, and analyze information, personnel, and/or materiel found during the conduct of operations. Also called **SE.** (JP 3-31)

situation report — A report giving the situation in the area of a reporting unit or formation. Also called **SITREP.** (JP 3-50)

situation template — A depiction of assumed adversary dispositions, based on that adversary's preferred method of operations and the impact of the operational environment if the adversary should adopt a particular course of action. See also **adversary template; course of action.** (JP 2-01.3)

small arms — Man portable, individual, and crew-served weapon systems used mainly against personnel and lightly armored or unarmored equipment.

small arms ammunition — Ammunition for small arms, i.e., all ammunition up to and including 20 millimeters (.787 inches).

sociocultural analysis — The analysis of adversaries and other relevant actors that integrates concepts, knowledge, and understanding of societies, populations, and other groups of people, including their activities, relationships, and perspectives across time and space at varying scales. Also called **SCA.** (JP 2-0)

sociocultural factors — The social, cultural, and behavioral factors characterizing the relationships and activities of the population of a specific region or operational environment. (JP 2-01.3)

solatium — Monetary compensation given in areas where it is culturally appropriate to alleviate grief, suffering, and anxiety resulting from injuries, death, and property loss with a monetary payment. (JP 1-06)

sortie — In air operations, an operational flight by one aircraft. (JP 3-30)

sortie allotment message — The means by which the joint force commander allots excess sorties to meet requirements of subordinate commanders that are expressed in their air employment and/or allocation plan. Also called **SORTIEALOT.** (JP 3-30)

source — 1. A person, thing, or activity from which information is obtained. 2. In clandestine activities, a person (agent), normally a foreign national, in the employ of an intelligence activity for intelligence purposes. 3. In interrogation activities, any person who furnishes information, either with or without the knowledge that the information is being used for intelligence purposes. See also **agent; collection agency.** (JP 2-01)

source management — The process of registering and monitoring the use of sources involved in counterintelligence and human intelligence operations to protect the security of the operations and avoid conflicts among operational elements. (JP 2-01.2)

source registry — A source record/catalogue of leads and sources acquired by collectors and centralized for management, coordination and deconfliction of source operations. (JP 2-01.2)

space asset — Equipment that is an individual part of a space system, which is or can be placed in space or directly supports space activity terrestrially. (JP 3-14)

space assignment — An assignment to the individual Military Departments/Services by the appropriate transportation operating agency of movement capability, which completely or partially satisfies the stated requirements of the Military Departments/Services for the operating month and that has been accepted by them without the necessity for referral to the Joint Transportation Board for allocation. (JP 4-01)

space capability — 1. The ability of a space asset to accomplish a mission. 2. The ability of a terrestrial-based asset to accomplish a mission in or through space. See also **space asset.** (JP 3-14)

space control — Operations to ensure freedom of action in space for the United States and its allies and, when directed, deny an adversary freedom of action in space. See also **combat service support; combat support; negation; space systems.** (JP 3-14)

space coordinating authority — A commander or individual assigned responsibility for planning, integrating, and coordinating space operations support in the operational area. Also called **SCA.** (JP 3-14)

space environment — The environment corresponding to the space domain, where electromagnetic radiation, charged particles, and electric and magnetic fields are the dominant physical influences, and that encompasses the earth's ionosphere and magnetosphere, interplanetary space, and the solar atmosphere. (JP 3-59)

space force application — Combat operations in, through, and from space to influence the course and outcome of conflict by holding terrestrial targets at risk. See also **ballistic missile; force protection.** (JP 3-14)

space force enhancement — Combat support operations and force-multiplying capabilities delivered from space systems to improve the effectiveness of military forces as well as support other intelligence, civil, and commercial users. See also **combat support .** (JP 3-14)

space forces — The space and terrestrial systems, equipment, facilities, organizations, and personnel necessary to access, use and, if directed, control space for national security. See also **national security; space systems.** (JP 3-14)

space power — The total strength of a nation's capabilities to conduct and influence activities to, in, through, and from space to achieve its objectives. (JP 3-14)

space situational awareness — Cognizance of the requisite current and predictive knowledge of the space environment and the operational environment upon which space operations depend. (JP 3-14)

space superiority — The degree of dominance in space of one force over any others that permits the conduct of its operations at a given time and place without prohibitive interference from space-based threats. . (JP 3-14)

space support — Launching and deploying space vehicles, maintaining and sustaining spacecraft on-orbit, rendezvous and proximity operations, disposing of (including deorbiting and recovering) space capabilities, and reconstitution of space forces, if required. See also **combat service support.** (JP 3-14)

space surveillance — The observation of space and of the activities occurring in space. See also **space control.** (JP 3-14)

space systems — All of the devices and organizations forming the space network. (JP 3-14)

space weather — The conditions and phenomena in space and specifically in the near-Earth environment that may affect space assets or space operations. See also **space asset.** (JP 3-59)

special access program — A sensitive program, approved in writing by a head of agency with original top secret classification authority, that imposes need-to-know and access controls beyond those normally provided for access to confidential, secret, or top secret information. The level of controls is based on the criticality of the program and the assessed hostile intelligence threat. The program may be an acquisition program, an intelligence program, or an operations and support program. Also called **SAP.** (JP 3-05.1)

special actions — Those functions that due to particular sensitivities, compartmentation, or caveats cannot be conducted in normal staff channels and therefore require extraordinary processes and procedures and may involve the use of sensitive capabilities. (JP 3-05.1)

special air operation — An air operation conducted in support of special operations and other clandestine, covert, and psychological activities. (JP 3-05.1)

special boat squadron — A permanent Navy echelon III major command to which two or more special boat units are assigned for some operational and all administrative purposes. The squadron is tasked with the training and deployment of these special boat units and may augment naval special warfare task groups and task units. Also called **SBS.** (JP 3-05.1)

special boat team — United States Navy forces organized, trained, and equipped to conduct or support special operations with combatant craft and other small craft. Also called **SBT.** (JP 3-05)

special cargo — Cargo that requires special handling or protection, such as pyrotechnics, detonators, watches, and precision instruments. (JP 4-01.5)

special event — An international or domestic event, contest, activity, or meeting, which by its very nature, or by specific statutory or regulatory authority, may warrant security, safety, and/or other logistical support or assistance from the Department of Defense. (DODD 3025.18)

special forces — US Army forces organized, trained, and equipped to conduct special operations with an emphasis on unconventional warfare capabilities. Also called **SF.** (JP 3-05)

special forces group — The largest Army combat element for special operations consisting of command and control, special forces battalions, and a support battalion capable of long duration missions. Also called **SFG.** (JP 3-05)

specialization — An arrangement within an alliance wherein a member or group of members most suited by virtue of technical skills, location, or other qualifications assume(s) greater responsibility for a specific task or significant portion thereof for one or more other members. (JP 3-16)

special mission unit — A generic term to represent a group of operations and support personnel from designated organizations that is task-organized to perform highly classified activities. Also called **SMU.** (JP 3-05.1)

special operations — Operations requiring unique modes of employment, tactical techniques, equipment and training often conducted in hostile, denied, or politically sensitive environments and characterized by one or more of the following: time sensitive, clandestine, low visibility, conducted with and/or through indigenous forces, requiring regional expertise, and/or a high degree of risk. Also called **SO.** (JP 3-05)

special operations combat control team — A team of Air Force personnel organized, trained, and equipped to conduct and support special operations. Under clandestine, covert, or low-visibility conditions, these teams establish and control air assault zones; assist aircraft by verbal control, positioning, and operating navigation aids; conduct limited offensive direct action and special reconnaissance operations; and assist in the insertion and extraction of special operations forces. Also called **SOCCT.** See also **combat control team.** (JP 3-05.1)

special operations command — A subordinate unified or other joint command established by a joint force commander to plan, coordinate, conduct, and support joint special operations within the joint force commander's assigned operational area. Also called **SOC.** See also **special operations.** (JP 3-05)

special operations command and control element — A special operations element that is the focal point for the synchronization of special operations forces activities with conventional forces activities. Also called **SOCCE.** See also **command and control; joint force special operations component commander; special operations; special operations forces.** (JP 3-05)

special operations forces — Those Active and Reserve Component forces of the Military Services designated by the Secretary of Defense and specifically organized, trained, and equipped to conduct and support special operations. Also called **SOF.** See also **Air Force special operations forces; Army special operations forces; naval special warfare forces.** (JP 3-05.1)

special operations liaison element — A special operations liaison team provided by the joint force special operations component commander to the joint force air component commander (if designated), or appropriate Service component air command and control organization, to coordinate, deconflict, and integrate special operations air, surface, and subsurface operations with conventional air operations. Also called **SOLE.** See also **joint force air component commander; joint force special operations component commander; special operations.** (JP 3-05)

special operations mission planning folder — The package that contains the materials required to execute a given special operations mission. It will include the mission tasking letter, mission tasking package, original feasibility assessment (as desired), initial assessment (as desired), target intelligence package, plan of execution, infiltration and exfiltration plan of execution, and other documentation as required or desired. Also called **SOMPF.** (JP 3-05.1)

special operations naval mobile environment team — A team of Navy personnel organized, trained, and equipped to support naval special warfare forces by providing weather, oceanographic, mapping, charting, and geodesy support. Also called **SONMET.** (JP 3-05.1)

special operations-peculiar — Equipment, material, supplies, and services required for special operations missions for which there is no Service-common requirement. Also called **SO-peculiar.** See also **Service-common; special operations.** (JP 3-05)

special operations weather team — A task organized team of Air Force personnel organized, trained, and equipped to collect critical environmental information from data sparse areas. Also called **SOWT.** (JP 3-05)

special operations wing — An Air Force special operations wing. Also called **SOW.** (JP 3-05.1)

special reconnaissance — Reconnaissance and surveillance actions conducted as a special operation in hostile, denied, or politically sensitive environments to collect or verify information of strategic or operational significance, employing military capabilities not normally found in conventional forces. Also called **SR.** (JP 3-05)

special tactics team — An Air Force task-organized element of special tactics that may include combat control, pararescue, tactical air control party, and special operations weather personnel. Also called **STT.** See also **combat search and rescue; special operations; special operations forces; terminal attack control.** (JP 3-05)

specified combatant command — A command, normally composed of forces from a single Military Department, that has a broad, continuing mission, normally functional, and is established and so designated by the President through the Secretary of Defense with the advice and assistance of the Chairman of the Joint Chiefs of Staff. (JP 1)

specified task — In the context of joint operation planning, a task that is specifically assigned to an organization by its higher headquarters. See also **essential task; implied task.** (JP 5-0)

split-mission oriented protective posture — The concept of maintaining heightened protective posture only in those areas (or zones) that are contaminated, allowing personnel in uncontaminated areas to continue to operate in a reduced posture. Also called **split-MOPP.** (JP 3-11)

spoke — The portion of the hub and spoke distribution system that refers to transportation mode operators responsible for scheduled delivery to a customer of the "hub". See also **distribution; distribution system; hub; hub and spoke distribution.** (JP 4-09)

spot — 1. To determine by observation, deviations of ordnance from the target for the purpose of supplying necessary information for the adjustment of fire. 2. To place in a proper location. 3. An approved shipboard helicopter landing site. See also **ordnance.** (JP 3-04)

spot net — Radio communication net used by a spotter in calling fire.

spot report — A concise narrative report of essential information covering events or conditions that may have an immediate and significant effect on current planning and operations that is afforded the most expeditious means of transmission consistent with requisite security. Also called **SPOTREP.** (Note: In reconnaissance and surveillance usage, spot report is not to be used.) (JP 3-09.3)

spotter — 1. An observer stationed for the purpose of observing and reporting results of naval gunfire to the firing agency and who also may be employed in designating targets. (JP 3-09) 2. In intelligence, an agent or illegal assigned to locate and assess individuals in positions of value to an intelligence service. (JP 2-01.2)

spotting — Parking aircraft in an approved shipboard landing site. (JP 3-04)

spreader bar — A device specially designed to permit the lifting and handling of containers or vehicles and breakbulk cargo. (JP 4-01.6)

squadron — 1. An organization consisting of two or more divisions of ships, or two or more divisions (Navy) or flights of aircraft. 2. The basic administrative aviation unit of the Army, Navy, Marine Corps, and Air Force. 3. Battalion-sized ground or aviation units. (JP 3-32)

stability operations — An overarching term encompassing various military missions, tasks, and activities conducted outside the United States in coordination with other instruments of national power to maintain or reestablish a safe and secure environment, provide essential governmental services, emergency infrastructure reconstruction, and humanitarian relief. (JP 3-0)

stabilized patient — A patient whose airway is secured, hemorrhage is controlled, shock treated, and fractures are immobilized. (JP 4-02)

stable patient — A patient for whom no inflight medical intervention is expected but the potential for medical intervention exists. (JP 4-02)

staff judge advocate — A judge advocate so designated in the Army, Air Force, or Marine Corps, and the principal legal advisor of a Navy, Coast Guard, or joint force command who is a judge advocate. Also called **SJA.** (JP 1-04)

staging — Assembling, holding, and organizing arriving personnel, equipment, and sustaining materiel in preparation for onward movement. See also **staging area.** (JP 3-35)

staging area — 1. **Amphibious or airborne** — A general locality between the mounting area and the objective of an amphibious or airborne expedition, through which the expedition or parts thereof pass after mounting, for refueling, regrouping of ships, and/or exercise, inspection, and redistribution of troops. 2. **Other movements** — A general locality established for the concentration of troop units and transient personnel between movements over the lines of communications. Also called **SA.** See also **airborne; marshalling; staging.** (JP 3-35)

staging base — 1. An advanced naval base for the anchoring, fueling, and refitting of transports and cargo ships as well as replenishment of mobile service squadrons. (JP 4-01.2) 2. A landing and takeoff area with minimum servicing, supply, and shelter provided for the temporary occupancy of military aircraft during the course of movement from one location to another. (JP 3-18)

standardization — The process by which the Department of Defense achieves the closest practicable cooperation among the Services and Department of Defense agencies for the most efficient use of research, development, and production resources, and agrees to adopt on the broadest possible basis the use of: a. common or compatible operational, administrative, and logistic procedures; b. common or compatible technical procedures and criteria; c. common, compatible, or interchangeable supplies, components, weapons, or equipment; and d. common or compatible tactical doctrine with corresponding organizational compatibility. (JP 4-02)

standard operating procedure — A set of instructions covering those features of operations which lend themselves to a definite or standardized procedure without loss of effectiveness. The procedure is applicable unless ordered otherwise. Also called **SOP.** (JP 3-31)

standard use Army aircraft flight route — Route established below the coordinating altitude to facilitate the movement of Army aviation assets. Route is normally located in the corps through brigade rear areas of operation and do not require approval by the airspace control authority. Also called **SAAFR.** (JP 3-52)

Standby Reserve — Those units and members of the Reserve Component (other than those in the Ready Reserve or Retired Reserve) who are liable for active duty only, as provided in Title 10, United States Code, Sections 10151, 12301, and 12306. See also **active duty; Ready Reserve; Reserve Component; Retired Reserve.** (JP 4-05)

standing joint force headquarters — A staff organization operating under a flag or general officer providing a combatant commander with a full-time, trained joint command and control element integrated into the combatant commander's staff whose focus is on contingency and crisis action planning. Also called **SJFHQ.** (JP 3-0)

standing operating procedure — See **standard operating procedure.** (JP 3-31)

standing rules for the use of force — Preapproved directives to guide United States forces on the use of force during various operations. Also called **SRUF.** (JP 3-28)

stateless person — A person who is not considered as a national by any state under the operation of its law. See also **dislocated civilian; displaced person; evacuee; refugee.** (JP 3-29)

station time — In air transport operations, the time at which crews, passengers, and cargo are to be on board and ready for the flight. (JP 3-17)

status-of-forces agreement — A bilateral or multilateral agreement that defines the legal position of a visiting military force deployed in the territory of a friendly state. Also called **SOFA.** (JP 3-16)

sterilizer — In mine warfare, a device included in mines to render the mine permanently inoperative on expiration of a pre-determined time after laying. (JP 3-15)

stockage objective — The maximum quantities of materiel to be maintained on hand to sustain current operations, which will consist of the sum of stocks represented by the operating level and the safety level. (JP 4-08)

Stock Number — See **national stock number.**

stockpile to target sequence — 1. The order of events involved in removing a nuclear weapon from storage and assembling, testing, transporting, and delivering it on the target. 2. A document that defines the logistic and employment concepts and related physical environments involved in the delivery of a nuclear weapon from the stockpile to the target. It may also define the logistic flow involved in moving nuclear weapons to and from the stockpile for quality assurance testing, modification and retrofit, and the recycling of limited life components.

stop-loss — Presidential authority under Title 10, United States Code, Section 12305 to suspend laws relating to promotion, retirement, or separation of any member of the Armed Forces determined essential to the national security of the United States ("laws relating to promotion" broadly includes, among others, grade tables, current general or flag officer authorizations, and E8 and 9 limits). This authority may be exercised by the President only if reservists are serving on active duty under Title 10, United States Code authorities for Presidential Reserve Call-up, partial mobilization, or full mobilization. See also **mobilization; partial mobilization; Presidential Reserve Call-up.** (JP 4-05)

stowage — The method of placing cargo into a single hold or compartment of a ship to prevent damage, shifting, etc. (JP 3-02)

stowage factor — The number that expresses the space, in cubic feet, occupied by a long ton of any commodity as prepared for shipment, including all crating or packaging. (JP 4-01.2)

stowage plan — A completed stowage diagram showing what materiel has been loaded and its stowage location in each hold, between-deck compartment, or other space in a ship, including deck space. (JP 4-01.5)

strafing — The delivery of automatic weapons fire by aircraft on ground targets.

strategic communication — Focused United States Government efforts to understand and engage key audiences to create, strengthen, or preserve conditions favorable for the advancement of United States Government interests, policies, and objectives through the use of coordinated programs, plans, themes, messages, and products synchronized with the actions of all instruments of national power. Also called **SC.** (JP 5-0)

strategic concept — The course of action accepted as the result of the estimate of the strategic situation which is a statement of what is to be done in broad terms. (JP 5-0)

strategic debriefing — Debriefing activity conducted to collect information or to verify previously collected information in response to national or theater level collection priorities. (JP 2-01.2)

strategic direction — The processes and products by which the President, Secretary of Defense, and Chairman of the Joint Chiefs of Staff provide strategic guidance to the Joint Staff, combatant commands, Services, and combat support agencies. (JP 5-0)

strategic estimate — The broad range of strategic factors that influence the commander's understanding of its operational environment and its determination of missions, objectives, and courses of action. See also **estimate; national intelligence estimate.** (JP 5-0)

strategic intelligence — Intelligence required for the formation of policy and military plans at national and international levels. Strategic intelligence and tactical intelligence differ primarily in level of application, but may also vary in terms of scope and detail. See also **intelligence; operational intelligence; tactical intelligence.** (JP 2-01.2)

strategic level of war — The level of war at which a nation, often as a member of a group of nations, determines national or multinational (alliance or coalition) strategic security objectives and guidance, then develops and uses national resources to achieve those objectives. See also **operational level of war; tactical level of war.** (JP 3-0)

strategic mobility — The capability to deploy and sustain military forces worldwide in support of national strategy. (JP 4-01)

strategic plan — A plan for the overall conduct of a war. (JP 5-0)

strategic sealift — The afloat pre-positioning and ocean movement of military materiel in support of US and multinational forces. (JP 4-01.5)

strategic sealift forces — Sealift forces composed of ships, cargo handling and delivery systems, and the necessary operating personnel. See also **force.** (JP 4-01.6)

strategic sealift shipping — Common-user ships of the Military Sealift Command force, including pre-positioned ships after their pre-positioning mission has been completed and they have been returned to the operational control of the Military Sealift Command. See also **Military Sealift Command; Military Sealift Command force.** (JP 4-01.2)

strategic warning — A warning prior to the initiation of a threatening act. See also **tactical warning.**

strategy — A prudent idea or set of ideas for employing the instruments of national power in a synchronized and integrated fashion to achieve theater, national, and/or multinational objectives. (JP 3-0)

strike — An attack to damage or destroy an objective or a capability. (JP 3-0)

strike coordination and reconnaissance — A mission flown for the purpose of detecting targets and coordinating or performing attack or reconnaissance on those targets. Also called **SCAR.** (JP 3-03)

stuffing — Packing of cargo into a container. See also **unstuffing.** (JP 4-09)

submarine operating authority — The naval commander exercising operational control of submarines. Also called **SUBOPAUTH.** (JP 3-32)

subordinate campaign plan — A combatant command prepared plan that satisfies the requirements under a Department of Defense campaign plan, which, depending upon the circumstances, transitions to a supported or supporting plan in execution. (JP 5-0)

subordinate command — A command consisting of the commander and all those individuals, units, detachments, organizations, or installations that have been placed under the command by the authority establishing the subordinate command. (JP 1)

subordinate unified command — A command established by commanders of unified commands, when so authorized by the Secretary of Defense through the Chairman of the Joint Chiefs of Staff, to conduct operations on a continuing basis in accordance with the criteria set forth for unified commands. See also **area command; functional component command; operational control; subordinate command; unified command.** (JP 1)

subsidiary landing — In an amphibious operation, a landing usually made outside the designated landing area, the purpose of which is to support the main landing. (JP 3-02)

subversion — Actions designed to undermine the military, economic, psychological, or political strength or morale of a governing authority. See also **unconventional warfare.** (JP 3-24)

summit — The highest altitude above mean sea level that a projectile reaches in its flight from the gun to the target; the algebraic sum of the maximum ordinate and the altitude of the gun.

sun-synchronous orbit — An orbit in which the satellite's orbital plane is at a fixed orientation to the sun, i.e., the orbit precesses about the earth at the same rate that the earth orbits the sun. (JP 3-14)

supercargo — Personnel that accompany cargo on board a ship for the purpose of accomplishing en route maintenance and security. (JP 4-01.5)

supplies — In logistics, all materiel and items used in the equipment, support, and maintenance of military forces. See also **component; equipment.** (JP 4-0)

supply — The procurement, distribution, maintenance while in storage, and salvage of supplies, including the determination of kind and quantity of supplies. a. **producer phase**—That phase of military supply that extends from determination of procurement schedules to acceptance of finished supplies by the Services. b. **consumer phase**—That phase of military supply that extends from receipt of finished supplies by the Services through issue for use or consumption. (JP 4-0)

supply chain — The linked activities associated with providing materiel from a raw materiel stage to an end user as a finished product. See also **supply; supply chain management.** (JP 4-09)

supply chain management — A cross-functional approach to procuring, producing, and delivering products and services to customers. The broad management scope includes subsuppliers, suppliers, internal information, and funds flow. See also **supply; supply chain.** (JP 4-09)

supply management — See **inventory control.**

supply support activity — Activities assigned a Department of Defense activity address code and that have a supply support mission, i.e., direct support supply units, missile support elements, and maintenance support units. Also called **SSA.** (JP 4-09)

support — 1. The action of a force that aids, protects, complements, or sustains another force in accordance with a directive requiring such action. 2. A unit that helps another unit in battle. 3. An element of a command that assists, protects, or supplies other forces in combat. See also **close support; direct support; general support; interdepartmental or agency support; inter-Service support; mutual support.** (JP 1)

supported commander — 1. The commander having primary responsibility for all aspects of a task assigned by the Joint Strategic Capabilities Plan or other joint operation planning authority. 2. In the context of joint operation planning, the commander who prepares operation plans or operation orders in response to requirements of the Chairman of the Joint Chiefs of Staff. 3. In the context of a support command relationship, the commander who receives assistance from another commander's force or capabilities, and who is responsible for ensuring that the supporting commander understands the assistance required. See also **support; supporting commander.** (JP 3-0)

supported unit — As related to contracted support, a supported unit is the organization that is the recipient, but not necessarily the requester of, contractor-provided support. See also **requiring activity.** (JP 4-10)

supporting arms — Weapons and weapons systems of all types employed to support forces by indirect or direct fire. (JP 3-02)

supporting arms coordination center — A single location on board an amphibious command ship in which all communication facilities incident to the coordination of fire support of the artillery, air, and naval gunfire are centralized. This is the naval counterpart to the fire support coordination center utilized by the landing force. Also called **SACC.** See also **fire support coordination center.** (JP 3-09.3)

supporting commander — 1. A commander who provides augmentation forces or other support to a supported commander or who develops a supporting plan. 2. In the context of a support command relationship, the commander who aids, protects, complements, or sustains another commander's force, and who is responsible for providing the assistance required by the supported commander. See also **support; supported commander.** (JP 3-0)

supporting fire — Fire delivered by supporting units to assist or protect a unit in combat. (JP 3-09)

supporting operations — In amphibious operations, those operations conducted by forces other than those conducted by the amphibious force. See also **amphibious force; amphibious operation.** (JP 3-02)

supporting plan — An operation plan prepared by a supporting commander, a subordinate commander, or an agency to satisfy the requests or requirements of the supported commander's plan. See also **supported commander; supporting commander.** (JP 5-0)

suppression — Temporary or transient degradation by an opposing force of the performance of a weapons system below the level needed to fulfill its mission objectives. (JP 3-01)

suppression of enemy air defenses — Activity that neutralizes, destroys, or temporarily degrades surface-based enemy air defenses by destructive and/or disruptive means. Also called **SEAD.** See also **electromagnetic spectrum; electronic warfare.** (JP 3-01)

surface action group — A temporary or standing organization of combatant ships, other than carriers, tailored for a specific tactical mission. Also called **SAG.** See **group; mission.** (JP 3-32)

surface combatant — A ship constructed and armed for combat use with the capability to conduct operations in multiple maritime roles against air, surface and subsurface threats, and land targets. (JP 3-32)

Surface Deployment and Distribution Command — A major command of the US Army, and the US Transportation Command's component command responsible for designated continental United States land transportation as well as common-user water terminal and traffic management service to deploy, employ, sustain, and redeploy US forces on a global basis. Also called **SDDC.** See also **transportation component command.** (JP 4-09)

surface-to-air missile site — A plot of ground prepared in such a manner that it will readily accept the hardware used in surface-to-air missile system. (JP 3-01)

surface-to-air weapon — A surface-launched weapon for use against airborne targets. Examples include missiles, rockets, and air defense guns. (JP 3-09.3)

surface warfare — That portion of maritime warfare in which operations are conducted to destroy or neutralize enemy naval surface forces and merchant vessels. Also called **SUW.** (JP 3-32)

surf line — The point offshore where waves and swells are affected by the underwater surface and become breakers. (JP 4-01.6)

surf zone — The area of water from the surf line to the beach. See also **surf line.** (JP 4-01.6)

surveillance — The systematic observation of aerospace, surface, or subsurface areas, places, persons, or things, by visual, aural, electronic, photographic, or other means. (JP 3-0)

surveillance approach — An instrument approach conducted in accordance with directions issued by a controller referring to the surveillance radar display.

survivability — All aspects of protecting personnel, weapons, and supplies while simultaneously deceiving the enemy. (JP 3-34)

survival, evasion, resistance, and escape — Actions performed by isolated personnel designed to ensure their health, mobility, safety, and honor in anticipation of or preparation for their return to friendly control. Also called **SERE.** (JP 3-50)

suspect — 1. In counterdrug operations, a track of interest where correlating information actually ties the track of interest to alleged illegal drug operations. See also **counterdrug operations; track of interest.** 2. An identity applied to a track that is potentially hostile because of its characteristics, behavior, origin, or nationality. See also **assumed friend; neutral; unknown**. (JP 3-07.4)

sustainment — The provision of logistics and personnel services required to maintain and prolong operations until successful mission accomplishment. (JP 3-0)

sustainment, restoration, and modernization — The fuels asset sustainment program within Defense Energy Support Center that provides a long-term process to cost-effectively sustain, restore, and modernize fuel facilities. Also called **S/RM.** (JP 4-03)

synchronization — 1. The arrangement of military actions in time, space, and purpose to produce maximum relative combat power at a decisive place and time. 2. In the intelligence context, application of intelligence sources and methods in concert with the operation plan to answer intelligence requirements in time to influence the decisions they support. (JP 2-0)

synchronized clock — A technique of timing the delivery of fires by placing all units on a common time. The synchronized clock uses a specific hour and minute based on either local or universal time. Local time is established using the local time zone. (JP 3-09.3)

synthesis — In intelligence usage, the examining and combining of processed information with other information and intelligence for final interpretation. (JP 2-0)

system — A functionally, physically, and/or behaviorally related group of regularly interacting or interdependent elements; that group of elements forming a unified whole. (JP 3-0)

systems support contract — A prearranged contract awarded by a Service acquisition program management office that provides technical support, maintenance and, in some cases, repair parts for selected military weapon and support systems. See also **external support contract; theater support contract.** (JP 4-10)

T

table of allowance — An equipment allowance document that prescribes basic allowances of organizational equipment, and provides the control to develop, revise, or change equipment authorization inventory data. Also called **TOA.** (JP 4-09)

TABOO frequencies — Any friendly frequency of such importance that it must never be deliberately jammed or interfered with by friendly forces including international distress, safety, and controller frequencies. See also **electronic warfare.** (JP 3-13.1)

tactical air command center — The principal US Marine Corps air command and control agency from which air operations and air defense warning functions are directed. It is the senior agency of the US Marine air command and control system that serves as the operational command post of the aviation combat element commander. It provides the facility from which the aviation combat element commander and his battle staff plan, supervise, coordinate, and execute all current and future air operations in support of the Marine air-ground task force. The tactical air command center can provide integration, coordination, and direction of joint and combined air operations. Also called **Marine TACC.** (JP 3-09.3)

tactical air control center — The principal air operations installation (ship-based) from which all aircraft and air warning functions of tactical air operations are controlled. Also called **Navy TACC.** (JP 3-09.3)

tactical air control party — A subordinate operational component of a tactical air control system designed to provide air liaison to land forces and for the control of aircraft. Also called **TACP.** (JP 3-09.3)

tactical air coordinator (airborne) — An officer who coordinates, from an aircraft, the actions of other aircraft engaged in air support of ground or sea forces. Also called **TAC(A).** See also **forward observer.** (JP 3-09.3)

tactical air direction center — An air operations installation under the overall control of the Navy tactical air control center or the Marine Corps tactical air command center, from which aircraft and air warning service functions of tactical air operations in support of amphibious operations are directed. Also called **TADC.** (JP 3-09.3)

tactical airfield fuel dispensing system — A tactical aircraft refueling system deployed by a Marine air-ground task force in support of air operations at an expeditionary airfield or a forward arming and refueling point. Also called **TAFDS.** (JP 4-03)

tactical air officer (afloat) — The officer (aviator) under the amphibious task force commander who coordinates planning of all phases of air participation of the amphibious operation and air operations of supporting forces en route to and in the objective area. Until control is passed ashore, this officer exercises control over all operations of the tactical air control center (afloat) and is charged with the following: a.

control of all aircraft in the objective area assigned for tactical air operations, including offensive and defensive air; b. control of all other aircraft entering or passing through the objective area; and c. control of all air warning facilities in the objective area. (JP 3-02)

tactical air operations center — The principal air control agency of the US Marine air command and control system responsible for airspace control and management. It provides real-time surveillance, direction, positive control, and navigational assistance for friendly aircraft. It performs real-time direction and control of all antiair warfare operations, to include manned interceptors and surface-to-air weapons. It is subordinate to the tactical air command center. Also called **TAOC.** (JP 3-09.3)

tactical air support element — An element of a US Army division, corps, or field army tactical operations center consisting of Army component intelligence staff officer and Army component operations staff officer air personnel who coordinate and integrate tactical air support with current tactical ground operations.

tactical assembly area — An area that is generally out of the reach of light artillery and the location where units make final preparations (pre-combat checks and inspections) and rest, prior to moving to the line of departure. See also **assembly area; line of departure.** (JP 3-35)

tactical combat casualty care — A set of trauma management guidelines customized for use on the battlefield that maintains a sharp focus on the most common causes of preventable deaths on the battlefield: external hemorrhage; tension pneumothorax; and airway obstruction. (JP 4-02)

tactical combat force — A combat unit, with appropriate combat support and combat service support assets, that is assigned the mission of defeating Level III threats. Also called **TCF.** (JP 3-10)

tactical control — The authority over forces that is limited to the detailed direction and control of movements or maneuvers within the operational area necessary to accomplish missions or tasks assigned. Also called **TACON.** See also **combatant command; combatant command (command authority); operational control.** (JP 1)

tactical data link — A Joint Staff-approved, standardized communication link suitable for transmission of digital information. Tactical digital information links interface two or more command and control or weapons systems via a single or multiple network architecture and multiple communication media for exchange of tactical information. Also called **TDL.** (JP 6-0)

tactical exploitation of national capabilities — Congressionally mandated program to improve the combat effectiveness of the Services through more effective military use of national programs. Also called **TENCAP.** (JP 2-01)

tactical intelligence — Intelligence required for the planning and conduct of tactical operations. See also **intelligence.** (JP 2-01.2)

tactical level of war — The level of war at which battles and engagements are planned and executed to achieve military objectives assigned to tactical units or task forces. See also **operational level of war; strategic level of war.** (JP 3-0)

tactical-logistical group — Representatives designated by troop commanders to assist Navy control officers aboard control ships in the ship-to-shore movement of troops, equipment, and supplies. Also called **TACLOG group.** (JP 3-02)

tactical minefield — A minefield that is employed to directly attack enemy maneuver as part of a formation obstacle plan and is laid to delay, channel, or break up an enemy advance, giving the defending element a positional advantage over the attacker. (JP 3-15)

tactical obstacles — Those obstacles employed to disrupt enemy formations, to turn them into a desired area, to fix them in position under direct and indirect fires, and to block enemy penetrations. (JP 3-15)

tactical questioning — Direct questioning by any Department of Defense personnel of a captured or detained person to obtain time-sensitive tactical intelligence information, at or near the point of capture or detention and consistent with applicable law. Also called **TQ.** (JP 3-63)

tactical recovery of aircraft and personnel — A Marine Corps mission performed by an assigned and briefed aircrew for the specific purpose of the recovery of personnel, equipment, and/or aircraft when the tactical situation precludes search and rescue assets from responding and when survivors and their location have been confirmed. Also called **TRAP.** (JP 3-50)

tactical reserve — A part of a force held under the control of the commander as a maneuvering force to influence future action.

tactical warning — 1. A warning after initiation of a threatening or hostile act based on an evaluation of information from all available sources. 2. In satellite and missile surveillance, a notification to operational command centers that a specific threat event is occurring. The component elements that describe threat events are as follows: a. **country of origin** — Country or countries initiating hostilities; b. **event type and size** — Identification of the type of event and determination of the size or number of weapons; c. **country under attack** — Determined by observing trajectory of an object and predicting its impact point; and d. **event time** — Time the hostile event occurred. See also **attack assessment; strategic warning.**

tactics — The employment and ordered arrangement of forces in relation to each other. See also **procedures; techniques.** (CJCSM 5120.01)

tare weight — The weight of a container deducted from gross weight to obtain net weight or the weight of an empty container. (JP 4-09)

target — 1. An entity or object that performs a function for the adversary considered for possible engagement or other action. 2. In intelligence usage, a country, area, installation, agency, or person against which intelligence operations are directed. 3. An area designated and numbered for future firing. 4. In gunfire support usage, an impact burst that hits the target. See also **objective area.** (JP 3-60)

target acquisition — The detection, identification, and location of a target in sufficient detail to permit the effective employment of weapons. Also called **TA.** See also **target analysis.** (JP 3-60)

target analysis — An examination of potential targets to determine military importance, priority of attack, and weapons required to obtain a desired level of damage or casualties. See also **target acquisition.** (JP 3-60)

target area of interest — The geographical area where high-value targets can be acquired and engaged by friendly forces. Not all target areas of interest will form part of the friendly course of action; only target areas of interest associated with high priority targets are of interest to the staff. These are identified during staff planning and wargaming. Target areas of interest differ from engagement areas in degree. Engagement areas plan for the use of all available weapons; target areas of interest might be engaged by a single weapon. Also called **TAI.** See also **area of interest; high-value target; target.** (JP 2-01.3)

target audience — An individual or group selected for influence. Also called **TA.** (JP 3-13)

target complex — A geographically integrated series of target concentrations. See also **target.** (JP 3-60)

target component — A set of targets within a target system performing a similar function. See also **target; target critical damage point.** (JP 3-60)

target critical damage point — The part of a target component that is most vital. Also called **critical node.** See also **target; target component.** (JP 3-05.1)

target development — The systematic examination of potential target systems - and their components, individual targets, and even elements of targets - to determine the necessary type and duration of the action that must be exerted on each target to create an effect that is consistent with the commander's specific objectives. (JP 3-60)

targeteer — An individual who has completed formal targeting training in an established Service or joint school and participates in the joint targeting cycle in their current duties. (JP 3-60)

target folder — A folder, hardcopy or electronic, containing target intelligence and related materials prepared for planning and executing action against a specific target. See also **target.** (JP 3-60)

target information center — The agency or activity responsible for collecting, displaying, evaluating, and disseminating information pertaining to potential targets. Also called **TIC.** See also **target.** (JP 3-02)

targeting — The process of selecting and prioritizing targets and matching the appropriate response to them, considering operational requirements and capabilities. See also **joint targeting coordination board; target.** (JP 3-0)

target intelligence — Intelligence that portrays and locates the components of a target or target complex and indicates its vulnerability and relative importance. See also **target; target complex.** (JP 3-60)

target location error — The difference between the coordinates generated for a target and the actual location of the target. Target location error is expressed primarily in terms of circular and vertical errors or infrequently, as spherical error. Also called **TLE.** (JP 3-09.3)

target materials — Graphic, textual, tabular, digital, video, or other presentations of target intelligence, primarily designed to support operations against designated targets by one or more weapon(s) systems. See also **Air Target Materials Program.** (JP 3-60)

target nomination list — A prioritized list of targets drawn from the joint target list and nominated by component commanders, appropriate agencies, or the joint force commander's staff for inclusion on the joint integrated prioritized target list. Also called **TNL.** See also **candidate target list; joint integrated prioritized target list; target.** (JP 3-60)

target of opportunity — 1. A target identified too late, or not selected for action in time, to be included in deliberate targeting that, when detected or located, meets criteria specific to achieving objectives and is processed using dynamic targeting. 2. A target visible to a surface or air sensor or observer, which is within range of available weapons and against which fire has not been scheduled or requested. See also **dynamic targeting; target; unplanned target; unanticipated target.** (JP 3-60)

target stress point — The weakest point (most vulnerable to damage) on the critical damage point. Also called **vulnerable node.** See also **target critical damage point.** (JP 3-05.1)

target system — 1. All the targets situated in a particular geographic area and functionally related. 2. A group of targets that are so related that their destruction will produce some particular effect desired by the attacker. See also **target; target complex.** (JP 3-60)

target system analysis — An all-source examination of potential target systems to determine relevance to stated objectives, military importance, and priority of attack. Also called **TSA.** (JP 3-60)

target system assessment — The broad assessment of the overall impact and effectiveness of the full spectrum of military force applied against the operation of an enemy target system, significant subdivisions of the system, or total combat effectiveness relative to the operational objectives established. See also **target system.** (JP 3-60)

target system component — A set of targets belonging to one or more groups of industries and basic utilities required to produce component parts of an end product, or one type of a series of interrelated commodities. (JP 3-60)

task — A clearly defined action or activity specifically assigned to an individual or organization that must be done as it is imposed by an appropriate authority. (JP 1)

task component — A subdivision of a fleet, task force, task group, or task unit, organized by the respective commander or by higher authority for the accomplishment of specific tasks. (JP 3-32)

task element — A component of a naval task unit organized by the commander of a task unit or higher authority. (JP 3-32)

task force — A component of a fleet organized by the commander of a task fleet or higher authority for the accomplishment of a specific task or tasks. Also called **TF.** (JP 3-32)

task force counterintelligence coordinating authority — An individual that affects the overall coordination of counterintelligence activities (in a joint force intelligence directorate counterintelligence and human intelligence staff element, joint task force configuration), with other supporting counterintelligence organizations, and supporting agencies to ensure full counterintelligence coverage of the task force operational area. Also called **TFCICA.** See also **counterintelligence; counterintelligence activities; joint task force.** (JP 2-01.2)

task group — A component of a naval task force organized by the commander of a task force or higher authority. Also called **TG.** (JP 3-32)

tasking order — A method used to task and to disseminate to components, subordinate units, and command and control agencies projected targets and specific missions. In addition, the tasking order provides specific instructions concerning the mission planning agent, targets, and other control agencies, as well as general instructions for accomplishment of the mission. Also called **TASKORD.** See also **mission; target.** (JP 3-05.1)

task order — Order for services placed against an established contract. See also **civil augmentation program; cost-plus award fee contract.** (JP 4-10)

task organization — An organization that assigns to responsible commanders the means with which to accomplish their assigned tasks in any planned action. (JP 3-33)

task unit — A component of a naval task group organized by the commander of a task group or higher authority. (JP 3-32)

T-day — See **times.**

tear line — A physical line on an intelligence message or document separating categories of information that have been approved for foreign disclosure and release. (JP 2-0)

technical analysis — In imagery interpretation, the precise description of details appearing on imagery. (JP 2-03)

technical assistance — The providing of advice, assistance, and training pertaining to the installation, operation, and maintenance of equipment. (JP 3-22)

technical documentation — Visual information documentation (with or without sound as an integral documentation component) of an actual event made for purposes of evaluation. Typically, technical documentation contributes to the study of human or mechanical factors, procedures, and processes in the fields of medicine, science, logistics, research, development, test and evaluation, intelligence, investigations, and armament delivery. Also called **TECDOC.** (JP 3-61)

technical escort — An individual technically qualified and properly equipped to accompany designated material requiring a high degree of safety or security during shipment. (JP 3-15.1)

technical evaluation — The study and investigations by a developing agency to determine the technical suitability of material, equipment, or a system for use in the Services. (JP 3-15.1)

technical intelligence — Intelligence derived from the collection, processing, analysis, and exploitation of data and information pertaining to foreign equipment and materiel for the purposes of preventing technological surprise, assessing foreign scientific and technical capabilities, and developing countermeasures designed to neutralize an adversary's technological advantages. Also called **TECHINT.** See also **exploitation; intelligence.** (JP 2-0)

technical nuclear forensics — The collection, analysis and evaluation of pre-detonation (intact) and post-detonation (exploded) radiological or nuclear materials, devices, and debris, as well as the immediate effects created by a nuclear detonation. (JP 3-41)

technical review authority — The organization tasked to provide specialized technical or administrative expertise to the primary review authority or coordinating review authority for joint publications. Also called TRA. See also **coordinating review authority; joint publication; primary review authority.** (CJCSM 5120.01)

technical surveillance countermeasures — Techniques and measures to detect and neutralize a wide variety of hostile penetration technologies that are used to obtain unauthorized access to classified and sensitive information. Technical penetrations include the employment of optical, electro-optical, electromagnetic, fluidic, and acoustic means as the sensor and transmission medium, or the use of various types of stimulation or modification to equipment or building components for the direct or indirect transmission of information meant to be protected. Also called **TSCM.** See also **counterintelligence.** (JP 2-01.2)

techniques — Non-prescriptive ways or methods used to perform missions, functions, or tasks. See also **procedures; tactics.** (CJCSM 5120.01)

telecommunications — Any transmission, emission, or reception of signs, signals, writings, images, sounds, or information of any nature by wire, radio, visual, or other electromagnetic systems. (JP 6-0)

telemedicine — Rapid access to shared and remote medical expertise by means of telecommunications and information technologies to deliver health services and exchange health information for the purpose of improving patient care. (JP 4-02)

tempest — An unclassified term referring to technical investigations for compromising emanations from electrically operated information processing equipment; these investigations are conducted in support of emanations and emissions security. See also **counterintelligence.** (JP 2-01.2)

temporary interment — A site for the purpose of: a. the interment of the remains if the circumstances permit; or b. the reburial of remains exhumed from an emergency interment. See also **mortuary affairs.** (JP 4-06)

terminal — A facility designed to transfer cargo from one means of conveyance to another. See also **facility.** (JP 4-01.6)

terminal attack control — The authority to control the maneuver of and grant weapons release clearance to attacking aircraft. See also **joint terminal attack controller.** (JP 3-09.3)

terminal control — 1. The authority to direct aircraft to maneuver into a position to deliver ordnance, passengers, or cargo to a specific location or target. Terminal control is a type of air control. 2. Any electronic, mechanical, or visual control given to aircraft to facilitate target acquisition and resolution. See also **terminal guidance.** (JP 3-09.3)

terminal control area — A control area or portion thereof normally situated at the confluence of air traffic service routes in the vicinity of one or more major airfields. See also **control area; controlled airspace; control zone.** (JP 3-52)

terminal guidance — 1. The guidance applied to a guided missile between midcourse guidance and arrival in the vicinity of the target. 2. Electronic, mechanical, visual, or other assistance given an aircraft pilot to facilitate arrival at, operation within or over, landing upon, or departure from an air landing or airdrop facility. See also **terminal control.** (JP 3-03)

terminal guidance operations — Those actions that provide electronic, mechanical, voice or visual communications that provide approaching aircraft and/or weapons additional information regarding a specific target location. Also called **TGO.** (JP 3-09)

terminal operations — The reception, processing, and staging of passengers; the receipt, transit, storage, and marshalling of cargo; the loading and unloading of modes of transport conveyances; and the manifesting and forwarding of cargo and passengers to destination. See also **operation; terminal.** (JP 4-01.5)

terminal phase — That portion of the flight of a ballistic missile that begins when the warhead or payload reenters the atmosphere and ends when the warhead or payload detonates, releases its submunitions, or impacts. See also **boost phase; midcourse phase.** (JP 3-01)

termination criteria — The specified standards approved by the President and/or the Secretary of Defense that must be met before a joint operation can be concluded. (JP 3-0)

terms of reference — 1. A mutual agreement under which a command, element, or unit exercises authority or undertakes specific missions or tasks relative to another command, element, or unit. 2. The directive providing the legitimacy and authority to undertake a mission, task, or endeavor. Also called **TORs.** (JP 3-0)

terrain analysis — The collection, analysis, evaluation, and interpretation of geographic information on the natural and man-made features of the terrain, combined with other relevant factors, to predict the effect of the terrain on military operations. (JP 2-03)

terrain avoidance system — A system which provides the pilot or navigator of an aircraft with a situation display of the ground or obstacles which project above either a horizontal plane through the aircraft or a plane parallel to it, so that the pilot can maneuver the aircraft to avoid the obstruction. (JP 3-50)

terrain flight — (*) Flight close to the Earth's surface during which airspeed, height, and/or altitude are adapted to the contours and cover of the ground in order to avoid enemy detection and fire. Also called **TERF.**

terrain intelligence — Intelligence on the military significance of natural and man-made characteristics of an area. (JP 3-15)

terrestrial environment — The Earth's land area, including its man-made and natural surface and sub-surface features, and its interfaces and interactions with the atmosphere and the oceans. (JP 3-59)

territorial airspace — Airspace above land territory and internal, archipelagic, and territorial waters. (JP 1)

territorial waters — A belt of ocean space adjacent to and measured from the coastal states baseline to a maximum width of 12 nautical miles. (JP 1)

terrorism — The unlawful use of violence or threat of violence to instill fear and coerce governments or societies. Terrorism is often motivated by religious, political, or other ideological beliefs and committed in the pursuit of goals that are usually political. See also **antiterrorism; combating terrorism; counterterrorism; force protection condition.** (JP 3-07.2)

terrorist threat level — An intelligence threat assessment of the level of terrorist threat faced by US personnel and interests in a foreign country. The assessment is based on a continuous intelligence analysis of a minimum of five elements: terrorist group existence, capability, history, trends, and targeting. There are four threat levels: **LOW, MODERATE, SIGNIFICANT,** and **HIGH.** Threat levels should not be confused with force protection conditions. Threat level assessments are provided to senior leaders to assist them in determining the appropriate local force protection condition. (The Department of State also makes threat assessments, which may differ from those determined by Department of Defense.) (JP 3-07.2)

theater — The geographical area for which a commander of a geographic combatant command has been assigned responsibility. (JP 1)

theater antisubmarine warfare commander — A Navy commander assigned to develop plans and direct assigned and attached assets for the conduct of antisubmarine warfare within an operational area. Also called **TASWC.** (JP 3-32)

theater-assigned transportation assets — Transportation assets that are assigned under the combatant command (command authority) of a geographic combatant commander. See also **combatant command (command authority); single manager for transportation.** (JP 4-01)

theater detainee reporting center — The field operating agency of the national detainee reporting center. It is the central tracing agency within the theater, responsible for maintaining information on all detainees and their personal property within a theater of operations or assigned area of operations. Also called **TDRC.** (JP 3-63)

theater distribution — The flow of personnel, equipment, and materiel within theater to meet the geographic combatant commander's missions. See also **distribution; theater; theater distribution system.** (JP 4-09)

theater distribution system — A distribution system comprised of four independent and mutually supported networks within theater to meet the geographic combatant commander's requirements: the physical network; the financial network; the information network; and the communications network. See also **distribution; distribution plan; distribution system; theater; theater distribution.** (JP 4-01)

theater event system — Architecture for reporting ballistic missile events, composed of three independent processing and reporting elements: the joint tactical ground stations, tactical detection and reporting, and the space-based infrared system mission control station. Also called **TES.** (JP 3-14)

theater hospitalization capability — Essential care and health service support capabilities to either return the patient to duty and/or stabilization to ensure the patient can tolerate evacuation to a definitive care facility outside the theater, which is known as Role 3 in North Atlantic Treaty Organization doctrine. (JP 4-02)

theater of operations — An operational area defined by the geographic combatant commander for the conduct or support of specific military operations. Also called **TO.** See also **theater of war.** (JP 3-0)

theater of war — Defined by the President, Secretary of Defense, or the geographic combatant commander as the area of air, land, and water that is, or may become, directly involved in the conduct of major operations and campaigns involving combat. See also **area of responsibility; theater of operations.** (JP 3-0)

theater patient movement requirements center — The activity responsible for intratheater patient movement management (medical regulating and aeromedical evacuation scheduling), the development of theater-level patient movement plans and schedules, the monitoring and execution in concert with the Global Patient Movement Requirements Center. Also called **TPMRC.** (JP 4-02)

theater special operations command — A subordinate unified command established by a combatant commander to plan, coordinate, conduct, and support joint special operations. Also called **TSOC.** See also **special operations.** (JP 3-05)

theater strategy — An overarching construct outlining a combatant commander's vision for integrating and synchronizing military activities and operations with the other instruments of national power in order to achieve national strategic objectives. See also **national military strategy; national security strategy; strategy.** (JP 3-0)

theater support contract — A type of contingency contract that is awarded by contracting officers in the operational area serving under the direct contracting authority of the Service component, special operations force command, or designated joint head of contracting activity for the designated contingency operation. See also **external support contract; systems support contract.** (JP 4-10)

thermal crossover — The natural phenomenon that normally occurs twice daily when temperature conditions are such that there is a loss of contrast between two adjacent objects on infrared imagery. (JP 3-09.3)

thermal radiation — 1. The heat and light produced by a nuclear explosion. 2. Electromagnetic radiations emitted from a heat or light source as a consequence of its temperature. (JP 3-41)

thorough decontamination — Decontamination carried out by a unit to reduce contamination on personnel, equipment, materiel, and/or working areas equal to natural background or to the lowest possible levels, to permit the partial or total removal of individual protective equipment and to maintain operations with minimum degradation. See also **immediate decontamination; operational decontamination.** (JP 3-11)

threat analysis — In antiterrorism, a continual process of compiling and examining all available information concerning potential terrorist activities by terrorist groups which could target a facility. A threat analysis will review the factors of a terrorist group's existence, capability, intentions, history, and targeting, as well as the security environment within which friendly forces operate. Threat analysis is an essential step in identifying probability of terrorist attack and results in a threat assessment. See also **antiterrorism.** (JP 3-07.2)

threat and vulnerability assessment — In antiterrorism, the pairing of a facility's threat analysis and vulnerability analysis. See also **antiterrorism.** (JP 3-07.2)

threat assessment — In antiterrorism, examining the capabilities, intentions, and activities, past and present, of terrorist organizations as well as the security environment within which friendly forces operate to determine the level of threat. Also called **TA.** (JP 3-07.2)

threat reduction cooperation — Activities undertaken with the consent and cooperation of host nation authorities in a permissive environment to enhance physical security, and to reduce, dismantle, redirect, and/or improve protection of a state's existing weapons of mass destruction program, stockpiles, and capabilities. Also called **TRC.** (JP 3-40)

threat warning — The urgent communication and acknowledgement of time-critical information essential for the preservation of life and/or vital resources. (JP 2-01)

throughput — 1. In transportation, the average quantity of cargo and passengers that can pass through a port on a daily basis from arrival at the port to loading onto a ship or plane, or from the discharge from a ship or plane to the exit (clearance) from the port complex. (JP 4-01.5) 2. In patient movement and care, the maximum number of patients (stable or stabilized) by category, that can be received at the airport, staged, transported, and received at the proper hospital within any 24-hour period. (JP 4-02)

throughput capacity — The estimated capacity of a port or an anchorage to clear cargo and/or passengers in 24 hours usually expressed in tons for cargo, but may be expressed in any agreed upon unit of measurement. See also **clearance capacity**. (JP 4-01.5)

time-definite delivery — The consistent delivery of requested logistic support at a time and destination specified by the receiving activity. See also **logistic support.** Also called **TDD.** (JP 4-09)

time of flight — In artillery, mortar, and naval gunfire support, the time in seconds from the instant a weapon is fired, launched, or released from the delivery vehicle or weapons system to the instant it strikes or detonates.

time on target — The actual time at which munitions impact the target. Also called **TOT.** (JP 3-09.3)

time-phased force and deployment data — The time-phased force data, non-unit cargo and personnel data, and movement data for the operation plan or operation order or ongoing rotation of forces. Also called **TPFDD.** See also **time-phased force and deployment list.** (JP 5-0)

time-phased force and deployment list — Appendix 1 to Annex A of the operation plan. It identifies types and/or actual units required to support the operation plan and indicates origin and ports of debarkation or ocean area. It may also be generated as a computer listing from the time-phased force and deployment data. Also called **TPFDL.** See also **Joint Operation Planning and Execution System; time-phased force and deployment data.** (JP 4-05)

times — The Chairman of the Joint Chiefs of Staff coordinates the proposed dates and times with the commanders of the appropriate unified and specified commands, as well as any recommended changes to when specified operations are to occur (C-, D-, M-days end at 2400 hours Universal Time [Zulu time] and are assumed to be 24 hours long for planning). (JP 5-0)

time-sensitive target — A joint force commander validated target or set of targets requiring immediate response because it is a highly lucrative, fleeting target of opportunity or it poses (or will soon pose) a danger to friendly forces. Also called **TST.** (JP 3-60)

time to target — The number of minutes and seconds to elapse before aircraft ordnance impacts on target. Also called **TTT**. (JP 3-09.3)

tophandler — A device specially designed to permit the lifting and handling of containers from the top with rough terrain container handlers. See also **container**. (JP 4-01.6)

topographic map — A map that presents the vertical position of features in measurable form as well as their horizontal positions. (JP 2-03)

top secret — Security classification that shall be applied to information, the unauthorized disclosure of which reasonably could be expected to cause exceptionally grave damage to the national security that the original classification authority is able to identify or describe. (EO 13526)

torture — As defined by Title 18, US Code, Section 2340, it is any act committed by a person acting under color of law specifically intended to inflict severe physical or mental pain or suffering (other than pain or suffering incidental to lawful sanctions) upon another person within his custody or physical control. "Severe mental pain or suffering" means the prolonged mental harm caused by or resulting from: (a) the intentional infliction or threatened infliction of severe physical pain or suffering; (b) the administration or application, or threatened administration or application, of mind-altering substances or other procedures calculated to disrupt profoundly the senses or personality; (c) the threat of imminent death; or (d) the threat that another person will imminently be subjected to death, severe physical pain or suffering, or the administration or application of mind-altering substances or other procedures calculated to disrupt profoundly the senses or personality. (JP 2-01.2)

total materiel requirement — The sum of the peacetime force material requirement and the war reserve material requirement.

total mobilization — See **mobilization**.

toxic industrial biological — Any biological material manufactured, used, transported, or stored by industrial, medical, or commercial processes which could pose an infectious or toxic threat. Also called **TIB**. (JP 3-11)

toxic industrial chemical — A chemical developed or manufactured for use in industrial operations or research by industry, government, or academia that poses a hazard. Also called **TIC**. (JP 3-11)

toxic industrial material — A generic term for toxic, chemical, biological, or radioactive substances in solid, liquid, aerosolized, or gaseous form that may be used, or stored for use, for industrial, commercial, medical, military, or domestic purposes. Also called **TIM**. (JP 3-11)

toxic industrial radiological — Any radiological material manufactured, used, transported, or stored by industrial, medical, or commercial processes. Also called **TIR.** (JP 3-11)

track — 1. A series of related contacts displayed on a data display console or other display device. 2. To display or record the successive positions of a moving object. 3. To lock onto a point of radiation and obtain guidance therefrom. 4. To keep a gun properly aimed, or to point continuously a target-locating instrument at a moving target. 5. The actual path of an aircraft above or a ship on the surface of the Earth. 6. One of the two endless belts on which a full-track or half-track vehicle runs. 7. A metal part forming a path for a moving object such as the track around the inside of a vehicle for moving a mounted machine gun. (JP 3-01)

track correlation — Correlating track information for identification purposes using all available data. (JP 3-01)

tracking — Precise and continuous position-finding of targets by radar, optical, or other means. (JP 3-07.4)

track management — Defined set of procedures whereby the commander ensures accurate friendly and enemy unit and/or platform locations, and a dissemination procedure for filtering, combining, and passing that information to higher, adjacent, and subordinate commanders. (JP 3-01)

track of interest — In counterdrug operations, contacts that meet the initial identification criteria applicable in the area where the contacts are detected. Also called **TOI.** See also **suspect.** (JP 3-07.4)

tradecraft — Specialized methods and equipment used in the organization and activity of intelligence organizations, especially techniques and methods for handling communications with agents. Operational practices and skills used in the performance of intelligence related duties. (JP 2-01.2)

traffic flow security — The protection resulting from features, inherent in some cryptoequipment, that conceal the presence of valid messages on a communications circuit, normally achieved by causing the circuit to appear busy at all times.

traffic management — The direction, control, and supervision of all functions incident to the procurement and use of freight and passenger transportation services. (JP 4-09)

traffic pattern — The traffic flow that is prescribed for aircraft landing at, taxiing on, and taking off from an airport. The usual components of a traffic pattern are upwind leg, crosswind leg, downwind leg, base leg, and final approach.

training aid — Any item developed or procured with the primary intent that it shall assist in training and the process of learning. (JP 1-06)

training and readiness oversight — The authority that combatant commanders may exercise over assigned Reserve Component forces when not on active duty or when on active duty for training. Also called **TRO.** See also **combatant commander.** (JP 1)

training unit — A unit established to provide military training to individual reservists or to Reserve Component units. (JP 4-05)

transient forces — Forces that pass or stage through, or base temporarily within, the operational area of another command but are not under its operational control. See also **force.** (JP 1)

transit zone — The path taken by either airborne or seaborne smugglers. See also **arrival zone.** (JP 3-07.4)

transmission security — The component of communications security that results from all measures designed to protect transmissions from interception and exploitation by means other than cryptanalysis. See also **communications security.** (JP 6-0)

transnational threat — Any activity, individual, or group not tied to a particular country or region that operates across international boundaries and threatens United States national security or interests. (JP 3-26)

transport area — In amphibious operations, an area assigned to a transport organization for the purpose of debarking troops and equipment. See also **inner transport area; outer transport area**. (JP 3-02)

transportation closure — The actual arrival date of a specified movement requirement at port of debarkation. (JP 3-35)

transportation component command — A major command of its parent Service under United States Transportation Command, which includes Air Force Air Mobility Command, Navy Military Sealift Command, and Army Military Surface Deployment and Distribution Command. Also called **TCC.** (JP 4-01.6)

transportation feasibility — A determination that the capability exists to move forces, equipment, and supplies from the point of origin to the final destination within the time required. See also **operation plan.** (JP 4-09)

transportation feasible — A determination made by the supported commander that a draft operation plan can be supported with the apportioned transportation assets. (JP 5-0)

transportation priorities — Indicators assigned to eligible traffic that establish its movement precedence. Appropriate priority systems apply to the movement of traffic by sea and air. In times of emergency, priorities may be applicable to continental United States movements by land, water, or air. (JP 4-09)

transportation system — All the land, water, and air routes and transportation assets engaged in the movement of United States forces and their supplies during military operations, involving both mature and immature theaters and at the strategic, operational, and tactical levels of war. (JP 4-01)

transport group — An element that directly deploys and supports the landing of the landing force, and is functionally designated as a transport group in the amphibious task force organization. A transport group provides for the embarkation, movement to the objective, landing, and logistic support of the landing force. Transport groups comprise all sealift and airlift in which the landing force is embarked. They are categorized as follows: a. airlifted groups; b. Navy amphibious ship transport groups; and c. strategic sealift shipping groups. (JP 3-02)

transshipment point — A location where material is transferred between vehicles. (JP 4-01.5)

troop space cargo — Cargo such as sea or barracks bags, bedding rolls or hammocks, locker trunks, and office equipment, normally stowed in an accessible place. This cargo will also include normal hand-carried combat equipment and weapons to be carried ashore by the assault troops.

turnaround — The length of time between arriving at a point and being ready to depart from that point. (JP 4-01.5)

turning movement — A variation of the envelopment in which the attacking force passes around or over the enemy's principal defensive positions to secure objectives deep in the enemy's rear to force the enemy to abandon his position or divert major forces to meet the threat. (JP 3-06)

two-person rule — A system designed to prohibit access by an individual to nuclear weapons and certain designated components by requiring the presence at all times of at least two authorized persons, each capable of detecting incorrect or unauthorized procedures with respect to the task to be performed.

Intentionally Blank

U

unanticipated target — A target of opportunity that was unknown or not expected to exist in the operational environment. See also **operational area; target; target of opportunity.** (JP 3-60)

unauthorized commitment — An agreement that is not binding solely because the United States Government representative who made it lacked the authority to enter into that agreement on behalf of the United States Government. See also **ratification.** (JP 4-10)

uncertain environment — Operational environment in which host government forces, whether opposed to or receptive to operations that a unit intends to conduct, do not have totally effective control of the territory and population in the intended operational area. (JP 3-0)

unconventional assisted recovery — Nonconventional assisted recovery conducted by special operations forces. Also called **UAR.** See also **authenticate; evader; recovery.** (JP 3-50)

unconventional assisted recovery coordination cell — A compartmented special operations forces facility, established by the joint force special operations component commander, staffed on a continuous basis by supervisory personnel and tactical planners to coordinate, synchronize, and de-conflict nonconventional assisted recovery operations within the operational area assigned to the joint force commander. Also called **UARCC.** See also **joint operations center; joint personnel recovery center; special operations forces; unconventional assisted recovery.** (JP 3-50)

unconventional warfare — Activities conducted to enable a resistance movement or insurgency to coerce, disrupt, or overthrow a government or occupying power by operating through or with an underground, auxiliary, and guerrilla force in a denied area. Also called **UW.** (JP 3-05)

undersea warfare — Military operations conducted to establish and maintain control of the undersea portion of the maritime domain. Also called **USW.** See also **antisubmarine warfare; mine warfare.** (JP 3-32)

underwater demolition — The destruction or neutralization of underwater obstacles that is normally accomplished by underwater demolition teams. (JP 3-34)

underwater demolition team — A group of officers and enlisted specially trained and equipped to accomplish the destruction or neutralization of underwater obstacles and associated tasks. Also called **UDT.** (JP 3-34)

unexploded explosive ordnance — Explosive ordnance which has been primed, fused, armed or otherwise prepared for action, and which has been fired, dropped, launched, projected, or placed in such a manner as to constitute a hazard to operations, installations, personnel, or material and remains unexploded either by malfunction or design or for any other cause. Also called **UXO.** See also **explosive ordnance.** (JP 3-15)

unified action — The synchronization, coordination, and/or integration of the activities of governmental and nongovernmental entities with military operations to achieve unity of effort. (JP 1)

unified combatant command — See **unified command.** (JP 1)

unified command — A command with a broad continuing mission under a single commander and composed of significant assigned components of two or more Military Departments that is established and so designated by the President, through the Secretary of Defense with the advice and assistance of the Chairman of the Joint Chiefs of Staff. Also called **unified combatant command.** See also **combatant command; subordinate unified command.** (JP 1)

Unified Command Plan — The document, approved by the President, that sets forth basic guidance to all unified combatant commanders; establishes their missions, responsibilities, and force structure; delineates the general geographical area of responsibility for geographic combatant commanders; and specifies functional responsibilities for functional combatant commanders. Also called **UCP.** See also **combatant command; combatant commander.** (JP 1)

uniformed services — The Army, Navy, Air Force, Marine Corps, Coast Guard, National Oceanic and Atmospheric Administration, and Public Health Services. See also **Military Department; Service.** (JP 1-0)

unit — 1. Any military element whose structure is prescribed by competent authority. 2. An organization title of a subdivision of a group in a task force. 3. A standard or basic quantity into which an item of supply is divided, issued, or used. Also called **unit of issue.** 4. With regard to Reserve Component of the Armed Forces, a selected reserve unit organized, equipped, and trained for mobilization to serve on active duty as a unit or to augment or be augmented by another unit. (JP 3-33)

unit aircraft — Those aircraft provided an aircraft unit for the performance of a flying mission. (JP 3-17)

United States — Includes the land area, internal waters, territorial sea, and airspace of the United States, including a. United States territories; and b. Other areas over which the United States Government has complete jurisdiction and control or has exclusive authority or defense responsibility. (JP 1)

United States Armed Forces — Used to denote collectively the Army, Marine Corps, Navy, Air Force, and Coast Guard. See also **Armed Forces of the United States.** (JP 1)

United States controlled shipping — That shipping under United States flag and selected ships under foreign flag considered to be under effective United States control. See also **effective United States controlled ships.** (JP 4-01.2)

United States message text format — A program designed to enhance joint and combined combat effectiveness through standardization of message formats, data elements, and information exchange procedures. Also called **USMTF.** (JP 3-50)

United States naval ship — A public vessel of the United States that is in the custody of the Navy and is: a. Operated by the Military Sealift Command and manned by a civil service crew; or b. Operated by a commercial company under contract to the Military Sealift Command and manned by a merchant marine crew. Also called **USNS.** See also **Military Sealift Command.** (JP 4-01.2)

United States Signals Intelligence System — The unified organization of signals intelligence activities under the direction of the Director, National Security Agency/Chief, Central Security Service. It consists of the National Security Agency/Central Security Service, the components of the Military Services authorized to conduct signals intelligence, and such other entities (other than the Federal Bureau of Investigation) authorized by the National Security Council or the Secretary of Defense to conduct signals intelligence activities. Also called **USSS.** See also **counterintelligence.** (JP 2-01.2)

unit identification code — A six-character, alphanumeric code that uniquely identifies each Active, Reserve, and National Guard unit of the Armed Forces. Also called **UIC.** (JP 1-0)

unitized load — A single item or a number of items packaged, packed, or arranged in a specified manner and capable of being handled as a unit. Unitization may be accomplished by placing the item or items in a container or by banding them securely together.

unit line number — A seven-character alphanumeric code that describes a unique increment of a unit deployment, i.e., advance party, main body, equipment by sea and air, reception team, or trail party, in the time-phased force and deployment data. Also called **ULN.** (JP 3-35)

unit loading — The loading of troop units with their equipment and supplies in the same vessels, aircraft, or land vehicles. (JP 4-01.5)

unit movement control center — A temporary organization activated by major subordinate commands and subordinate units during deployment to control and manage marshalling and movement. Also called **UMCC.** See also **deployment; marshaling; unit.** (JP 3-35)

unit movement data — A unit equipment and/or supply listing containing corresponding transportability data. Tailored unit movement data has been modified to reflect a specific movement requirement. Also called **UMD**. (JP 3-35)

unit of issue — In its special storage meaning, refers to the quantity of an item; as each number, dozen, gallon, pair, pound, ream, set, yard. Usually termed unit of issue to distinguish from "unit price." See also **unit.**

unit personnel and tonnage table — A table included in the loading plan of a combat-loaded ship as a recapitulation of totals of personnel and cargo by type, listing cubic measurements and weight. Also called **UP&TT**. (3-02.1)

unit type code — A Joint Chiefs of Staff developed and assigned code, consisting of five characters that uniquely identify a "type unit." Also called **UTC**. (JP 4-02)

unity of command — The operation of all forces under a single responsible commander who has the requisite authority to direct and employ those forces in pursuit of a common purpose. (JP 3-0)

unity of effort — Coordination and cooperation toward common objectives, even if the participants are not necessarily part of the same command or organization, which is the product of successful unified action. (JP 1)

Universal Joint Task List — A menu of capabilities that may be selected by a joint force commander to accomplish the assigned mission. Also called **UJTL**. (JP 3-33)

universal polar stereographic grid — A military grid prescribed for joint use in operations in limited areas and used for operations requiring precise position reporting. It covers areas between the 80 degree parallels and the poles. (JP 2-03)

Universal Time — A measure of time that conforms, within a close approximation, to the mean diurnal rotation of the Earth and serves as the basis of civil timekeeping. Also called **ZULU time**. (Formerly called Greenwich Mean Time.) (JP 5-0)

unknown — 1. A code meaning "information not available." 2. An unidentified target. An aircraft or ship that has not been determined to be hostile, friendly, or neutral using identification friend or foe and other techniques, but that must be tracked by air defense or naval engagement systems. 3. An identity applied to an evaluated track that has not been identified. See also **assumed friend; friend; neutral; suspect.** (JP 3-01)

unmanned aircraft — An aircraft or balloon that does not carry a human operator and is capable of flight under remote control or autonomous programming. Also called **UA**. (JP 3-52)

unmanned aircraft system — That system whose components include the necessary equipment, network, and personnel to control an unmanned aircraft. Also called **UAS.** (JP 3-52)

unplanned target — A target of opportunity that is known to exist in the operational environment. See also **operational area; target; target of opportunity.** (JP 3-60)

unrestricted reporting — A process that a Service member uses to disclose, without requesting confidentiality or restricted reporting, that he or she is the victim of a sexual assault. (JP 1-0)

unstable patient — A patient whose physiological status is in fluctuation and for whom emergent, treatment, and/or surgical intervention are anticipated during treatment or evacuation; and the patient's rapidly changing status and requirements are beyond the standard en route care capability and requires medical/surgical augmentation. (JP 4-02)

unstuffing — The removal of cargo from a container. Also called **stripping.** (JP 4-09)

use of force policy — Policy guidance issued by the Commandant, US Coast Guard, on the use of force and weapons. (JP 3-03)

US forces — All Armed Forces (including the Coast Guard) of the United States, any person in the Armed Forces of the United States, and all equipment of any description that either belongs to the US Armed Forces or is being used (including Type I and II Military Sealift Command vessels), escorted, or conveyed by the US Armed Forces. (JP 1)

US national — US citizen and US permanent and temporary legal resident aliens. (JP 1)

US person — For intelligence purposes, a US person is defined as one of the following: (1) a US citizen; (2) an alien known by the intelligence agency concerned to be a permanent resident alien; (3) an unincorporated association substantially composed of US citizens or permanent resident aliens; or (4) a corporation incorporated in the United States, except for those directed and controlled by a foreign government or governments. (JP 2-01.2)

Intentionally Blank

V

validate — Execution procedure used by combatant command components, supporting combatant commanders, and providing organizations to confirm to the supported commander and United States Transportation Command that all the information records in a time-phased force and deployment data not only are error-free for automation purposes, but also accurately reflect the current status, attributes, and availability of units and requirements. (JP 5-0)

validation — 1. A process associated with the collection and production of intelligence that confirms that an intelligence collection or production requirement is sufficiently important to justify the dedication of intelligence resources, does not duplicate an existing requirement, and has not been previously satisfied. (JP 2-01) 2. A part of target development that ensures all vetted targets meet the objectives and criteria outlined in the commander's guidance and ensures compliance with the law war and rules of engagement. (JP 3-60) 3. In computer modeling and simulation, the process of determining the degree to which a model or simulation is an accurate representation of the real world from the perspective of the intended uses of the model or simulation. (JP 3-35) 4. Execution procedure whereby all the information records in a time-phased force and deployment data are confirmed error free and accurately reflect the current status, attributes, and availability of units and requirements. See also **time-phased force and deployment data; verification.** (JP 3-35)

vehicle-borne improvised explosive device — A device placed or fabricated in an improvised manner on a vehicle incorporating destructive, lethal, noxious, pyrotechnic, or incendiary chemicals and designed to destroy, incapacitate, harass, or distract. Otherwise known as a car bomb. Also called **VBIED.** (JP 3-10)

vehicle cargo — Wheeled or tracked equipment, including weapons, that require certain deck space, head room, and other definite clearance. (JP 4-01.2)

vehicle summary and priority table — A table listing all vehicles by priority of debarkation from a combat-loaded ship. It includes the nomenclature, dimensions, square feet, cubic feet, weight, and stowage location of each vehicle; the cargo loaded in each vehicle; and the name of the unit to which the vehicle belongs. Also called **VS&PT.** (JP 3-02.1)

verification — 1. In arms control, any action, including inspection, detection, and identification, taken to ascertain compliance with agreed measures. (JP 3-41) 2. In computer modeling and simulation, the process of determining that a model or simulation implementation accurately represents the developer's conceptual description and specifications. See also **configuration management; validation.** (JP 3-13.1)

vertex height — See **maximum ordinate.**

vertical and/or short takeoff and landing — Vertical and/or short takeoff and landing capability for aircraft.

vertical envelopment — A tactical maneuver in which troops that are air-dropped, air-landed, or inserted via air assault, attack the rear and flanks of a force, in effect cutting off or encircling the force.. (JP 3-18)

vertical landing zone — A specified ground area for landing vertical takeoff and landing aircraft to embark or disembark troops and/or cargo. A landing zone may contain one or more landing sites. Also called **VLZ**. See also **landing zone; vertical takeoff and landing aircraft.** (JP 3-02)

vertical replenishment — The use of a helicopter for the transfer of materiel to or from a ship. Also called **VERTREP.** (JP 3-04)

vertical stowage — A method of stowage in depth within a single compartment by which loaded items are continually accessible for unloading, and the unloading can be completed without corresponding changes or prior unloading of other cargo. (JP 3-02.1)

vertical takeoff and landing aircraft — Fixed-wing aircraft and helicopters capable of taking off or landing vertically. Also called **VTOL aircraft.** See also **vertical landing zone.** (JP 3-02)

vetting — A part of target development that assesses the accuracy of the supporting intelligence to targeting. (JP 3-60)

visual information — Various visual media with or without sound. Generally, visual information includes still and motion photography, audio video recording, graphic arts, visual aids, models, display, and visual presentations. Also called **VI.** (JP 3-61)

visual meteorological conditions — Weather conditions in which visual flight rules apply; expressed in terms of visibility, ceiling height, and aircraft clearance from clouds along the path of flight. Also called **VMC.** See also **instrument meteorological conditions.** (JP 3-04)

Voluntary Intermodal Sealift Agreement — An agreement that provides the Department of Defense with assured access to United States flag assets, both vessel capacity and intermodal systems, to meet Department of Defense contingency requirements. Also called **VISA.** See also **intermodal** (JP 4-01.2)

voluntary tanker agreement — An agreement established by the Maritime Administration to provide for United States commercial tanker owners and operators to voluntarily make their vessels available to satisfy the Department of Defense to meet contingency or war requirements for point-to-point petroleum, oils, and lubricants movements. Also called **VTA.** (JP 4-01.2)

vulnerability — 1. The susceptibility of a nation or military force to any action by any means through which its war potential or combat effectiveness may be reduced or its will to fight diminished. (JP 3-01) 2. The characteristics of a system that cause it to suffer a definite degradation (incapability to perform the designated mission) as a result of having been subjected to a certain level of effects in an unnatural (man-made) hostile environment. (JP 3-60) 3. In information operations, a weakness in information system security design, procedures, implementation, or internal controls that could be exploited to gain unauthorized access to information or an information system. See also **information operations.** (JP 3-13)

vulnerability assessment — A Department of Defense, command, or unit-level evaluation (assessment) to determine the vulnerability of a terrorist attack against an installation, unit, exercise, port, ship, residence, facility, or other site. Identifies areas of improvement to withstand, mitigate, or deter acts of violence or terrorism. Also called **VA.** (JP 3-07.2)

Intentionally Blank

W

walk-in — An unsolicited contact who provides information. (JP 2-01.2)

warden system — An informal method of communication used to pass information to US citizens during emergencies. See also **noncombatant evacuation operations.** (JP 3-68)

warning intelligence — Those intelligence activities intended to detect and report time-sensitive intelligence information on foreign developments that forewarn of hostile actions or intention against United States entities, partners, or interests. (JP 2-0)

warning order — 1. A preliminary notice of an order or action that is to follow. 2. A planning directive that initiates the development and evaluation of military courses of action by a supported commander and requests that the supported commander submit a commander's estimate. 3. A planning directive that describes the situation, allocates forces and resources, establishes command relationships, provides other initial planning guidance, and initiates subordinate unit mission planning. Also called **WARNORD.** (JP 5-0)

war reserve materiel requirement — That portion of the war materiel requirement required to be on hand on D-day. This level consists of the war materiel requirement less the sum of the peacetime assets assumed to be available on D-day and the war materiel procurement capability. (JP 4-02)

war reserves — Stocks of materiel amassed in peacetime to meet the increase in military requirements consequent upon an outbreak of war. (JP 4-01.5)

war reserve stock — That portion of total materiel assets designated to satisfy the war reserve materiel requirement. Also called **WRS.** See also **reserve; war reserve materiel requirement; war reserves.** (JP 2-03)

wartime reserve modes — Characteristics and operating procedures of sensor, communications, navigation aids, threat recognition, weapons, and countermeasures systems that will contribute to military effectiveness if unknown to or misunderstood by opposing commanders before they are used, but could be exploited or neutralized if known in advance. Also called **WARM.** (JP 3-13.1)

Washington Liaison Group — An interagency committee and/or joint monitoring body, chaired by the Department of State with representation from the Department of Defense, established to coordinate the preparation and implementation of plans for evacuation of United States citizens abroad in emergencies. Also called **WLG.** (JP 3-68)

waterspace management — The allocation of waterspace in terms of antisubmarine warfare attack procedures to permit the rapid and effective engagement of hostile submarines while preventing inadvertent attacks on friendly submarines. Also called **WSM.** (JP 3-32)

wave — A formation of forces, including ships, craft, amphibious vehicles or aircraft, required to beach or land about the same time. Waves can be classified by function: scheduled, on-call, or non-scheduled. Waves can also be classified by type of craft, e.g., assault, helicopter, or landing craft. (JP 3-02)

W-day — See **times.**

weaponeer — An individual who has completed requisite training to determine the quantity and type of lethal or nonlethal means required to create a desired effect on a given target. (JP 3-60)

weaponeering — The process of determining the quantity of a specific type of lethal or nonlethal means required to create a desired effect on a given target. (JP 3-60)

weapon engagement zone — In air defense, airspace of defined dimensions within which the responsibility for engagement of air threats normally rests with a particular weapon system. Also called **WEZ.** (JP 3-01)

weapons control status — An air defense control measure declared for a particular area and time by an area air defense commander, or delegated subordinate commander, based on the rules of engagement designed to establish the freedom for fighters and surface air defense weapons to engage threats. Also call **WCS.** (JP 3-01)

weapons free zone — An air defense zone established for the protection of key assets or facilities, other than air bases, where weapon systems may be fired at any target not positively recognized as friendly. (JP 3-01)

weapons of mass destruction — Chemical, biological, radiological, or nuclear weapons capable of a high order of destruction or causing mass casualties and exclude the means of transporting or propelling the weapon where such means is a separable and divisible part from the weapon. Also called **WMD.** See also **special operations.** (JP 3-40)

weapons of mass destruction active defense — Active measures to defeat an attack with chemical, biological, radiological, or nuclear weapons by employing actions to divert, neutralize, or destroy those weapons or their means of delivery while en route to their target. Also called **WMD active defense.** (JP 3-40)

weapons of mass destruction consequence management — Actions authorized by the Secretary of Defense to mitigate the effects of a weapon of mass destruction attack or event and, if necessary, provide temporary essential operations and services at home and abroad. Also called **WMD CM.** (JP 3-40)

weapons of mass destruction elimination — Actions undertaken in a hostile or uncertain environment to systematically locate, characterize, secure, and disable, or destroy weapons of mass destruction programs and related capabilities. Also called **WMD elimination.** (JP 3-40)

weapons of mass destruction interdiction — Operations to track, intercept, search, divert, seize, or otherwise stop the transit of weapons of mass destruction, its delivery systems, or related materials, technologies, and expertise. Also called **WMD interdiction.** (JP 3-40)

weapons of mass destruction offensive operations — Actions to disrupt, neutralize, or destroy a weapon of mass destruction threat before it can be used, or to deter subsequent use of such weapons. Also called **WMD offensive operations.** (JP 3-40)

weapons of mass destruction security cooperation and partner activities — Activities to improve or promote defense relationships and capacity of allied and partner nations to execute or support the other military mission areas to combat weapons of mass destruction through military-to-military contact, burden sharing arrangements, combined military activities, and support to international activities. Also called **WMD security cooperation.** (JP 3-40)

weapons readiness state — The degree of readiness of air defense weapons which can become airborne or be launched to carry out an assigned task and normally expressed in numbers of weapons and numbers of minutes. (JP 3-01)

weapons release authority — The authority originating from the President to engage or direct engagement of ballistic missile threats using ground-based interceptors of the ground-based midcourse defense. Also call **WRA.** (JP 3-01)

weapons technical intelligence — A category of intelligence and processes derived from the technical and forensic collection and exploitation of improvised explosive devices, associated components, improvised weapons, and other weapon systems. Also called **WTI.** (JP 3-15.1)

weapon system — A combination of one or more weapons with all related equipment, materials, services, personnel, and means of delivery and deployment (if applicable) required for self-sufficiency. (JP 3-0)

wellness — Force health protection program that consolidates and incorporates physical and mental fitness, health promotion, and environmental and occupational health. See also **force health protection.** (JP 4-02)

wharf — A structure built of open rather than solid construction along a shore or a bank that provides cargo-handling facilities. See also **quay.** (JP 4-01.5)

wing — 1. An Air Force unit composed normally of one primary mission group and the necessary supporting organizations, i.e., organizations designed to render supply, maintenance, hospitalization, and other services required by the primary mission groups. Primary mission groups may be functional, such as combat, training, transport, or service. 2. A fleet air wing is the basic organizational and administrative unit for naval-, land-, and tender-based aviation. Such wings are mobile units to which are assigned aircraft squadrons and tenders for administrative organization control. 3. A balanced Marine Corps task organization of aircraft groups and squadrons, together with appropriate command, air control, administrative, service, and maintenance units. A standard Marine Corps aircraft wing contains the aviation elements normally required for the air support of a Marine division. 4. A flank unit; that part of a military force to the right or left of the main body.

withdrawal operation — A planned retrograde operation in which a force in contact disengages from an enemy force and moves in a direction away from the enemy. (JP 3-17)

witting — A term of intelligence art that indicates that one is not only aware of a fact or piece of information but also aware of its connection to intelligence activities. (JP 2-01.2)

working capital fund — A revolving fund established to finance inventories of supplies and other stores, or to provide working capital for industrial-type activities. (JP 1-06)

working group — An enduring or ad hoc organization within a joint force commander's headquarters consisting of a core functional group and other staff and component representatives whose purpose is to provide analysis on the specific function to users. Also called **WG.** (JP 3-33)

X

Intentionally Blank

Y

Intentionally Blank

Z

zone of action — A tactical subdivision of a larger area, the responsibility for which is assigned to a tactical unit; generally applied to offensive action. (JP 3-09)

zone of fire — An area into which a designated ground unit or fire support ship delivers, or is prepared to deliver, fire support. Fire may or may not be observed. Also called ZF. (JP 3-09)

ZULU time — See **Universal Time.**

Intentionally Blank

APPENDIX A
ABBREVIATIONS AND ACRONYMS

A

A	analog
A&P	administrative and personnel; analysis and production
A-1	director of manpower, personnel, and services (Air Force)
A2C2	Army airspace command and control
A-2	intelligence staff officer (Air Force)
A-3	operations directorate (COMAFFOR staff); operations staff officer (Air Force)
A-4	director of logistics (Air Force)
A-5	plans directorate (COMAFFOR staff)
A-6	communications staff officer (Air Force)
A-7	director of installations and mission support (Air Force)
AA	assessment agent; avenue of approach
AA&E	arms, ammunition, and explosives
AAA	antiaircraft artillery; arrival and assembly area; assign alternate area
AABB	American Association of Blood Banks
AABWS	amphibious assault bulk water system
AAC	activity address code
AACG	arrival airfield control group
AADC	area air defense commander
AADP	area air defense plan
AAEC	aeromedical evacuation control team
AAFES	Army and Air Force Exchange Service
AAFIF	automated air facility information file
AAFS	amphibious assault fuel system
AAFSF	amphibious assault fuel supply facility
AAG	aeronautical assignment group
AAGS	Army air-ground system
AAI	air-to-air interface
AAM	air-to-air missile
AAMDC	US Army Air and Missile Defense Command
AAOE	arrival and assembly operations element
AAOG	arrival and assembly operations group
AAP	Allied administrative publication; assign alternate parent
AAR	after action report; after action review; air-to-air refueling area
AAS	amphibious assault ship
AAST	aeromedical evacuation administrative support team
AAT	automatic analog test; aviation advisory team
AATCC	amphibious air traffic control center
AAU	analog applique unit

AAV	amphibious assault vehicle
AAW	antiair warfare
AB	air base
ABCA	American, British, Canadian, Australian, and New Zealand
ABCS	Army Battle Command System
ABD	airbase defense
ABFC	advanced base functional component
ABFDS	aerial bulk fuel delivery system
ABFS	amphibious bulk fuel system
ABGD	air base ground defense
ABIS	Automated Biometric Identification System
ABL	airborne laser
ABLTS	amphibious bulk liquid transfer system
ABM	antiballistic missile
ABN	airborne
ABNCP	Airborne Command Post
ABO	air base operability; blood typing system
ABP	air battle plan
A/C	aircraft
AC	Active Component; alternating current
AC2	airspace command and control
AC-130	Hercules
ACA	air clearance authority; airlift clearance authority; airspace control authority; airspace coordination area
ACAA	automatic chemical agent alarm
ACAPS	area communications electronics capabilities
ACAT	aeromedical evacuation command augmentation team
ACB	amphibious construction battalion
ACC	Air Combat Command; air component commander; area coordination center; Army Contracting Command
ACCE	air component coordination element
ACCON	acoustic condition
ACCS	air command and control system
ACCSA	Allied Communications and Computer Security Agency
ACD	automated cargo documentation
ACDO	assistant command duty officer
ACE	airborne command element (USAF); air combat element (NATO); Allied Command Europe; aviation combat element (USMC)
ACEOI	automated communications-electronics operating instructions
ACF	air contingency force
ACI	assign call inhibit
ACIC	Army Counterintelligence Center
ACINT	acoustic intelligence
ACK	acknowledgement

ACL	access control list; allowable cabin load
ACLANT	Allied Command Atlantic
ACLP	affiliated contingency load planning
ACM	advanced conventional munitions; advanced cruise missile; air combat maneuver; air contingency Marine air-ground task force (MAGTF); airspace coordinating measure
ACMREQ	airspace control means request; airspace coordination measures request
ACN	assign commercial network
ACO	administrative contracting officer; airspace control order
ACOA	adaptive course of action
ACOC	area communications operations center
ACOCC	air combat operations command center
ACOS	assistant chief of staff
ACP	access control point; air commander's pointer; airspace control plan; Allied communications publication; assign common pool
ACR	armored cavalry regiment (Army); assign channel reassignment
ACS	agile combat support; air-capable ship; airspace control system; auxiliary crane ship
ACSA	acquisition and cross-servicing agreement; Allied Communications Security Agency
AC/S, C4I	Assistant Chief of Staff, Command, Control, Communications, Computers, and Intelligence (USMC)
ACT	activity; advance civilian team; Allied Command Transformation
ACU	assault craft unit
ACV	aircraft cockpit video; armored combat vehicle
ACW	advanced conventional weapons
A/D	analog-to-digital
AD	active duty; advanced deployability; air defense; automatic distribution; priority add-on
ADA	aerial damage assessment; air defense artillery
A/DACG	arrival/departure airfield control group
ADAFCO	air defense artillery fire control officer
ADAL	authorized dental allowance list
ADAM	air defense airspace management
ADAM/BAE	air defense airspace management/brigade aviation element
ADAMS	Allied Deployment and Movement System
ADANS	Air Mobility Command Deployment Analysis System
ADC	air defense commander; area damage control
ADCAP	advanced capability
A/DCG	arrival/departure control group

ADCI/MS	Associate Director of Central Intelligence for Military Support
ADCON	administrative control
ADD	assign on-line diagnostic
ADDO	Assistant Deputy Director for Operations
ADDO(MS)	Assistant Deputy Director for Operations/Military Support
ADE	airdrop damage estimate; assign digit editing
ADF	automatic direction finding
ADIZ	air defense identification zone
ADKC/RCU	Automatic Key Distribution Center/Rekeying Control Unit
ADL	advanced distributed learning; assign XX (SL) routing
ADMIN	administration
ADN	Allied Command Europe desired ground zero number
ADNET	anti-drug network
ADOC	air defense operations center
ADP	air defense plan; automated data processing
ADPE	automated data processing equipment
ADPS	automatic data processing system
ADR	accident data recorder; aircraft damage repair; airfield damage repair
ADRA	Adventist Development and Relief Agency
ADRP	Army doctrine reference publication
ADS	air defense section; air defense sector; amphibian discharge site
ADSIA	Allied Data Systems Interoperability Agency
ADSW	active duty for special work
ADT	active duty for training; assign digital transmission group; automatic digital tester
ADUSD(TP)	Assistant Deputy Under Secretary of Defense, Transportation Policy
ADVON	advanced echelon
ADW	air defense warnings
ADWC	air defense warning condition
ADZ	amphibious defense zone
A/E	ammunition/explosives
AE	aeromedical evacuation; assault echelon; attenuation equalizer
AEC	aeromedical evacuation crew
AECA	Arms Export Control Act
AECM	aeromedical evacuation crew member
AECS	aeromedical evacuation command squadron
AECT	aeromedical evacuation control team
AEG	air expeditionary group
AEHF	advanced extremely high frequency
AELT	aeromedical evacuation liaison team

AEOS	aeromedical evacuation operations squadron
AEOT	aeromedical evacuation operations team
AEPS	aircrew escape propulsion system
AEPST	aeromedical evacuation plans and strategy team
AES	aeromedical evacuation squadron
AESC	aeromedical evacuation support cell
AET	airport emergency team
AETC	Air Education and Training Command
AETF	air and space expeditionary task force
A/ETF	automated/electronic target folder
AEU	assign essential user bypass
AEW	airborne early warning
AF	Air Force; Air Force (form); amphibious force
AF/A2	Air Force Director of Intelligence, Surveillance, and Reconnaissance
AFAARS	Air Force After Action Reporting System
AFARN	Air Force air request net
AFATDS	Advanced Field Artillery Tactical Data System
AFB	Air Force base
AFC	area frequency coordinator; automatic frequency control
AFCA	Air Force Communications Agency
AFCAP	Air Force contract augmentation program; Armed Forces contract augmentation program
AFCB	Armed Forces Chaplains Board
AFCC	Air Force Component Commander
AFCCC	Air Force Combat Climatology Center
AFCEE	Air Force Center for Engineering and the Environment
AFCENT	Allied Forces Central Europe (NATO)
AFCERT	Air Force computer emergency response team
AFCESA	Air Force Civil Engineering Support Agency
AFD	assign fixed directory
AFDA	Air Force doctrine annex
AFDC	Air Force Doctrine Center
AFDD	Air Force doctrine document
AFDIGS	Air Force digital graphics system
AFDIL	Armed Forces DNA Identification Laboratory
AFDIS	Air Force Weather Agency Dial In Subsystem
AF/DP	Deputy Chief of Staff for Personnel, United States Air Force
AFE	Armed Forces Entertainment
AFEES	Armed Forces Examining and Entrance Station
AFFIS	Air Facilities File Information System
AFFMA	Air Force Frequency Management Agency
AFFOR	Air Force forces
AFH	Air Force handbook
AFHSC	Armed Forces Health Surveillance Center
AFI	Air Force instruction

AFIAA	Air Force Intelligence Analysis Agency
AFID	anti-fratricide identification device
AF/IL	Deputy Chief of Staff for Installations and Logistics, USAF
AFIP	Armed Forces Institute of Pathology
AFIRB	Armed Forces Identification Review Board
AFIS	American Forces Information Service
AFISRA	Air Force Intelligence, Surveillance, and Reconnaissance Agency
AFIWC	Air Force Information Warfare Center
AFJI	Air Force joint instruction
AFJMAN	Air Force Joint Manual
AFLC	Air Force Logistics Command
AFLE	Air Force liaison element
AFLNO	Air Force liaison officer
AFMAN	Air Force manual
AFMC	Air Force Materiel Command
AFMD	Air Force Mission Directive
AFME	Armed Forces Medical Examiner
AFMES	Armed Forces Medical Examiner System
AFMIC	Armed Forces Medical Intelligence Center
AFMLO	Air Force Medical Logistics Office
AFMS	Air Force Medical Service
AFNORTH	Air Force North; Allied Forces Northern Europe (NATO)
AFNORTHWEST	Allied Forces North West Europe (NATO)
AFNSEP	Air Force National Security and Emergency Preparedness Agency
AFOA	Air Force Operations Activity
AFOE	assault follow-on echelon
AFOSI	Air Force Office of Special Investigations
AFPAM	Air Force pamphlet
AFPC	Air Force Personnel Center
AFPD	Air Force policy directive
AFPEO	Armed Forces Professional Entertainment Overseas
AFR	Air Force Reserve; assign frequency for network reporting
AFRAT	Air Force Radiation Assessment Team
AFRC	Air Force Reserve Command; Armed Forces Recreation Center
AFRCC	Air Force rescue coordination center
AFRL	Air Force Research Laboratory
AFRRI	Armed Forces Radiobiology Research Institute
AFRTS	American Forces Radio and Television Service
AFS	aeronautical fixed service
AFSATCOM	Air Force satellite communications (system)
AFSB	Army field support brigade
AFSC	Armed Forces Staff College; Army Field Support Center; United States Air Force specialty code

AFSCN	Air Force Satellite Control Network
AFSMO	Air Force Spectrum Management Office
AFSOC	Air Force Special Operations Command; Air Force special operations component
AFSOCC	Air Force special operations control center
AFSOD	Air Force special operations detachment
AFSOE	Air Force special operations element
AFSOF	Air Force special operations forces
AFSOUTH	Allied Forces, South (NATO)
AFSPACE	United States Space Command Air Force
AFSPC	Air Force Space Command
AFSPOC	Air Force Space Operations Center
AFSTRAT	Air Forces Strategic
AFTAC	Air Force Technical Applications Center
AFTH	Air Force Theater Hospital
AFTN	Aeronautical Fixed Telecommunications Network
AFTO	Air Force technical order
AFTRANS	Air Force Transportation Component
AFTTP	Air Force tactics, techniques, and procedures; Air Force technical training publication
AFTTP(I)	Air Force tactics, techniques, and procedures (instruction)
AFW	Air Force Weather
AFWA	Air Force Weather Agency
AFWCF	Air Force working capital fund
AFWIN	Air Force Weather Information Network
AF/XO	Deputy Chief of Staff for Plans and Operations, United States Air Force
AF/XOI	Air Force Director of Intelligence, Surveillance, and Reconnaissance
AF/XOO	Director of Operations, United States Air Force
A/G	air to ground
AG	adjutant general (Army)
AGARD	Advisory Group for Aerospace Research and Development
AGE	aerospace ground equipment
AGIL	airborne general illumination lightself
AGL	above ground level
AGM-28A	Hound Dog
AGM-65	Maverick
AGM-69	short range attack missile
AGR	Active Guard and Reserve
AGS	aviation ground support
AHA	alert holding area
AHD	antihandling device
AI	airborne interceptor; air interdiction; area of interest
AIA	Air Intelligence Agency
AIASA	annual integrated assessment for security assistance

AIC	air intercept controller; assign individual compressed dial; Atlantic Intelligence Command
AICF/USA	Action Internationale Contre La Faim (International Action Against Hunger)
AIDS	acquired immunodeficiency syndrome
AIF	automated installation intelligence file
AIFA	AAFES Imprest Fund Activity
AIG	addressee indicator group
AIIRS	automated intelligence information reporting system
AIK	assistance in kind
AIM	Airman's Information Manual
AIM-7	Sparrow
AIM-9	Sidewinder
AIM-54A	Phoenix
AIMD	aircraft intermediate maintenance department
AIMT	air interdiction of maritime targets
AIP	aeronautical information publication
AIQC	antiterrorism instructor qualification course
AIRBAT	Airborne Intelligence, Surveillance, and Reconnaissance Requirements-Based Allocation Tool
AIRCENT	Allied Air Forces Central Europe (NATO)
AIRES	advanced imagery requirements exploitation system
AIREVACCONFIRM	air evacuation confirmation
AIREVACREQ	air evacuation request
AIREVACRESP	air evacuation response
AIRNORTHWEST	Allied Air Forces North West Europe (NATO)
AIRREQRECON	air request reconnaissance
AIRSOUTH	Allied Air Forces Southern Europe (NATO)
AIRSUPREQ	air support request
AIS	automated information system
AIT	aeromedical isolation team; automated identification technology; automated information technology
AIU	Automatic Digital Network Interface Unit
AJ	anti-jam
AJBPO	area joint blood program office
AJCC	alternate joint communications center
AJ/CM	anti-jam control modem
AJF	allied joint force
AJFP	adaptive joint force packaging
AJMRO	area joint medical regulating office
AJNPE	airborne joint nuclear planning element
AJP	allied joint publication
AK	commercial cargo ship
AKNLDG	acknowledge message
ALARA	as low as reasonably achievable
ALCC	airlift control center

ALCE	airlift control element
ALCF	airlift control flight
ALCG	analog line conditioning group
ALCM	air-launched cruise missile
ALCOM	United States Alaskan Command
ALCON	all concerned
ALCS	airlift control squadron
ALCT	airlift control team
ALD	airborne laser designator; available-to-load date
ALE	airlift liaison element
ALEP	amphibious lift enhancement program
ALERFA	alert phase (ICAO)
ALERT	attack and launch early reporting to theater
ALERTORD	alert order
ALLOREQ	air allocation request; allocation request
ALLTV	all light level television
ALMSNSCD	airlift mission schedule
ALN	ammunition lot number
ALNOT	alert notice
ALO	air liaison officer
ALOC	air line of communications
ALORD	alert launch order
ALP	allied logistic publication
ALSA	Air Land Sea Application (Center)
ALSE	aviation life support equipment
ALSS	advanced logistic support site
ALT	acquisition, logistics, and technology
ALTD	airborne laser target designator
ALTRV	altitude reservation
ALTTSC	alternate Tomahawk strike coordinator
A/M	approach and moor
AM	amplitude modulation
AMAL	authorized medical allowance list
AMB	air mobility branch; ambassador
AMBUS	ambulance bus
AMC	airborne mission coordinator; Air Mobility Command; Army Materiel Command: midpoint compromise search area
AMCC	allied movement coordination center; alternate military command center
AMCIT	American citizen
AMCM	airborne mine countermeasures
AMC/SGXM	Air Mobility Command/Command Surgeon's Office
AMCT	air mobility control team
AMD	air and missile defense; air mobility division
AMDC	air and missile defense commander

AME	antenna mounted electronics
AMEDD	Army Medical Department
AMEDDCS	U.S. Army Medical Department Center and School
AMedP	allied medical publication
AMEMB	American Embassy
AMETL	agency mission-essential task list
AMF(L)	ACE Mobile Force (Land) (NATO)
AMH	automated message handler
AMHS	automated message handling system
AMIO	alien migrant interdiction operations
AMLO	air mobility liaison officer
AMMO	ammunition
AMOC	Air and Marine Operations Center (DHS)
AMOCC	air mobility operations control center
AMOG	air mobility operations group
AMOPES	Army Mobilization and Operations Planning and Execution System
AMOPS	Army mobilization and operations planning system; Army mobilization operations system
AMOS	air mobility operations squadron
AMOSS	Air and Marine Operations Surveillance System
AMOW	air mobility operations wing
AMP	amplifier; analysis of mobility platform
AMPE	automated message processing exchange
AMPN	amplification
AMP-PAT	analysis of mobility platform suite of port analysis tools
AMPSSO	Automated Message Processing System Security Office (or Officer)
AMRAAM	advanced medium-range air-to-air missile
AMS	Aerial Measuring System (DOE); air mobility squadron; Army management structure; Asset Management System
AMSS	air mobility support squadron
AMT	aerial mail terminal
amu	atomic mass unit
AMVER	automated mutual-assistance vessel rescue system
AMW	air mobility wing; amphibious warfare
AMX	air mobility express
AN	alphanumeric; analog nonsecure
ANCA	Allied Naval Communications Agency
ANDVT	advanced narrowband digital voice terminal
ANG	Air National Guard
ANGLICO	air-naval gunfire liaison company
ANGUS	Air National Guard of the United States
A/NM	administrative/network management
ANMCC	Alternate National Military Command Center

ANN	assign NNX routing
ANR	Alaskan North American Aerospace Defense Command Region
ANSI	American National Standards Institute
ANX	assign NNXX routing
ANY	assign NYX routing
ANZUS	Australia-New Zealand-United States Treaty
AO	action officer; administration officer; air officer; area of operations; aviation ordnance person
AO&M	administration, operation, and maintenance
AOA	amphibious objective area
AOB	advanced operations base; aviation operations branch
AOC	air and space operations center (USAF); air operations center; Army operations center
AOCC	air operations control center
AOC-E	Aviation Operations Center-East (USCS)
AOCU	analog orderwire control unit
AOC-W	Aviation Operations Center-West (USCS)
AOD	air operations directive; on-line diagnostic
AOF	azimuth of fire
AOG	Army Operations Group
AOI	area of interest
AOP	air operations plan; area of probability
AOR	area of responsibility
AOSS	aviation ordnance safety supervisor
AOTR	Aviation Operational Threat Response
AP	allied publication; antipersonnel; average power
APA	Army pre-positioned afloat
APAN	All Partners Access Network; Asia-Pacific Area Network
APC	aerial port commander; assign preprogrammed conference list
APCC	alternate processing and correlation center
APE	airfield pavement evaluation
APES	Automated Patient Evacuation System
APEX	Adaptive Planning and Execution
APF	afloat pre-positioning force
APG	aimpoint graphic
APHIS	Animal and Plant Health Inspection Service (USDA)
APIC	allied press information center
APL	antipersonnel land mine
APO	afloat pre-positioning operations; Air Force post office; Army post office
APOD	aerial port of debarkation
APOE	aerial port of embarkation
APORT	aerial port
APORTSREP	air operations bases report

APP	allied procedural publication
APPS	analytical photogrammetric positioning system
APR	assign primary zone routing
APS	aerial port squadron; Army pre-positioned stocks
APS-3	afloat pre-positioning stocks; Army pre-positioned stocks-3
APU	auxiliary power unit
AR	air refueling; Army regulation; Army reserve
ARB	alternate recovery base; assign receive bypass lists
ARBS	angle rate bombing system
ARC	air Reserve Components; American Red Cross
ARCENT	United States Army Central Command
ARCT	air refueling control team
ARDF	automatic radio direction finding
AREC	air resource element coordinator
ARFOR	Army forces
ARG	Accident Response Group (DOE); amphibious ready group
ARGO	automatic ranging grid overlay
ARINC	Aeronautical Radio Incorporated
ARL-M	airborne reconnaissance low-multifunction
ARM	antiradiation missile
ARNG	Army National Guard
ARNGUS	Army National Guard of the United States
ARP	air refueling point
ARPERCEN	United States Army Reserve Personnel Center
ARQ	automatic request-repeat
ARRC	Allied Command Europe Rapid Reaction Corps (NATO)
ARRDATE	arrival date
ARS	acute radiation syndrome; air rescue service
ARSOA	Army special operations aviation
ARSOC	Army special operations component
ARSOF	Army special operations forces
ARSOTF	Army special operations task force
ARSPACE	Army Space Command
ARSPOC	Army space operations center
ART	air reserve technician
ARTCC	air route traffic control center
ARTS III	Automated Radar Tracking System
ARTYMET	artillery meteorological
AS	analog secure
A/S	anti-spoofing
ASA	automatic spectrum analyzer
ASA(ALT)	Assistant Secretary of the Army for Acquisition, Logistics, and Technology
ASAP	as soon as possible

ASARS	Advanced Synthetic Aperture Radar System
ASAS	All Source Analysis System
ASAT	antisatellite weapon
ASB	naval advanced support base
ASBP	Armed Services Blood Program
ASBPO	Armed Services Blood Program Office
ASC	acting service chief; Aeronautical Systems Center; Air Systems Command; Army Sustainment Command; assign switch classmark; Automatic Digital Network switching center
ASCC	Air Standardization Coordinating Committee; Army Service component command; Army Service component commander
ASCIET	all Services combat identification evaluation team
ASCII	American Standard Code for Information Interchange
ASCOPE	areas, structures, capabilities, organizations, people, and events
ASCS	air support control section; air support coordination section
ASD	Assistant Secretary of Defense
ASD(A&L)	Assistant Secretary of Defense (Acquisition and Logistics)
ASD(C)	Assistant Secretary of Defense (Comptroller)
ASD(C3I)	Assistant Secretary of Defense (Command, Control, Communications, and Intelligence)
ASD(FM&P)	Assistant Secretary of Defense (Force Management and Personnel)
ASD(FMP)	Assistant Secretary of Defense (Force Management Policy)
ASD(GSA)	Assistant Secretary of Defense for Global Strategic Affairs
ASD(HA)	Assistant Secretary of Defense (Health Affairs)
ASD(HD)	Assistant Secretary of Defense (Homeland Defense)
ASD(HD&ASA)	Assistant Secretary of Defense (Homeland Defense and Americas' Security Affairs)
ASDI	analog simple data interface
ASDIA	All-Source Document Index
ASD(ISA)	Assistant Secretary of Defense (International Security Affairs)
ASD(ISP)	Assistant Secretary of Defense (International Security Policy)
ASD(LA)	Assistant Secretary of Defense (Legislative Affairs)
ASD(L&MR)	Assistant Secretary of Defense for Logistics and Materiel Readiness
ASD(NII)	Assistant Secretary of Defense (Networks and Information Integration)
ASD(OEPP)	Assistant Secretary of Defense for Operational Energy Plans and Programs

ASD(P&L)	Assistant Secretary of Defense (Production and Logistics)
ASD(PA)	Assistant Secretary of Defense (Public Affairs)
ASD(PA&E)	Assistant Secretary of Defense (Program Analysis and Evaluation)
ASD(RA)	Assistant Secretary of Defense (Reserve Affairs)
ASD(RSA)	Assistant Secretary of Defense (Regional Security Affairs)
ASD(S&R)	Assistant Secretary of Defense (Strategy and Requirements)
ASD(SO/LIC)	Assistant Secretary of Defense (Special Operations and Low-Intensity Conflict)
ASD(SO/LIC&IC)	Assistant Secretary of Defense for Special Operations and Low-Intensity Conflict and Interdependent Capabilities
ASE	aircraft survivability equipment; automated stabilization equipment
ASF	aeromedical staging facility
ASG	Allied System for Geospatial Intelligence; area support group
ASH	Assistant Administrator for Security and Hazardous Materials; Assistant Secretary for Health (DHHS)
ASI	assign and display switch initialization
ASIC	Air and Space Interoperability Council
ASIF	Airlift Support Industrial Fund
ASL	allowable supply list; archipelagic sea lane; assign switch locator (SL) routing; authorized stockage list (Army)
ASM	air-to-surface missile; armored scout mission; Army Spectrum Manager; automated scheduling message
ASMD	antiship missile defense
ASMO	Army Spectrum Management Office
ASN(RD&A)	Assistant Secretary of the Navy for Research, Development and Acquisition
ASO	advanced special operations; air support operations
ASOC	air support operations center
ASOFDTG	as of date/time group
ASP	ammunition supply point
ASPA	American Service-Members' Protection Act
A-Space	Analytic Space
ASPP	acquisition systems protection program
ASPPO	Armed Service Production Planning Office
ASPR	Office of Assistant Secretary for Preparedness and Response (DHHS)
ASR	air support request; available supply rate
ASSETREP	transportation assets report
AST	assign secondary traffic channels
ASTS	aeromedical staging squadron

ASW	antisubmarine warfare; average surface wind
ASWBPL	Armed Services Whole Blood Processing Laboratories
ASWC	antisubmarine warfare commander
AT	annual training; antitank; antiterrorism
At	total attainable search area
ATA	Airlift Tanker Association
ATAC	antiterrorism alert center (Navy)
ATACC	advanced tactical air command center
ATACMS	Army Tactical Missile System
ATACO	air tactical actions control officer
ATACS	Army Tactical Communications System
ATAF	Allied Tactical Air Force (NATO)
ATBM	antitactical ballistic missile
ATC	Air Threat Conference; air traffic control; air transportable clinic (USAF)
ATCA	Allied Tactical Communications Agency
ATCAA	air traffic control assigned airspace
ATCC	Antiterrorism Coordinating Committee
ATCC-SSG	Antiterrorism Coordinating Committee-Senior Steering Group
ATCRBS	Air Traffic Control Radar Beacon System
ATCS	air traffic control section
ATDLS	Advanced Tactical Data Link System
ATDM	adaptive time division multiplexer
ATDS	airborne tactical data system
ATEP	Antiterrorism Enterprise Portal
ATF	Advanced Targeting FLIR; amphibious task force; Bureau of Alcohol, Tobacco, Firearms, and Explosives (DOJ)
AT/FP	antiterrorism/force protection
ATG	amphibious task group; assign trunk group cluster
ATGM	antitank guided missile; antitank guided munition
ATH	air transportable hospital; assign thresholds
ATHS	Airborne Target Handover System
ATI	asset target interaction
ATM	advanced trauma management; air target material; assign traffic metering
ATMCT	air terminal movement control team
ATMP	Air Target Materials Program
ATN	assign thresholds
AtN	attack the network
ATO	air tasking order; antiterrorism officer
ATOC	air tactical operations center; air terminal operations center
ATP	advance targeting pod; allied tactical publication; Army tactical publication
ATR	attrition reserve
ATS	air traffic service; assign terminal service

ATSD(AE)	Assistant to the Secretary of Defense (Atomic Energy)
ATSD(IO)	Assistant to the Secretary of Defense (Intelligence Oversight)
ATSD(NCB)	Assistant to the Secretary of Defense for Nuclear and Chemical and Biological Defense Programs
ATT	assign terminal type
ATTP	Army tactics, techniques, and procedures
ATTU	air transportable treatment unit
ATWG	antiterrorism working group
AU	African Union
AUEL	automated unit equipment list
AUF	airborne use of force
AUG	application user group
AUIC	active duty unit identification code
AUSCANNZUKUS	Australian, Canadian, New Zealand, United Kingdom, United States
AUTODIN	Automatic Digital Network
AUX	auxiliary
AV	air vehicle; asset visibility
AV-8	Harrier
AVCAL	aviation consolidated allowance list
AVDTG	analog via digital trunk group
AVGAS	aviation gasoline
AVIM	aviation intermediate maintenance
AVL	anti-vehicle land mine; assign variable location
AVOU	analog voice orderwire unit
AVOW	analog voice orderwire
AVS	asset validation system; asset visibility system; audiovisual squadron
AVUM	aviation unit maintenance
AV/VI	audiovisual/visual information
AW	air warfare
AWACS	Airborne Warning and Control System
AWC	air warfare commander
AWCAP	airborne weapons corrective action program
AWDS	automated weather distribution system
AWG	Asymmetric Warfare Group (Army)
AWN	Automated Weather Network
AWOL	absent without leave
AWS	Air Weather Service
AWSE	armament weapons support equipment
AWSIM	air warfare simulation model
AWSR	Air Weather Service regulation
AXO	abandoned explosive ordnance
AXX	assign XXX routing
AZR	assign zone restriction lists

B

B	cross-over barrier pattern
B-52	Stratofortress
B&A	boat and aircraft
BAE	brigade aviation element
BAF	backup alert force
BAG	baggage
BAH	basic allowance for housing
BAI	backup aircraft inventory; battlefield air interdiction
BALO	battalion air liaison officer
BALS	berthing and loading schedule
BAS	basic allowance for subsistence; battalion aid station
BATF	Bureau of Alcohol, Tobacco, and Firearms
B/B	baseband
BB	breakbulk
bbl	barrel (42 US gallons)
BC	bottom current
BCA	border crossing authority
BCAT	beddown capability assessment tool
BCC	battle control center
BCD	battlefield coordination detachment
BCG	beach control group
BCI	bit count integrity
BCL	battlefield coordination line
BCN	beacon
BCOC	base cluster operations center
BCR	baseline change request
BCT	brigade combat team
BCTP	battle command training program
BCU	beach clearance unit
BD	barge derrick
BDA	battle damage assessment
BDAREP	battle damage assessment report
BDC	blood donor center
BDE	brigade
BDL	beach discharge lighter
BDOC	base defense operations center
BDR	battle damage repair
BDRP	Biological Defense Research Program
BDZ	base defense zone
BE	basic encyclopedia
BEAR	base expeditionary airfield resources
BEE	bioenvironmental engineering officer
BEI	biometrics-enabled intelligence

BEN	base encyclopedia number
BE number	basic encyclopedia number
BER	bit error ratio
BES	budget estimate submission
BEST	border enforcement security task force
BfV	*Bundesamt für Verfassungsschutz (*federal office for defending the Constitution)
BGC	boat group commander
BHR	Bureau of Humanitarian Response
BI	battlefield injury; battle injury
BIA	behavioral influences analysis; Bureau of Indian Affairs
BIAR	biometric intelligence analysis report
BIAS	Battlefield Illumination Assistance System
BICES	battlefield information collection and exploitation system (NATO)
BIDDS	Base Information Digital Distribution System
BIDE	basic identity data element
BIFC	Boise Interagency Fire Center
BIFS	Border Intelligence Fusion Section
BIH	International Time Bureau (Bureau International d'l'Heure)
BII	base information infrastructure
BIMA	Biometrics Identity Management Agency
BINM	Bureau of International Narcotics Matters
BIO	biological; Bureau of International Organizations
BIS	Bureau of Industry and Security
BISS	base installation security system
BIT	built-in test
BITE	built-in test equipment
BIU	beach interface unit
BKA	*Bundeskriminalamt* (federal criminal office)
BL	biocontainment level
BLCP	beach lighterage control point
BLDREP	blood report
BLDSHIPREP	blood shipment report
BLM	Bureau of Land Management
BLOS	beyond line-of-sight
BLT	battalion landing team
BM	ballistic missile; battle management; beach module
BMC4I	Battle Management Command, Control, Communications, Computers, and Intelligence
BMCT	begin morning civil twilight
BMD	ballistic missile defense
BMDO	Ballistic Missile Defense Organization
BMDS	ballistic missile defense system
BMET	biomedical electronics technician

BMNT	begin morning nautical twilight
BMU	beachmaster unit
BN	battalion
BND	*Bundesnachrichtendienst* (federal intelligence service)
BOA	basic ordering agreement
BOC	base operations center; bomb on coordinate
BOCCA	Bureau of Coordination of Civil Aircraft (NATO)
BOG	beach operations group
BOH	bottom of hill
BORFIC	Border Patrol Field Intelligence Center
BOS	base operating support; battlefield operating system
BOSG	base operations support group
BOS-I	base operating support-integrator
BOSS	base operating support service
BOT	bomb on target
BP	battle position; block parity
BPA	blanket purchase agreement
BPC	building partnership capacity
BPD	blood products depot
BPG	beach party group
BPI	bits per inch
BPLAN	base plan
BPO	blood program office
BPPBS	bi-annual planning, programming, and budget system
bps	bits per second
BPSK	biphase shift keying
BPT	beach party team
BPWRR	bulk petroleum war reserve requirement
BPWRS	bulk petroleum war reserve stocks
BR	budget review
BRAC	base realignment and closure
BRACE	Base Resource and Capability Estimator
BRC	base recovery course
BS	battle staff; broadcast source
BSA	beach support area; brigade support area
BSB	brigade support battalion
BSC	black station clock
BSC ro	black station clock receive out
BSCT	behavioral science consultation team
BSD	blood supply detachment
BSI	base support installation
BSP	base support plan
BSRP	bureau strategic resource plan
BSSG	brigade service support group
BSU	blood supply unit
BSZ	base security zone

BT	bathythermograph
BTB	believed-to-be
BTC	blood transshipment center
BTG	basic target graphic
BTOC	battalion tactical operations center
BTS	Border and Transportation Security (DHS)
BTU	beach termination unit
BULK	bulk cargo
BUMEDINST	Bureau of Medicine and Surgery instruction
BVR	beyond visual range
BW	biological warfare
BWC	Biological Weapons Convention
BZ	buffer zone

C

C	Celsius; centigrade; clock; compromise band; coverage factor; creeping line pattern
C&A	certification and accreditation
C&E	communications and electronics
C&LAT	cargo and loading analysis table
C2	command and control
C2-attack	an offensive form of command and control warfare
C2CRE	command and control chemical, biological, radiological, and nuclear response element
C2E	command and control element
C2IP	Command and Control Initiatives Program
C2IPS	Command and Control Information Processing System
C2P	command and control protection
C2-protect	a defensive form of command and control warfare
C2S	command and control support
C-2X	coalition Intelligence Directorate counterintelligence and human intelligence staff element
C3	command, control, and communications
C3AG	Command, Control, and Communications Advisory Group
C3CM	command, control, and communications countermeasures
C3I	command, control, communications, and intelligence
C3IC	coalition coordination, communications, and integration center
C3SMP	Command, Control, and Communications Systems Master Plan
C4CM	command, control, communications, and computer countermeasures
C4I	command, control, communications, computers, and intelligence
C4IFTW	command, control, communications, computers, and intelligence for the Warrior

C4S	command, control, communications, and computer systems
C4 systems	command, control, communications, and computer systems
C-5	Galaxy
C-17	Globemaster III
C-21	Learjet
C-27	Spartan
C-130	Hercules
C-141	Starlifter
CA	chaplain assistant; civil administration; civil affairs; combat assessment; coordinating altitude; credibility assessment; criticality assessment
C/A	course acquisition
CAA	civil air augmentation; combat aviation advisors; command arrangement agreement
CAAF	contractor personnel authorized to accompany the force
CAB	combat aviation brigade
CAC	common access card; current actions center
CACOM	civil affairs command
CACTIS	community automated intelligence system
CAD	Canadian air division; cartridge actuated device; collective address designator
CADRS	concern and deficiency reporting system
CADS	containerized ammunition distribution system
CAF	Canadian Air Force; combat air forces; commander, airborne/air assault force
CAFMS	computer-assisted force management system
CAG	carrier air group; civil affairs group; collective address group
CAGO	contractor acquired government owned
CAIMS	conventional ammunition integrated management system
CAINS	carrier aircraft inertial navigation system
CAIS	civil authority information support
CAISE	civil authority information support element
CAL	caliber; critical asset list
CALA	Community Airborne Library Architecture
CALCM	conventional air-launched cruise missile
CALICS	communication, authentication, location, intentions, condition, and situation
CALMS	computer-aided load manifesting system
CAM	chemical agent monitor; crisis action module
CAMOC	Caribbean Air and Marine Operations Center
CAMPS	Consolidated Air Mobility Planning System
CAMT	countering air and missile threats
CANA	convulsant antidote for nerve agent
CANADA COM	Canada Command

CANR	Canadian North American Aerospace Defense Command Region
CANUS	Canada-United States
CANUS BDD	Canada-United States Basic Defense Document
CANUS CDP	Canada-United States Combined Defense Plan
CAO	chief administrative officer; civil affairs operations; counterair operation
CAOC	combat air operations center; combined air operations center
CAO SOP	standing operating procedures for coordination of atomic operations
CAP	Civil Air Patrol; civil augmentation program; combat air patrol; configuration and alarm panel; Consolidated Appeals Process (UN); crisis action planning
CAPT	civil affairs planning team
CAR	Chief of the Army Reserve
CARA	chemical, biological, radiological, nuclear, and high-yield explosives analytical and remediation activity
CARDA	continental United States airborne reconnaissance for damage assessment; continental United States area reconnaissance for damage assessment
CARE	Cooperative for Assistance and Relief Everywhere (CAREUSA)
CARIBROC	Caribbean Regional Operations Center
CARP	computed air release point; contingency alternate route plan
CARS	combat arms regimental system
CARVER	criticality, accessibility, recuperability, vulnerability, effect, and recognizability
CAS	casualty; civil aviation security; close air support
CASEVAC	casualty evacuation
CASF	contingency aeromedical staging facility
CASP	computer-aided search planning
CASPER	contact area summary position report
CASREP	casualty report
CASREQ	close air support request
CAT	category; civil affairs team; crisis action team
CATCC	carrier air traffic control center
CATF	commander, amphibious task force
CAU	crypto ancillary unit; cryptographic auxiliary unit
CAVU	ceiling and visibility unlimited
CAW	carrier air wing
CAW/ESS	crisis action weather and environmental support system
CAX	computer-assisted exercise
C-B	chemical-biological
CB	chemical-biological; construction battalion (SEABEES)

CBBLS	hundreds of barrels
CBCP	Customs and Border Clearance Program (DOD)
CBD	chemical, biological defense
CBFS	cesium beam frequency standard
CBG	coalition building guide
CBIRF	Chemical-Biological Incident Response Force
CBLTU	common battery line terminal unit
CBMR	capabilities-based munitions requirements
CBMU	construction battalion maintenance unit
CBP	capabilities-based planning; Customs and Border Protection (DHS)
CBPO	Consolidated Base Personnel Office
CBPS	chemical biological protective shelter
CBR	chemical, biological, and radiological
CBRN	Caribbean Basin Radar Network; chemical, biological, radiological, and nuclear
CBRN CM	chemical, biological, radiological, and nuclear consequence management
CBRNE	chemical, biological, radiological, nuclear, and high-yield explosives
CBRN hazard	chemical, biological, radiological, and nuclear hazard
CBRN passive defense	chemical, biological, radiological, and nuclear passive defense
CBRT	chemical-biological response team
CBS	common battery signaling
CBSA	Canadian Border Services Agency
CBT	common battery terminal
CbT	combating terrorism
CbT-RIF	Combating Terrorism Readiness Initiatives Fund
CBTZ	combat zone
CBU	conference bridge unit; construction battalion unit
CBW	chemical and biological warfare
C/C	cabin cruiser; cast off and clear
CC	component command (NATO); component commander; critical capability
CC&D	camouflage, concealment, and deception
CCA	carrier-controlled approach; central contracting authority; circuit card assembly; combat cargo assistant; container control activity; contamination control area; contingency capabilities assessment; contract construction agent
CCAP	combatant command AFRTS planner
CCAS	contingency contract administration services
CCAS-C	contingency contract administration services commander
CCATT	critical care air transport team

CCB	Community Counterterrorism Board; configuration control board
CCC	coalition coordination cell; coalition coordination center; crisis coordination center; critical control circuit; cross-cultural communications course
CCD	camouflage, concealment, and deception
CCDB	consolidated counterdrug database
CCDR	combatant commander
CCE	container control element; continuing criminal enterprise
CCEB	Combined Communications-Electronics Board
CCF	collection coordination facility
CCG	combat communications group; crisis coordination group
CCGD	commander, Coast Guard district
CCIB	command center integration branch
CCICA	command counterintelligence coordinating authority
CCIF	Combatant Commander Initiative Fund
CCIP	continuously computed impact point
CCIR	commander's critical information requirement; International Radio Consultative Committee
CCIS	common channel interswitch signaling
CCITT	International Telegraph and Telephone Consultative Committee
CCIU	CEF control interface unit
CCJTF	commander, combined joint task force
CCL	communications/computer link
CCLI	computer control list item
CCMD	combatant command
CCO	central control officer; combat cargo officer; command and control office; contingency contracting officer
CCOI	critical contact of interest
CCP	casualty collection point; consolidated cryptologic program; consolidation and containerization point
CCPDS	command center processing and display system
CCR	closed circuit refueling
CCRD	combatant commander's required delivery date
C-CS	communication and computer systems
CCS	central control ship; commander's communication synchronization; container control site
CCSA	combatant command support agent
CCSD	command communications service designator; control communications service designator
CCT	collaborative contingency targeting; combat control team
CCTI	Chairman of the Joint Chiefs of Staff commended training issue
CCTV	closed circuit television

CCW	1980 United Nations Convention on Conventional Weapons; continuous carrier wave
CD	channel designator; compact disc; counterdrug; customer direct
C-day	unnamed day on which a deployment operation begins
CDC	Centers for Disease Control and Prevention
CDD	chemical decontamination detachment
CDE	collateral damage estimation
CDERA	Caribbean Disaster Emergency Response Agency
CDF	combined distribution frame
CDI	cargo disposition instructions; conditioned diphase
C di	conditioned diphase
CDHAM	Center for Disaster and Humanitarian Assistance Medicine
CDIP	combined defense improvement project
CDIPO	counterdrug intelligence preparation for operations
CDLMS	common data link management system
CDM	cable driver modem
CDMGB	cable driver modem group buffer
CDN	compressed dial number
CDO	command duty officer; commander, detainee operations
CDOC	counterdrug operations center
CDOPS	counterdrug operations
CDP	commander's dissemination policy; landing craft air cushion departure point
CDR	commander; continuous data recording
CDRAFNORTH	Commander, Air Force North
CDRAFSOF	commander, Air Force special operations forces
CDRCFCOM	Commander, Combined Forces Command
CDRESC	commander, electronic security command
CDREUDAC	Commander, European Command Defense Analysis Center (ELINT) or European Data Analysis Center
CDRFORSCOM	Commander, Forces Command
CDRG	catastrophic disaster response group (FEMA)
CDRJSOTF	commander, joint special operations task force
CDRL	contract data requirements list
CDRMTMC	Commander, Military Traffic Management Command
CDRNORAD	Commander, North American Aerospace Defense Command
CD-ROM	compact disc read-only memory
CDRTSOC	commander, theater special operations command
CDRUNC	Commander, United Nations Command
CDRUSAFRICOM	Commander, United States Africa Command
CDRUSAINSCOM	Commander, United States Army Intelligence and Security Command
CDRUSARNORTH	Commander, United States Army, North
CDRUSCENTCOM	Commander, United States Central Command

CDRUSCYBERCOM	Commander, United States Cyber Command
CDRUSELEMNORAD	Commander, United States Element, North American Aerospace Defense Command
CDRUSEUCOM	Commander, United States European Command
CDRUSJFCOM	Commander, United States Joint Forces Command
CDRUSNAVEUR	Commander, United States Naval Forces, Europe
CDRUSNORTHCOM	Commander, United States Northern Command
CDRUSPACOM	Commander, United States Pacific Command
CDRUSSOCOM	Commander, United States Special Operations Command
CDRUSSOUTHCOM	Commander, United States Southern Command
CDRUSSTRATCOM	Commander, United States Strategic Command
CDRUSTRANSCOM	Commander, United States Transportation Command
CDS	Chief of Defence Staff (Canada); container delivery system
CDSSC	continuity of operations plan designated successor service chief
CDU	counterdrug update
C-E	communications-electronics
CE	casualty estimation; circular error; command element (MAGTF); communications-electronics; core element; counterespionage; critical element
CEA	captured enemy ammunition
CEB	combat engineer battalion
CEC	civil engineer corps
CECOM	communications-electronics command
CEDI	commercial electronic data interface
CEDREP	communications-electronics deployment report
CEE	captured enemy equipment
CEF	civil engineering file; common equipment facility
CEG	common equipment group
CEHC	Counter Explosive Hazards Center (Army)
CEI	critical employment indicator
CELLEX	cellular exploitation
CEM	combined effects munition
CEMC	communications-electronics management center
CEMIRT	civil engineer maintenance, inspection, and repair team
CENTRIXS	Combined Enterprise Regional Information Exchange System
CEOI	communications-electronics operating instructions
CEP	cable entrance panel; Chairman's Exercise Program
CEPOD	communications-electronics post-deployment report
CERF	Central Emergency Revolving Fund (UN)
CERFP	chemical, biological, radiological, nuclear, and high-yield explosives enhanced response force package
CERP	Commanders' Emergency Response Program
CERT	computer emergency response team; contingency engineering response team

CERTSUB	certain submarine
CES	coast earth station
CESE	civil engineering support equipment; communications equipment support element
CESG	communications equipment support group
CESO	civil engineer support office
CESPG	civil engineering support plan group; civil engineering support planning generator
CEXC	combined explosives exploitation cell
CEW	Civilian Expeditionary Workforce
CF	Canadian forces; carrier furnished; causeway ferry; conventional forces; drift error confidence factor
CFA	Committee on Food Aid Policies and Programmes (UN); critical factors analysis
CFACC	combined force air component commander
CFB	Canadian forces base
CFC	Combined Forces Command, Korea
CF-COP	counterfire common operational picture
CFL	Contingency Planning Facilities List; coordinated fire line
CFLCC	coalition forces land component commander
CFM	cubic feet per minute
CFO	chief financial officer
CFPM	causeway ferry power module
CFR	Code of Federal Regulations
CFS	CI force protection source
CFSO	counterintelligence force protection source operations
CFST	coalition forces support team
CG	Chairman's guidance; Coast Guard; commanding general; Comptroller General
CG-652	Coast Guard Spectrum Management and Telecommunications Policy Division
CGAS	Coast Guard Air Station
CGAUX	Coast Guard Auxiliary
CGC	Coast Guard Cutter
CGCAP	Coast Guard capabilities plan
CGCG	Coast Guard Cryptologic Group
CGCIS	Coast Guard Counterintelligence Service
CGDEFOR	Coast Guard defense force
CGFMFLANT	Commanding General, Fleet Marine Forces, Atlantic
CGFMFPAC	Commanding General, Fleet Marine Forces, Pacific
CGIS	United States Coast Guard Investigative Service
CGLSMP	Coast Guard logistic support and mobilization plan
CGP	Coast Guard publication
CGRS	common geographic reference system
CGS	common ground station; continental United States ground station

CGUSAREUR	Commanding General, United States Army, Europe
CH	channel; contingency hospital
CH-53	Sea Stallion
CHAMPUS	Civilian Health and Medical Program for the Uniformed Services
CHARC	counterintelligence and human intelligence analysis and requirements cell
CHB	cargo-handling battalion
CHCS	composite health care system
CHCSS	Chief, Central Security Service
CHE	cargo-handling equipment; container-handling equipment
CHET	customs high endurance tracker
CHOP	change of operational control
CHPPM	US Army Center for Health Promotion and Preventive Medicine
CHRIS	chemical hazard response information system
CHRP	contaminated human remains pouch
CHSTR	characteristics of transportation resources
CHSTREP	characteristics of transportation resources report
CI	civilian internee; counterintelligence
CIA	Central Intelligence Agency
CIAP	Central Intelligence Agency program; central intelligence architecture plan; command intelligence architecture plan; command intelligence architecture program
CIAS	counterintelligence analysis section
CIAT	counterintelligence analytic team
CIB	combined information bureau; controlled image base
CIC	combat information center; combat intelligence center (Marine Corps); combined intelligence center; communications interface controller; content indicator code; counterintelligence center
CICA	counterintelligence coordinating authority
CICAD	Inter-American Drug Abuse Control Commission
CICC	counterintelligence coordination cell
CICR	counterintelligence collection requirement
CID	combat identification; combat intelligence division; criminal investigation division
CIDB	common intelligence database
CIDC	Criminal Investigation Division Command
CIDNE	Combined Information Data Network Exchange
CIE	collaborative information environment
CIEA	classification, identification, and engagement area
C-IED	counter-improvised explosive device
CIEG/CIEL	common information exchange glossary and language
CIFA	counterintelligence field activity
CIG	communications interface group

CIHO	counterintelligence/human intelligence officer
CIIR	counterintelligence information report
CI/KR	critical infrastructure and key resources
CIL	command information library; critical information list; critical item list
CILO	counterintelligence liaison officer
CIM	civil information management; compartmented information management
CIMIC	civil-military cooperation
CIN	cargo increment number
CIO	chief information officer; command intelligence officer
CIOC	counterintelligence operations cell
CIOTA	counterintelligence operational tasking authority
CIP	communications interface processor; critical infrastructure protection
CIPSU	communications interface processor pseudo line
CIR	continuing intelligence requirement
CIRM	International Radio-Medical Center
CIRV	common interswitch rekeying variable
CIRVIS	communications instructions for reporting vital intelligence sightings
CIS	common item support; Commonwealth of Independent States; communications interface shelter
CISAR	catastrophic incident search and rescue
CISD	critical incident stress debriefing
CISO	counterintelligence staff office; counterintelligence support officer
CITE	computer intrusion technical exploitation
CITP	counter-improvised explosive device targeting program
CIV	civilian
CIVMAR	civil service mariner
CIVPOL	civilian police
CIWG	communications interoperability working group
CJ-4	combined-joint logistics officer
CJATF	commander, joint amphibious task force
CJB	Congressional Justification Book
CJCS	Chairman of the Joint Chiefs of Staff
CJCSAN	Chairman of the Joint Chiefs of Staff Alerting Network
CJCSI	Chairman of the Joint Chiefs of Staff instruction
CJCSM	Chairman of the Joint Chiefs of Staff manual
CJDA	critical joint duty assignment
CJE	component joint data networks operations officer equivalent
CJLOTS	combined joint logistics over-the-shore
CJMAB	Central Joint Mortuary Affairs Board

CJMAO	Central Joint Mortuary Affairs Office; Chief, joint mortuary affairs office
CJMISTF	combined joint military information support task force
CJMTF	combined joint military information support operations task force
CJOC	Canada Joint Operations Command
CJSART	Criminal Justice Sector Assessment Rating Tool
CJSMPT	Coalition Joint Spectrum Management Planning Tool
CJTF	combined joint task force (NATO); commander, joint task force
CJTF-CS	Commander, Joint Task Force - Civil Support
CJTF-NCR	Commander, Joint Task Force - National Capital Region
C-JWICS	Containerized Joint Worldwide Intelligence Communications System
CKT	circuit
CL	class
CLA	landing craft air cushion launch area
CLASSRON	class squadron
CLB	combat logistics battalion
CLD	compact laser designator
CLDP	Commercial Law Development Program
CLEA	civilian law enforcement agency
C-level	category level
CLF	combat logistics force; commander, landing force
CLG	combat logistics group
CLGP	cannon-launched guided projectile
CLIA	Clinical Laboratory Improvement Amendments
CLIPS	communications link interface planning system
CLPSB	combatant commander logistic procurement support board
CLPT	contingency load planning team
CLR	combat logistics regiment
CLS	contracted logistic support
CLSS	combat logistic support squadron
CLT	civil liaison team; combat lasing team
CLZ	craft landing zone; cushion landing zone; landing craft air cushion landing zone
CM	Chairman's memorandum; collection manager; combination module; configuration management; control modem; countermine; cruise missile
Cm	mean coverage factor
cm	centimeter
CMA	collection management authority
CMAA	cooperative military airlift agreement
CMAH	commander of a combatant command's Mobile Alternate Headquarters
CM&D	collection management & dissemination

CMAOC	Casualty and Mortuary Affairs Operations Center
CMAT	consequence management advisory team
CMC	Commandant of the Marine Corps; crew management cell; Office of Civilian Military Cooperation (USAID)
Cmc	midpoint compromise coverage factor
CMCB	civil-military coordination board
CMCC	combined movement coordination center
CMD	command; cruise missile defense
CMDO	command military deception officer
CME	civil-military engagement
CMHT	consequence management home team
CMM	Office of Conflict Management and Mitigation (USAID)
CMMA	collection management mission application
CMO	Central Measurement and Signature Intelligence (MASINT) Organization; chief medical officer; chief military observer; civil-military operations; collection management office(r); configuration management office
CMOC	Cheyenne Mountain Operations Center; civil-military operations center
CMOS	cargo movement operations system; Cargo Movement Operations System (USAF); complementary metal-oxide semiconductor
CMP	communications message processor; contractor management plan
CMPF	commander, maritime pre-positioned force
CMPT	consequence management planning team
CM R&A	consequence management response and assessment
CMRT	consequence management response team
CMS	cockpit management system; command management system; community management staff; community security materiel system; contingency mutual support; crisis management system
CMSE	civil-military support element
CMST	consequence management support team
CMTS	comments
CMTU	cartridge magnetic tape unit
CMV	commercial motor vehicle
CMX	crisis management exercise
CNAC	Customs National Aviation Center (USCS)
C-NAF	component numbered air force
CNASP	chairman's net assessment for strategic planning
CNBG	commander, naval beach group
CNC	Crime and Narcotics Center (CIA)
CNCE	communications nodal control element
CNCI	Comprehensive National Cybersecurity Initiative
CND	counternarcotics division

CNE	computer network exploitation; Counter Narcotics Enforcement
CNGB	Chief, National Guard Bureau
CNIC	Commander, Navy Installations Command
CNM	classified notice to mariners
CNMOC	Commander, Naval Meteorology and Oceanography Command
CNO	Chief of Naval Operations
CNOG	Chairman, Nuclear Operations Group
CNRF	Commander, Naval Reserve Forces
CNSG	Commander, Naval Security Group
CNTY	country
CNWDI	critical nuclear weapons design information
CO	commanding officer; cyberspace operations
COA	course of action
COAA	course-of-action analysis
COAMPS	Coupled Ocean Atmosphere Mesoscale Prediction System
COB	collocated operating base; contingency operating base
COBOL	common business-oriented language
COC	combat operations center
CoC	Code of Conduct
COCOM	combatant command (command authority)
COD	carrier onboard delivery; combat operations division
COE	Army Corps of Engineers; common operating environment; concept of employment
COEDMHA	Center for Excellence in Disaster Management and Humanitarian Assistance
COF	chief of fires; conduct of fire
COFC	container on flatcar
COG	center of gravity; continuity of government
COGARD	Coast Guard
COI	community of interest; contact of interest
COIC	counter-improvised explosive device operations integration center
COIN	counterinsurgency
COLDS	cargo offload and discharge system
COLISEUM	community on-line intelligence system for end-users and managers
COLPRO	collective protection
COLS	concept of logistic support
COLT	combat observation and lasing team
COM	chief of mission; collection operations management; command; commander
COMACC	Commander, Air Combat Command
COMAFFOR	commander, Air Force forces
COMAFSOC	Commander, Air Force Special Operations Command

COMAJF	commander, allied joint force
COMALF	commander airlift forces
COMALOC	commercial air line of communications
COMARFOR	commander, Army forces
COMCAM	combat camera
COMCARGRU	commander, carrier group
COMCRUDESGRU	commander, cruiser destroyer group
COMDCAEUR	Commander, Defense Communications Agency Europe
COMDESRON	commander destroyer squadron
COMDT COGARD	Commandant, United States Coast Guard
COMDTINST	Commandant of the United States Coast Guard instruction
COMFLTCYBERCOM	Commander, Fleet Cyber Command
COMICEDEFOR	Commander, United States Forces, Iceland
COMIDEASTFOR	Commander, Middle East Forces
COMINEWARCOM	Commander, Mine Warfare Command
COMINT	communications intelligence
COMJCSE	Commander, Joint Communications Support Element
COMJIC	Commander, Joint Intelligence Center
COMLANDFOR	commander, land forces
COMLANTAREACOGARD	Commander, Coast Guard Atlantic Area
COMLOGGRU	combat logistics group
COMM	communications
COMMARFOR	commander, Marine Corps forces
COMMARFORNORTH	Commander, Marine Corps Forces North
COMMDZ	Commander, Maritime Defense Zone
COMNAV	Committee for European Airspace Coordination Working Group on Communications and Navigation Aids
COMNAVAIRLANT	Commander, Naval Air Force, Atlantic
COMNAVAIRPAC	Commander, Naval Air Force, Pacific
COMNAVAIRSYSCOM	Commander, Naval Air Systems Command
COMNAVCOMTELCOM	Commander, Naval Computer and Telecommunications Command
COMNAVELSG	Commander, Navy Expeditionary Logistics Support Group
COMNAVFOR	commander, Navy forces
COMNAVMETOCCOM	Commander, Naval Meteorology and Oceanography Command
COMNAVSEASYSCOM	Commander, Naval Sea Systems Command
COMNAVSECGRP	Commander, United States Navy Security Group
COMNAVSURFLANT	Commander, Naval Surface Force, Atlantic
COMNAVSURFPAC	Commander, Naval Surface Force, Pacific
COMNET	communications network
COMP	component
COMPACAF	Commander, Pacific Air Forces
COMPACAREACOGARD	Commander, Coast Guard Pacific Area
COMPACFLT	Commander, Pacific Fleet

COMPASS	common operational modeling, planning, and simulation strategy; Computerized Movement Planning and Status System
COMPES	contingency operations mobility planning and execution system
COMPLAN	communications plan
COMPUSEC	computer security
COMSAT	communications satellite
COMSC	Commander, Military Sealift Command
COMSCINST	Commander, Military Sealift Command instruction
COMSEC	communications security
COMSOC	Commander, Special Operations Command
COMSOCCENT	Commander, Special Operations Command, United States Central Command
COMSOCEUR	Commander, Special Operations Command, United States European Command
COMSOCPAC	Commander, Special Operations Command, Pacific
COMSOCSOUTH	Commander Special Operations Command, United States Southern Command
COMSOF	commander, special operations forces
COMSTAT	communications status
COMSUBLANT	Commander Submarine Force, United States Atlantic Fleet
COMSUBPAC	Commander Submarine Force, United States Pacific Fleet
COMSUPNAVFOR	commander, supporting naval forces
COMTAC	tactical communications
COMTENTHFLT	Commander, Tenth Fleet
COMUSAFE	Commander, United States Air Force in Europe
COMUSARCENT	Commander, United States Army Forces, Central Command
COMUSCENTAF	Commander, United States Air Force, Central Command
COMUSFLTFORCOM	Commander, United States Fleet Forces Command
COMUSFORAZ	Commander, United States Forces, Azores
COMUSJ	Commander, United States Forces, Japan
COMUSK	Commander, United States Forces, Korea
COMUSMARCENT	Commander, United States Marine Forces, Central Command
COMUSNAVCENT	Commander, United States Navy, Central Command
COMUSPACFLT	Commander, United States Pacific Fleet
COMUSSOCJFCOM	Commander Special Operations Command, United States Joint Forces Command
CONCAP	construction capabilities contract (Navy); Construction Capabilities Contract Process; construction capabilities contract program
CONEX	container express
CONEXPLAN	contingency and exercise plan
CONOPS	concept of operations

CONPLAN	concept plan; operation plan in concept format
CONR	continental United States North American Aerospace Defense Command Region
CONUS	continental United States
CONUSA	Continental United States Army
COOP	continuity of operations
COP	common operational picture
COP-CSE	common operational picture-combat support enabled
COPG	chairman, operations planners group
COPPERHEAD	name for cannon-launched guided projectile
COPS	communications operational planning system
COR	contracting officer representative
CORE	contingency response program
COS	chief of staff; chief of station; critical occupational specialty
COSMIC	North Atlantic Treaty Organization (NATO) security category
COSPAS	*cosmicheskaya sistyema poiska avariynch sudov* - space system for search of distressed vessels (Russian satellite system)
COSR	combat and operational stress reactions
COT	commanding officer of troops; crisis operations team
COTP	captain of the port
COTS	cargo offload and transfer system; commercial off-the-shelf; container offloading and transfer system
COU	cable orderwire unit
counter C3	counter command, control, and communications
COVCOM	covert communications
CP	check point; collection point; command post; contact point; control point; counterproliferation
CP&I	coastal patrol and interdiction
CPA	closest point of approach
CPD	combat plans division
CPE	customer premise equipment
CPFL	contingency planning facilities list
CPG	central processor group; Commander, Amphibious Group; Contingency Planning Guidance
CPI	crash position indicator
CPIC	coalition press information center
CPM	civilian personnel manual
CPO	chief petty officer; complete provisions only
CPR	cardiopulmonary resuscitation
CPRC	coalition personnel recovery center
CPS	characters per second; collective protective shelter
CPT	common procedural terminology
CPU	central processing unit
CPX	command post exercise

CR	civil reconnaissance; critical requirement
CRA	command relationships agreement; continuing resolution authority; coordinating review authority
CRAF	Civil Reserve Air Fleet
CRAM	control random access memory
CRB	configuration review board
CRC	circuit routing chart; Civilian Response Corps (DOS); civil response corps; coastal riverine company; control and reporting center; CONUS replacement center; COOP response cell; crisis reaction center; cyclic redundancy rate
CRD	capstone requirements document; chemical reconnaissance detachment
CRE	contingency response element; control reporting element
CREAPER	Communications and Radar Electronic Attack Planning Effectiveness Reference
CREST	casualty and resource estimation support tool
CREW	counter radio-controlled improvised explosive device electronic warfare
CRF	channel reassignment function; coastal riverine force
CRG	contingency response force; contingency response group
CRI	collective routing indicator
CRIF	cargo routing information file
CRITIC	critical information; critical intelligence communication; critical message (intelligence)
CRITICOMM	critical intelligence communications system
CRM	collection requirements management; comment resolution matrix; crew resource management
CrM	crisis management
CRO	combat rescue officer
CROP	common relevant operational picture
CRP	control and reporting post
CRRC	combat rubber raiding craft
CRS	Catholic Relief Services; Chairman's readiness system; coastal radio station; community relations service; container recovery system; Coordinator for Reconstruction and Stabilization
CRSG	country reconstruction and stabilization group
CRSP	centralized receiving and shipping point
CRT	cathode ray tube; chemical, biological, radiological, nuclear, and high-yield explosives response team; contingency response team
CRTS	casualty receiving and treatment ship
CR-UAV	close-range unmanned aerial vehicle
CRW	contingency response wing
CRYPTO	cryptographic

CS	call sign; Chaplain Service (Air Force); circuit switch; coastal station; combat service; combat support; content staging; controlled space; creeping line single-unit; critical source
CSA	Chief of Staff, United States Army; combat support agency; container stuffing activity
CSAAS	combat support agency assessment system
CSADR	combat support agency director's report
CSAF	Chief of Staff, United States Air Force
CSAM	computer security for acquisition managers
CSAR	combat search and rescue
CSAR3	combat support agency responsiveness and readiness report
CSARTE	combat search and rescue task element
CSARTF	combat search and rescue task force
CSB	contracting support brigade
CSB (MEB)	combat support brigade (maneuver enhancement brigade)
CSC	combat support center; community support center; convoy support center; creeping line single-unit coordinated; International Convention for Safe Containers
CSCC	coastal sea control commander
CSE	client server environment; combat support enhanced; combat support equipment; contingency support element; cyberspace support element
CSEL	circuit switch select line; combat survivor evader locator; command senior enlisted leader
CSEP	Chairman of the Joint Chiefs of Staff-sponsored exercise program
CSG	carrier strike group; Chairman's Staff Group; coordinating subgroup; cryptologic services group; Cryptologic Support Group
CSGN	coordinating subgroup for narcotics
CSH	combat support hospital
CSIF	communications service industrial fund
CSIP	contract support integration plan
CSIPG	circuit switch interface planning guide
CSL	combat stores list; cooperative security location
C-SMPP	Consolidated Satellite Communications Management Policies and Procedures
CSNP	causeway section, nonpowered
CSNP(BE)	causeway section, nonpowered (beach end)
CSNP(I)	causeway section, nonpowered (intermediate)
CSNP(SE)	causeway section, nonpowered (sea end)
CSO	Center for Special Operations (USSOCOM); communications support organization; controlled source operation
CSOA	combined special operations area

CSOB	command systems operations branch
CSOD	command systems operation division
CSP	call service position; career sea pay; causeway section, powered; commence search point; contracting support plan; crisis staffing procedures (JCS); cryptologic support package
CSPAR	combatant commander's preparedness assessment report
CSR	central source registry; combatant commander's summary report; commander's summary report; controlled supply rate
CSRF	common source route file
CSS	Central Security Service (NSA); combat service support; communications subsystem; coordinator surface search
CSSA	combat service support area
CSSAMO	combat service support automation management office
CSSB	combat sustainment support battalion
CSSC	coded switch set controller
C-SSE	consolidated satellite communications system expert
CSST	combat service support team
CSSU	combat service support unit
CST	contingency support team; customer service team
CSW	compartment stowage worksheet; coordinate seeking weapons
CT	computed tomography; control telemetry; counterterrorism; country team
CTA	common table of allowance
CTAF	counterterrorism analytical framework
CTAPS	contingency Theater Air Control System automated planning system
CTC	cargo transfer company (USA); counterterrorist center
CTDB	combating terrorism database
CTEP	combined training and education plan
CTF	combined task force; commander, task force; counter threat finance
CTF IAMD	commander, task force integrated air and missile defense
CTFP	Combating Terrorism Fellowship Program
CTG	commander, task group
CTID	communications transmission identifier
CTKB	combating terrorism knowledge base
CTL	candidate target list
CTM	core target material
CTOC	corps tactical operations center
CTP	common tactical picture
CTR	cooperative threat reduction
CTRIF	Combating Terrorism Readiness Initiative Fund

CTS	commodity tracking system; Contingency Tracking System; controlled technical services
CTSS	central targeting support staff
CTU	commander, task unit
CU	cubic capacity; common unit
CUI	controlled unclassified information
CUL	common-user logistics
CULT	common-user land transportation
CV	aircraft carrier; carrier; critical vulnerability; curriculum vitae
CVAMP	Core Vulnerability Assessment Management Program
CVN	aircraft carrier, nuclear
CVR	cockpit voice recorder
CVS	commercial vendor services
CVSD	continuous variable slope delta
CVT	criticality-vulnerability-threat
CVW	carrier air wing; cryptovariable weekly (GPS)
CVWC	carrier strike group air wing commander
CW	carrier wave; chemical warfare; continuous wave
CWA	chemical warfare agent
CWC	Chemical Weapons Convention; composite warfare commander
CWDE	chemical warfare defense equipment
CWMD	countering weapons of mass destruction
CWO	communications watch officer
CWP	causeway pier
CWPD	Conventional War Plans Division, Joint Staff (J-7)
CWR	calm water ramp
CWT	combat weather team; customer wait time
CY	calendar year

D

D	total drift, data
d	surface drift
D&D	denial and deception
D&F	determinations and findings
D&M	detection and monitoring
D&R	debrief and reintegrate
D-2X	Department of Defense-level counterintelligence and human intelligence staff element
D3A	decide, detect, deliver, and assess
D/A	digital-to-analog
DA	data adapter aerospace drift; data administrator; Department of the Army; Development Assistance; direct action; Directorate for Mission Services (DIA); double agent

Da	aerospace drift
DA&M	Director of Administration and Management
DAA	designated approving authority; display alternate area routing lists
DAADC	deputy area air defense commander
DAADC(AMD)	deputy area air defense commander for air and missile defense
DAAS	defense automatic addressing system
DAASO	defense automatic addressing system office
DAB	Defense Acquisition Board
DAC	Defense Intelligence Agency (DIA) counterintelligence and security activity; Department of Army civilians; Development Assistance Committee (OECD)
DACB	data adapter control block
DACG	departure airfield control group
DACM	data adapter control mode
DADCAP	dawn and dusk combat air patrol
DAF	Department of the Air Force
DAFL	directive authority for logistics
DAICC	domestic air interdiction coordinator center
DAL	defended asset list
DALIS	Disaster Assistance Logistics Information System
DALS	downed aviator locator system
DAMA	demand assigned multiple access
DAMES	defense automatic addressing system (DAAS) automated message exchange system
DAN	Diver's Alert Network
DAO	defense attaché office; defense attaché officer; department/agency/organization
DAP	designated acquisition program
DAR	Defense Acquisition Regulation; distortion adaptive receiver
DARO	Defense Airborne Reconnaissance Office
DARPA	Defense Advanced Research Projects Agency
DART	disaster assistance response team; downed aircraft recovery team; dynamic analysis and replanning tool
DAS	deep air support (USMC); defense attaché system; direct access subscriber; direct air support
DAS3	decentralized automated service support system
DASA	Department of the Army (DA) staff agencies
DASC	direct air support center
DASC(A)	direct air support center (airborne)
DASD	Deputy Assistant Secretary of Defense
DASD-CN	Deputy Assistant Secretary of Defense for Counternarcotics
DASD(H&RA)	Deputy Assistant Secretary of Defense (Humanitarian & Refugee Affairs)

DASD(I)	Deputy Assistant Secretary of Defense (Intelligence)
DASD(PK/HA)	Deputy Assistant Secretary of Defense (Peacekeeping and Humanitarian Affairs)
DASD(S&IO)	Deputy Assistant Secretary of Defense (Security and Information Operations)
DASSS	decentralized automated service support system
DAT	deployment action team
DATT	defense attaché
DATU	data adapter termination unit
dB	decibel
DBA	database administrator
DBDB	digital bathymetric database
DBDB-V	digital bathymetric database variable
DBG	database generation
DBI	defense budget issue
DBIDS	Defense Biometric Identification System
DBMS	database management system; Defense-Business Management System
DBSMC	Defense Business Systems Management Committee
DBSS	Defense Blood Standard System
DBT	design basis threat
DC3	Department of Defense Cyber Crime Center
D/C	downconverter
DC	Deputies Committee; direct current; Directorate of Counterintelligence (DIA); dislocated civilian
DCA	Defense Communications Agency; Defense Cooperation Agreements; defensive counterair; dual-capable aircraft
DCAA	Defense Contract Audit Agency
DCAM	Defense Medical Logistics Standard Support (DMLSS) customer assistance module
DCAPES	Deliberate and Crisis Action Planning and Execution Segments
DCC	damage control center; deployment control center
DCCC	defense collection coordination center
DCCEP	developing country combined exercise program
DCD	data collection device
DCE	defense coordinating element
D-cell	deployment cell
DCGS	distributed common ground/surface system
DCHA	Bureau for Democracy, Conflict, and Humanitarian Assistance (USAID)
DCHC	Defense Counterintelligence and Human Intelligence Center
DCHE	Defense Counterintelligence and Human Intelligence Enterprise
DCI	defense critical infrastructure; Director of Central Intelligence; dual channel interchange

D/CI&SP	Director, Counterintelligence and Security Programs
D/CIA	Director, Central Intelligence Agency
DCID	Director of Central Intelligence directive
DCIIS	Defense Counterintelligence Information System
DCIO	defense criminal investigative organization
DCIP	Defense Critical Infrastructure Program
DCIS	Defense Criminal Investigative Services
DCISE	Defense Industrial Base Collaborative Information Sharing Environment
DCJTF	deputy commander, joint task force
DCM	data channel multiplexer; deputy chief of mission
DCMA	Defense Contract Management Agency
DCMC	Office of Deputy Chairman, Military Committee
DCMO	deputy chief military observer
DCNO	Deputy Chief of Naval Operations
DCO	debarkation control officer; Defense Connect Online; defense coordinating officer; defensive cyberspace operations; dial central office
DCO-RA	defensive cyberspace operations response actions
DCP	Defense Continuity Program; detainee collection point
DCPA	Defense Civil Preparedness Agency
DCPG	digital clock pulse generator
DCR	doctrine, organization, training, materiel, leadership and education, personnel, and facilities change recommendation
DCRF	defense chemical, biological, radiological, and nuclear response force
DCS	Defense Courier Service; deputy chief of staff; digital computer system
DCSCU	dual capability servo control unit
DC/S for RA	Deputy Chief of Staff for Reserve Affairs
DCSINT	Deputy Chief of Staff for Intelligence
DCSLOG	Deputy Chief of Staff for Logistics, US Army
DCSOPS	Deputy Chief of Staff for Operations and Plans, United States Army
DCSPER	Deputy Chief of Staff for Personnel, United States Army
DCST	Defense Logistics Agency (DLA) contingency support team
DCTS	Defense Collaboration Tool Suite
DCW	Defense Collection Watch (DIA)
DD	Department of Defense form; destroyer (Navy ship)
DDA	Deputy Director for Administration (CIA); designated development activity
D-day	unnamed day on which operations commence or are scheduled to commence
DDC	data distribution center; defense distribution center
DDCI	Deputy Director of Central Intelligence (CIA)

DDCI/CM	Deputy Director of Central Intelligence for Community Management
DDED	defense distribution expeditionary depot
DDG	guided missile destroyer
DDI	Deputy Director of Intelligence (CIA); Director of Defense Intelligence
DDL	digital data link
DDM	Defense Logistics Agency Distribution Mapping; digital data modem
DDMA	Defense Distribution Mapping Activity
DDMS	Deputy Director for Military Support (NRO)
DDO	Deputy Director of Operations (CIA)
DDOC	Deployment and Distribution Operations Center (USTRANSCOM)
DDP	detailed deployment plan
DDR	disarmament, demobilization, and reintegration
DDR&E	director of defense research and engineering
DDRRR	disarmament, demobilization, repatriation, reintegration, and resettlement
DDS	defense dissemination system; Deployable Disbursing System; dry deck shelter
DDS&T	Deputy Director for Science & Technology (CIA)
DDSBn	deployment and distribution support battalion
DDSM	Defense Distinguished Service Medal
DDST	deployment and distribution support team
DDWSO	Deputy Director for Wargaming, Simulation, and Operations
DE	damage expectancy; decedent effects; delay equalizer; directed energy
De	total drift error
de	individual drift error
DEA	Drug Enforcement Administration (DOJ)
dea	aerospace drift error
DEACN	Drug Enforcement Administration Communications Network
DEAR	disease and environmental alert report
DEARAS	Department of Defense (DOD) Emergency Authorities Retrieval and Analysis System
DeCA	Defense Commissary Agency
DEERS	Defense Enrollment Eligibility Reporting System
DEFSMAC	Defense Special Missile and Aerospace Center
DEL	deployable equipment list
DEMARC	demarcation
de max	maximum drift error
DEMIL	demilitarization

de min	minimum drift error
de minimax	minimax drift error
DeMS	deployment management system
DEMUX	demultiplex
DEP	Delayed Entry Program; deployed
DEP&S	Drug Enforcement Plans and Support
DEPCJTF	deputy commander, joint task force
DEPID	deployment indicator code
DEPMEDS	deployable medical systems
DepOpsDeps	Service deputy operations deputies
DEPORD	deployment order
DESC	Defense Energy Support Center
DESCOM	Depot System Command (Army)
DESIGAREA	designated area message
DEST	destination; domestic emergency support team
DET	detainee
DETRESFA	distress phase (ICAO)
DEW	directed-energy warfare
DF	direction finding; dispersion factor; disposition form
DFARS	Department of Defense Federal Acquisition Regulation Supplement
DFAS	Defense Finance and Accounting Service
DFAS-DE	Defense Finance and Accounting Service-Denver
DFC	deputy force commander; detention facility commander
DFE	Defense Intelligence Agency forward element; Defense Joint Intelligence Operations Center forward element; division force equivalent
DFM	deterrent force module
DFO	disaster field office (FEMA)
DFR	Defense Fuel Region
DFR/E	Defense Fuel Region, Europe
DFRIF	Defense Freight Railway Interchange Fleet
DFR/ME	Defense Fuel Region, Middle East
DFSC	Defense Fuel Supply Center
DFSP	defense fuel support point
DFT	deployment for training
DG	defense guidance
DGIAP	Defense General Intelligence and Applications Program
DGM	digital group multiplex
DH	death due to hostilities; Directorate for Human Intelligence (DIA)
DHA	detainee holding area
DHB	Defense Health Board
DHE	defense human intelligence executor
DHHS	Department of Health and Human Services
DHM	Department of Defense human intelligence manager

DHMO	Department of Defense human intelligence management office
DHP	Defense Health Program
DHS	Defense Human Intelligence (HUMINT) Service; Department of Homeland Security; Director of Health Services
DI	Defense Intelligence Agency (DIA) Directorate for Analysis; DIA Directorate for Intelligence Production; discrete identifier; dynamic interface
DI&E	data integration and exploitation
DIA	Defense Intelligence Agency
DIAC	Defense Intelligence Analysis Center
DIA/DHX	Defense Intelligence Agency, Directorate of Human Intelligence, Office of Document and Media Operations
DIAM	Defense Intelligence Agency manual; Defense Intelligence Agency memorandum
DIAP	Defense Intelligence Analysis Program; Drug Interdiction Assistance Program
DIAR	Defense Intelligence Agency (DIA) regulation
DIB	defense industrial base
DIBITS	digital in-band interswitch trunk signaling
DIBRS	defense incident-based reporting system
DIBTS	digital in-band trunk signaling
DICO	Data Information Coordination Office
DID	Defense Intelligence Digest
DIDHS	Deployable Intelligence Data Handling System
DIDO	designated intelligence disclosure official
DIDS	Defense Intelligence Dissemination System
DIEB	Defense Intelligence Executive Board
DIEPS	Digital Imagery Exploitation Production System
DIG	digital
DIGO	Defence Imagery and Geospatial Organisation (Australia)
DII-COE	defense information infrastructure-common operating environment
DIILS	Defense Institute of International Legal Studies
DIJE	Defense Intelligence Joint Environment (United Kingdom)
DILPA	diphase loop modem-A
DIMA	drilling individual mobilization augmentee
DIMOC	Defense Imagery Management Operations Center
DIN	defense intelligence notice
DINET	Defense Industrial Net
DINFOS	Defense Information School
DIOC	drug interdiction operations center
DIOCC	Defense Intelligence Operations Coordination Center
DIPC	defense industrial plant equipment center

DIPF	defense intelligence priorities framework
DIPFAC	diplomatic facility
DIPGM	diphase supergroup modem
DIRINT	Director of Intelligence (USMC)
DIRJIATF	director, joint inter-agency task force
DIRLAUTH	direct liaison authorized
DIRM	Directorate for Information and Resource Management
DIRMOBFOR	director of mobility forces
DIRNSA	Director, National Security Agency
DIRSPACEFOR	director of space forces (USAF)
DIS	defense information system; Defense Investigative Service; distributed interactive simulation
DISA	Defense Information Systems Agency
DISA-LO	Defense Information Systems Agency - liaison officer
DISANMOC	Defense Information Systems Agency Network Management and Operations Center
DisasterAWARE	Disaster All-Hazard Warnings, Analysis, and Risk Evaluation System
DISCOM	division support command (Army)
DISGM	diphase supergroup
DISN	Defense Information Systems Network
DISN-E	Defense Information Systems Network – Europe
DISO	deception in support of operations security; defense intelligence support office
DISP	drug investigation support program (FAA)
DITDS	defense information threat data system; defense intelligence threat data system
DITSUM	defense intelligence terrorist summary
DJIOC	Defense Joint Intelligence Operations Center
DJS	Director, Joint Staff
DJSM	Director, Joint Staff memorandum
DJTFS	deputy joint task force surgeon
DLA	Defense Logistics Agency
DLAM	Defense Logistics Agency manual
DLAR	Defense Logistics Agency regulation
DLEA	drug law enforcement agency
DLED	dedicated loop encryption device
DLD	digital liaison detachment
DLIS	Defense Logistics Information Service
DLMS	Department Logistics Management System
DLP	data link processor
DLPMA	diphase loop modem A
DLQ	deck landing qualification
DLR	depot-level repairable
DLSA	Defense Legal Services Agency
DLSS	Defense Logistics Standard Systems

DLTM	digital line termination module
DLTU	digital line termination unit
DM	detection and monitoring; docking module
DMA	Defense Media Activity
dmax	maximum drift distance
DMB	datum marker buoy
DMC	data mode control
DMD	digital message device
DMDC	defense management data center; defense manpower data center
DME	distance measuring equipment
DMHS	Defense Message Handling System
DMI	director military intelligence
DMIGS	Domestic Mobile Integrated Geospatial-Intelligence System
dmin	minimum drift distance
DML	data manipulation language
DMLSS	Defense Medical Logistics Standard Support
DMMPO	Defense Medical Materiel Program Office
DMO	directory maintenance official
DMORT	Disaster Mortuary Operational Response Team
DMOS	duty military occupational specialty
DMPI	designated mean point of impact
DMRD	defense management resource decision
DMRIS	defense medical regulating information system
DMS	defense message system; defense meteorological system; director of military support
DMSB	Defense Medical Standardization Board
DMSM	Defense Meritorious Service Medal
DMSO	Defense Modeling and Simulation Office; director of major staff office; Division Medical Supply Office
DMSP	Defense Meteorological Satellite Program
DMSSC	defense medical systems support center
DMT	disaster management team (UN)
DMU	disk memory unit
DMZ	demilitarized zone
DN	digital nonsecure
DNA	Defense Nuclear Agency; deoxyribonucleic acid
DNAT	defense nuclear advisory team
DNBI	disease and nonbattle injury
DNC	digital nautical chart
DND	Department of National Defence (Canada)
DNDO	Domestic Nuclear Detection Office (DHS)
DNGA	Director of National Geospatial-Intelligence Agency
DNI	Director of National Intelligence; Director of Naval Intelligence
DNIF	duty not involving flying

DNMSP	driftnet monitoring support program
DNSO	Defense Network Systems Organization
DNVT	digital nonsecure voice terminal
DNY	display area code (NYX) routing
DOA	days of ammunition; dead on arrival; director of administration
DOB	date of birth; dispersal operating base
DOC	Department of Commerce; designed operational capability
DOCC	deep operations coordination cell
DOCDIV	documents division
DOCEX	document exploitation
DOCNET	Doctrine Networked Education and Training
DOD	Department of Defense
DODAAC	Department of Defense activity address code
DODAAD	Department of Defense Activity Address Directory
DODAC DOD	ammunition code
DODD	Department of Defense directive
DODDS	Department of Defense Dependent Schools
DODEX	Department of Defense intelligence system information system extension
DODFMR	Department of Defense Financial Management Regulation
DODI	Department of Defense instruction
DODIC	Department of Defense identification code
DODID	Department of Defense Intelligence Digest
DODIIS	Department of Defense Intelligence Information System
DODIN	Department of Defense information networks
DODIPC	Department of Defense intelligence production community
DODIPP	Department of Defense Intelligence Production Program
DOD-JIC	Department of Defense Joint Intelligence Center
DODM	data orderwire diphase modem; Department of Defense manual
DOE	Department of Energy
DOEHRS	Defense Occupational and Environmental Health Reporting System
DOF	degree of freedom
DOI	Defense Special Security Communications System (DSSCS) Operating Instructions; Department of the Interior
DOJ	Department of Justice
DOL	Department of Labor
DOM	day of month
DOMEX	document and media exploitation
DOMS	director of military support
DON	Department of the Navy
DOPMA	Defense Officer Personnel Management Act
DOR	date of rank

DOS	date of separation; days of supply; denial of service; Department of State; disk operating system
DOT	Department of Transportation
DOTEO	Department of Transportation emergency organization
DOTMLPF	doctrine, organization, training, materiel, leadership and education, personnel, and facilities
DOW	data orderwire; died of wounds
DOX-T	direct operational exchange-tactical
DOY	day of year
DP	Air Force component plans officer (staff); decisive point; Directorate for Policy Support (DIA); displaced person
dp	parachute drift
DPA	danger pay allowance; Defense Production Act
DPAP	Defense Procurement and Acquisition Policy
DPAS	Defense Priorities and Allocation System
DPC	deception planning cell; Defense Planning Committee (NATO)
DPEC	displaced person exploitation cell
DPI	desired point of impact
dpi	dots per inch
DPICM	dual purpose improved conventional munitions
DPKO	Department of Peacekeeping Operations (UN)
DPLSM	dipulse group modem
DPM	dissemination program manager
DPMO	Defense Prisoner of War/Missing Personnel Office
DPO	distribution process owner
DPP	data patch panel; distributed production program
DPPDB	digital point positioning database
DPQ	defense planning questionnaire (NATO)
DPR	display non-nodal routing
DPRB	Defense Planning and Resources Board
DPRE	displaced persons, refugees, and evacuees
DPS	data processing system
DPSC	Defense Personnel Support Center
DPSK	differential phase shift keying
DR	dead reckoning; digital receiver; disaster relief
DRB	Defense Resources Board
DRe	dead reckoning error
DRL	Bureau of Democracy, Human Rights, and Labor (DOS)
DRMD	deployments requirements manning document
DRMO	Defense Reutilization and Marketing Office
DRMS	Defense Reutilization and Marketing Service; distance root-mean-square
DRN	Disaster Response Network
DRO	departmental requirements office; departmental requirements officer

DRRS	Defense Readiness Reporting System
DRS	detainee reporting system
DRSN	Defense Red Switched Network
DRT	dead reckoning tracer; decontamination and reconnaissance team
DRTC	designated reporting technical control
DS	Directorate for Management and Chief Information Officer (DIA); direct support; doctrine sponsor
DSA	defense special assessment (DIA)
DSAA	Defense Security Assistance Agency
DSAID	defense sexual assault incident database
DSAR	Defense Supply Agency regulation
DSB	digital in-band trunk signaling (DIBTS) signaling buffer
DSC	defensive space control; digital selective calling; dual-status commander
DSCA	Defense Security Cooperation Agency; defense support of civil authorities
DSCP	Defense Supply Center Philadelphia
DSCR	Defense Supply Center Richmond
DSCS	Defense Satellite Communications System
DSCSOC	Defense Satellite Communications System operations center
DSDI	digital simple data interface
DSE	direct support element
DSF	District Stability Framework (USAID)
DSG	digital signal generator
DSI	defense simulation internet
DSL	display switch locator (SL) routing
DSM	decision support matrix
DSMAC	digital scene-matching area correlation
DSN	Defense Switched Network
DSNET	Defense Secure Network
DSNET-2	Defense Secure Network-2
DSO	Defense Spectrum Organization; defensive systems officer
DSOE	deployment schedule of events
DSP	Defense Satellite Program; Defense Support Program
DSPD	defense support to public diplomacy
DSPL	display system programming language
DSPS	Director, Security Plans and Service
DSR	defense source registry
DSS	Defense Security Service; Distribution Standard System
DSS/ALOC	direct support system/air line of communications
DSSCS	Defense Special Security Communications System
DSSM	Defense Superior Service Medal

DSSO	data system support organization; defense sensitive support office; defense systems support organization
DSSR	Department of State Standardized Regulation
DST	decision support template; Defense Logistics Agency support team; deployment support team; district support team
DSTP	Director of Strategic Target Planning
DSTR	destroy
DSTS-G	Defense Information Systems Network (DISN) Satellite Transmission Services - Global
DSVL	doppler sonar velocity log
DSVT	digital subscriber voice terminal
DT	Directorate for MASINT and Technical Collection (DIA)
DTA	Defense Threat Assessment; dynamic threat assessment
DTAM	defense terrorism awareness message
DTCI	Defense Transportation Coordination Initiative
DTD	detailed troop decontamination
DTE	data terminal equipment; developmental test and evaluation
DTED	digital terrain elevation data
DTG	date-time group; digital trunk group (digital transmission group)
D/T/ID	detect/track/identify
DTIP	Disruptive Technology Innovations Partnership (DIA)
DTL	designator target line
DTMF	dual tone multi-frequency
DTMR	defense traffic management regulation
DTO	division transportation office; drug trafficking organization
DTOC	division tactical operations center
DTR	defense transportation regulation
DTRA	Defense Threat Reduction Agency
DTRACS	Defense Transportation Reporting and Control System
DTRATCA	Defense Threat Reduction and Treaty Compliance Agency
DTS	Defense Transportation System; Defense Travel System; diplomatic telecommunications service
DTTS	Defense Transportation Tracking System
DTWR	defense terrorism warning report
DU	depleted uranium
DUSD	deputy under Secretary of Defense
DUSD (CI&S)	Deputy Under Secretary of Defense for Counterintelligence and Security
DUSDL	Deputy Under Secretary of Defense for Logistics
DUSD(L&MR)	Deputy Under Secretary of Defense for Logistics and Materiel Readiness
DUSDP	Deputy Under Secretary of Defense for Policy
DV	distinguished visitor

DVA	Department of Veterans Affairs
DVD	digital video device; digital video disc
DVITS	Digital Video Imagery Transmission System
DVOW	digital voice orderwire
DVT	deployment visualization tool
DWAS	Defense Working Capital Accounting System
DWMCF	double-wide modular causeway ferry
DWT	deadweight tonnage
DWTS	Digital Wideband Transmission System
DX	Directorate for External Relations (DIA)
DZ	drop zone
DZC	drop zone controller
DZCO	drop zone control officer
DZSO	drop zone safety officer
DZST	drop zone support team
DZSTL	drop zone support team leader

E

E	total probable error
E&DCP	evaluation and data collection plan
E&E	emergency and extraordinary expense authority
E&EE	emergency and extraordinary expense
E&I	engineering and installation
E&M	ear and mouth; special signaling leads
E1	Echelon 1
E2	Echelon 2
E3	Echelon 3; electromagnetic environmental effects
E4	Echelon 4
E5	Echelon 5
E-8C	joint surveillance, target attack radar system (JSTARS) aircraft
EA	electronic attack; emergency action; evaluation agent; executive agent; executive assistant
ea	each
EAC	echelons above corps (Army); emergency action; emergency action committee
EACS	expeditionary aeromedical evacuation crew member support
EACT	expeditionary aeromedical evacuation coordination team
EAD	earliest arrival date; echelons above division (Army); extended active duty
EADRU	Euro-Atlantic disaster response unit
EADS	Eastern Air Defense Sector
EAES	expeditionary aeromedical evacuation squadron
EAF	expeditionary aerospace forces; expeditionary airfield
EAI	executive agent instruction

EALT	earliest anticipated launch time
EAM	emergency action message
EAP	emergency action plan; emergency action procedures
EAPC	Euro-Atlantic Partnership Council
EAP-CJCS	emergency action procedures of the Chairman of the Joint Chiefs of Staff
EARLY	evasion and recovery supplemental data report
E-ARTS	en route automated radar tracking system
EASF	expeditionary aeromedical staging facility
EAST	expeditionary aeromedical evacuation staging team
EASTPAC	eastern Pacific Ocean
EBCDIC	extended binary coded decimal interchange code
EBS	environmental baseline survey
EC	electronic combat; enemy combatant; error control; European Community
ECAC	Electromagnetic Compatibility Analysis Center
ECB	echelons corps and below (Army)
ECC	engineer coordination cell; evacuation control center
ECES	expeditionary civil engineer squadron
ECHA	Executive Committee for Humanitarian Affairs (UN)
ECHO	European Community Humanitarian Aid Department
ECM	electronic countermeasures
ECN	electronic change notice; Minimum Essential Emergency Communications Network
ECO	electronic combat officer
ECOSOC	Economic and Social Council (UN)
ECP	emergency command precedence; engineering change proposal; entry control point
ECS	expeditionary combat support
ECU	environmental control unit
ED	envelope delay; evaluation directive
EDA	Economic Development Administration (DOC); excess defense articles
EDC	estimated date of completion
EDD	earliest delivery date
EDI	electronic data interchange
EDSS	equipment deployment and storage system
EE	emergency establishment
EEA	environmental executive agent
EEBD	emergency escape breathing device
EECT	end evening civil twilight
EED	electro-explosive device; emergency-essential designation
EEDAC	emergency essential Department of the Army civilian
EEE	emergency and extraordinary expense
EEFI	essential element of friendly information

EEI	essential element of information
EELV	evolved expendable launch vehicle
EEO	equal employment opportunity
EEPROM	electronic erasable programmable read-only memory
EER	enlisted employee review; extended echo ranging
EEZ	exclusive economic zone
EFA	engineering field activity
EFAC	emergency family assistance center
EFD	engineering field division
EFP	explosively formed projectile
EFST	essential fire support task
EFT	electronic funds transfer
EFTO	encrypt for transmission only
EGAT	Bureau of Economic Growth, Agriculture, and Trade (USAID)
EGM	Earth Gravity Model
EGS	Earth ground station
EH	explosive hazard
EHCC	explosive hazards coordination cell
EHDB	explosive hazard database
EHF	extremely high frequency
EHO	environmental health officer
EHRA	environmental health risk assessment
EHSA	environmental health site assessment
EHT	explosive hazard team
EI	environmental information; exercise item
EIA	Electronic Industries Association
EID	electrically initiated device
EIS	Environmental Impact Statement
eJMAPS	electronic Joint Manpower and Personnel System
EJPME	enlisted joint professional military education
ELBA	emergency locator beacon
ELCAS	elevated causeway system
ELCAS(M)	elevated causeway system (modular)
ELCAS(NL)	elevated causeway system (Navy lighterage)
ELD	emitter locating data
ELINT	electronic intelligence
ELIST	enhanced logistics intratheater support tool
ELOS	extended line of sight
ELPP	equal level patch panel
ELR	extra-long-range aircraft
ELSEC	electronics security
ELT	emergency locator transmitter
ELV	expendable launch vehicle
ELVA	emergency low-visibility approach
EM	electromagnetic; emergency management; executive manager

EMAC	emergency management assistance compact
E-mail	electronic mail
EMALL	electronic mall
EMBM	electromagnetic battle management
EMC	electromagnetic compatibility
EMCON	emission control
EMD	effective miss distance; expeditionary military information support detachment
EME	electromagnetic environment
EMEDS	Expeditionary Medical Support
EMF	expeditionary medical facility
EMI	electromagnetic interface; electromagnetic interference
EMIO	expanded maritime interception operations
EMOE	electromagnetic operational environment
EMP	electromagnetic pulse
EMR hazards	electromagnetic radiation hazards
EMS	electromagnetic spectrum; emergency medical services
EMSC	electromagnetic spectrum control
EMSEC	emanations security
EMT	emergency medical technician; emergency medical treatment; expeditionary military information support team
EMTF	expeditionary mobility task force
EMV	electromagnetic vulnerability
ENCOM	engineer command (Army)
ENDEX	exercise termination
ENL	enlisted
ENSCE	enemy situation correlation element
ENWGS	Enhanced Naval Warfare Gaming System
EO	electro-optical; end office; equal opportunity; executive order; eyes only
EOB	electromagnetic order of battle; electronic order of battle; enemy order of battle
EOC	early operational capability; emergency operating center; emergency operations center
EOCA	explosive ordnance clearance agent
EOD	explosive ordnance disposal
EODMU-1	explosive ordnance disposal mobile unit one
EOI	electro-optic(al) imagery
EO-IR	electro-optical-infrared
EO-IR CM	electro-optical-infrared countermeasure
EOL	end of link
EOM	end of message
EOP	emergency operating procedures
E-O TDA	electro-optical tactical decision aid
EOW	engineering orderwire

EP	electronic protection; emergency preparedness; emergency procedures; execution planning
EPA	Environmental Protection Agency; evasion plan of action
EPBX	electronic private branch exchange
EPC	Emergency Procurement Committee
EPF	enhanced palletized load system (PLS) flatrack
EPH	emergency planning handbook
EPIC	El Paso Intelligence Center
EPIRB	emergency position-indicating radio beacon
EPLO	emergency preparedness liaison officer
EPROM	erasable programmable read-only memory
EPU	expeditionary port unit
EPW	enemy prisoner of war
EPW/CI	enemy prisoner of war/civilian internee
ERC	exercise related construction
ERDC	Engineer Research and Development Center
ERG	Emergency Response Guidebook
ERGM	extended range guided munitions
ERO	engine running on or offload
ERPSS	En Route Patient Staging System
ERRO	Emergency Response and Recovery Office
ERSD	estimated return to service date
ERT	emergency response team (FEMA); engineer reconnaissance team
ERT-A	emergency response team - advance element
ERU	emergency response unit
ERW	explosive remnants of war
ES	electronic warfare support; executive secretariat
ESAC	Electromagnetic-Space Analysis Center (NSA)
ESB	engineer support battalion
ESC	Electronics Systems Center; expeditionary sustainment command
ESF	Economic Support Fund; emergency support function
ESG	executive steering group; expeditionary strike group
ESGN	electrically suspended gyro navigation
ESI	extremely sensitive information
ESK	electronic staff weather officer kit
ESM	expeditionary site mapping
ESO	embarkation staff officer; environmental science officer; Expeditionary Support Organization (DFAS)
ESOC	Emergency Supply Operations Center
ESOH	environmental, safety, and occupational health
ESORTS	Enhanced Status of Resources and Training System
ESP	engineer support plan
E-Space	Electromagnetic-Space
ESR	external supported recovery

EST	embarked security team; emergency service team; emergency support team (FEMA); en route support team
ET	electronics technician
ETA	estimated time of arrival
ETAC	emergency tactical air control
ETD	estimated time of departure
ETF	electronic target folder
ETI	estimated time of intercept
ETIC	estimated time for completion; estimated time in commission
ETM	electronic transmission; essential tasks matrix
ETO	Emergency Transportation Operations (DOT)
ETPL	endorsed TEMPEST products list
ETR	export traffic release
ETS	European telephone system
ETSS	extended training service specialist
ETX	end of text
EU	European Union
E-UAV	endurance unmanned aerial vehicle
EUB	essential user bypass
EURV	essential user rekeying variable
EUSA	Eighth US Army
EUSC	effective United States control; effective United States controlled
EUSCS	effective United States-controlled ships
EVC	evasion chart
EVE	equal value exchange
EW	early warning; electronic warfare
EWC	electronic warfare cell; electronic warfare coordinator
EWCA	electronic warfare control authority
EWCC	electronic warfare coordination cell
EWCS	electronic warfare control ship
EWD	Electronic Warfare Directorate (USSTRATCOM)
EWE	electronic warfare element
EWG	electronic warfare group
EW/GCI	early warning/ground-controlled intercept
EWIR	electronic warfare integrated reprogramming
EWO	electronic warfare officer
EXCIMS	Executive Council for Modeling and Simulations
ExCom	executive committee
EXDIR	Executive Director (CIA)
EXDIR/ICA	Executive Director for Intelligence Community Affairs (USG)
EXECSEC	executive secretary
EXER	exercise

EXORD	execute order
EXPLAN	exercise plan
EZ	exchange zone
EZCO	extraction zone control officer
EZM	engagement zone manager

F

F	Fahrenheit; flare patterns; flash
F2T2EA	find, fix, track, target, engage, and assess
F3EAD	find, fix, finish, exploit, analyze, and disseminate
F&ES	fire and emergency services
FA	feasibility assessment; field artillery
FAA	Federal Aviation Administration (DOT); Foreign Assistance Act
FAAO	Federal Aviation Administration order
FAAR	facilitated after-action review
FAC	forward air controller
FAC(A)	forward air controller (airborne)
FACE	forward aviation combat engineering
FACSFAC	fleet area control and surveillance facility
FACT	field advance civilian team; field assessment and coordination team
FAD	feasible arrival date
F/AD	force/activity designator
FAE	fuel air explosive
FAH	foreign affairs handbook
FALD	Field Administration and Logistics Division
FALOP	Forward Area Limited Observing Program
FAM	functional area manager
FAMP	forward area minefield planning
FAO	Food and Agriculture Organization (UN); foreign area officer
FAPES	Force Augmentation Planning and Execution System
FAR	Federal Acquisition Regulation; Federal Aviation Regulation; formal assessment report
FARC	Revolutionary Armed Forces of Colombia
FARP	forward arming and refueling point
FAS	Foreign Agricultural Service (USDA); frequency assignment subcommittee; fueling at sea; functional account symbol
FASCAM	family of scatterable mines
FAST	field assessment surveillance team; fleet antiterrorism security team
FAX	facsimile
FB	forward boundary

FBI	Federal Bureau of Investigation (DOJ)
FBIS	Foreign Broadcast Information Service
FBO	faith-based organization
FC	field circular; final coordination; fires cell (Army); floating causeway; floating craft; force commander
FCA	Foreign Claims Act; functional configuration audit
FCC	Federal Communications Commission; Federal coordinating center; functional combatant commander
FCE	forward command element
FCG	foreign clearance guide
FCM	foreign consequence management
FCO	federal coordinating officer
FCP	fire control party
FCT	firepower control team
FD	first draft; from temporary duty
FDA	Food and Drug Administration
FDBM	functional database manager
FDC	fire direction center
FDESC	force description
FDL	fast deployment logistics
FDLP	flight deck landing practice
FDM	frequency division multiplexing
FDO	fire direction officer; flexible deterrent option; flight deck officer; foreign disclosure officer
FDR/FA	flight data recorder/fault analyzer
FDS	fault detection system
FDSL	fixed directory subscriber list
FDSS	fault detection subsystem
FDSSS	flight deck status and signaling system
FDT	forward distribution team
FDUL	fixed directory unit list
FDX	full duplex
FE	facilities engineering
FEA	front-end analysis
FEBA	forward edge of the battle area
FEC	facilities engineering command; forward error correction
FECC	fires and effects coordination cell
FED-STD	federal standard
FEI	forensic-enabled intelligence
FEK	frequency exchange keying
FEMA	Federal Emergency Management Agency (DHS)
FEP	fleet satellite (FLTSAT) extremely high frequency (EHF) Package; foreign excess property
FEPP	federal excess personal property; foreign excess personal property
FES	fire emergency services

FEST	foreign emergency support team; forward engineer support team
FET	facility engineer team
FEU	forty-foot equivalent unit
FEZ	fighter engagement zone
FF	navy fast frigate
Ff	fatigue correction factor
FFA	free-fire area
FFC	force fires coordinator
FFCC	flight ferry control center; force fires coordination center (USMC)
FFD	foundation feature data
FFE	field force engineering; flame field expedients
FFG	guided missile frigate
FFH	fast frequency hopping
FFH-net	fast-frequency-hopping net
FFHT-net	fast-frequency-hopping training net
FFIR	friendly force information requirement
FFP	Food for Peace; fresh frozen plasma
FFT	friendly force tracking
FFTU	forward freight terminal unit
FG	fighter group
FGMDSS	Future Global Maritime Distress and Safety System
FGS	final governing standard; Force Generation Service (UN)
FH	fleet hospital
FHA	Bureau for Food and Humanitarian Assistance; foreign humanitarian assistance
FHC	family help center
F-hour	effective time of announcement by the Secretary of Defense to the Military Departments of a decision to mobilize Reserve units
FHP	force health protection
FHWA	Federal Highway Administration (DOT)
FI	foreign intelligence
FIA	functional interoperability architecture
FIC	force indicator code
FID	foreign internal defense
FIDAF	foreign internal defense augmentation force
FIE	foreign intelligence entity
FIFO	first-in-first-out
FinCEN	Financial Crimes Enforcement Network
FIR	first-impressions report; flight information region
FIRCAP	foreign intelligence requirements capabilities and priorities
1st IOC	1st Information Operations Command (Land)
FIS	flight information service; Foreign Intelligence Service
FISC	fleet and industrial supply center

FISINT	foreign instrumentation signals intelligence
FISS	foreign intelligence and security services
FIST	fire support team; fleet imagery support terminal; fleet intelligence support team
FIWC	fleet information warfare center
FIXe	navigational fix error
FLAR	forward-looking airborne radar
FLENUMMETOCCEN	Fleet Numerical Meteorology and Oceanography Center
FLENUMMETOCDET	Fleet Numerical Meteorological and Oceanographic Detachment
FLETC	Federal Law Enforcement Training Center (DHS)
FLIP	flight information publication; flight instruction procedures
FLIR	forward-looking infrared
FLITE	federal legal information through electronics
FLO/FLO	float-on/float-off
FLOLS	fresnel lens optical landing system
FLOT	forward line of own troops
FLP	force level planning
FLS	forward logistic site
FLSG	force logistic support group
FLTCYBERCOM	Fleet Cyber Command (Navy)
FLTSAT	fleet satellite
FLTSATCOM	fleet satellite communications
FM	field manual (Army); financial management; flare multiunit; force module; frequency management; frequency modulation; functional manager
FM/A	functional manager for analysis
FMA-net	frequency management A-net
FMAS	foreign media analysis subsystem
FMAT	financial management augmentation team
FMC	force movement characteristics; full mission-capable
FMCH	fleet multichannel
FMCR	Fleet Marine Corps Reserve
FMCSA	Federal Motor Carrier Safety Administration
FMI	field manual-interim
FMF	Fleet Marine Force
FMFP	foreign military financing program
FMID	force module identifier
FMO	frequency management office; functional manager office
FMP	foreign materiel program
FMS	force module subsystem; foreign military sales
FMSC	frequency management sub-committee
FMT-net	frequency management training net
FMV	full motion video
FN	foreign nation
FNMOC	Fleet Numerical Meteorology and Oceanography Center

FNMOD	Fleet Numerical Meteorological and Oceanographic Detachment
FNOC	Fleet Numerical Oceanographic Command
FNS	foreign nation support
FO	fiber optic; flash override; forward observer
FOB	forward operating base; forward operations base
FOC	full operational capability; future operations cell
FOD	field operations division; foreign object damage
FOFW	fiber optic field wire
FOG	Field Operations Guide for Disaster Assessment and Response
FOI	fault detection isolation
FOIA	Freedom of Information Act
FOIU	fiber optic interface unit
FOL	fiber optic link
FON	freedom of navigation (operations)
FOO	field ordering officer
FORCE	fuels operational readiness capability equipment (Air Force)
FORMICA	foreign military intelligence collection activities
FORSCOM	United States Army Forces Command
FORSTAT	force status and identity report
FOS	forward operating site; full operational status
FOT	follow-on operational test
FOUO	for official use only
FP	force protection; frequency panel
FPA	foreign policy advisor
FPC	final planning conference; future plans cell
FPCON	force protection condition
FPD	force protection detachment; foreign post differential
FPF	final protective fire
FPM	Federal personnel manual
FPO	fleet post office
FPOC	focal point operations center
FPS	Federal Protective Service; force protection source
FPTAS	flight path threat analysis simulation
FPTS	forward propagation by tropospheric scatter
FPWG	force protection working group
FR	final report; frequency response
FRA	Federal Railroad Administration (DOT)
FRAG	fragmentation code
FRAGORD	fragmentary order
FRC	federal resource coordinator; forward resuscitative care
FRD	formerly restricted data
FREQ	frequency
FRERP	Federal Radiological Emergency Response Plan
FRF	fragment retention film

FRMAC	Federal Radiological Monitoring and Assessment Center (DOE)
FRN	force requirement number
FRO	flexible response option
FROG	free rocket over ground
FRP	Federal response plan (USG)
FRRS	frequency resource record system
FRTP	fleet response training plan
FS	fighter squadron; file separator; file server; flare single-unit
fs	search radius safety factor
FSA	fire support area
FSB	fire support base; forward staging base; forward support base; forward support battalion
FSC	fire support cell; fire support coordinator (USMC)
FSCC	fire support coordination center (USMC)
FSCL	fire support coordination line
FSCM	fire support coordination measure
FSCOORD	fire support coordinator (Army)
FSE	fire support element
FSEM	fire support execution matrix
FSF	foreign security forces
FSI	Foreign Service Institute
FSK	frequency shift key
FSN	foreign service national
FSO	fire support officer; flight safety officer; foreign service officer
FSP	functional support plan
FSR	field service representative
FSRT	fatality search and recovery team
FSS	fast sealift ship; fire support station; flight service station
FSSG	force service support group (USMC)
FSST	forward space support to theater
FST	fleet surgical team
FSU	former Soviet Union; forward support unit
FSW	feet of seawater
ft	feet; foot
ft3	cubic feet
FTC	Federal Trade Commission
FTCA	Foreign Tort Claims Act
FTN	force tracking number
FTP	file transfer protocol
FTRG	fleet tactical readiness group
FTS	Federal Telecommunications System; Federal telephone service; file transfer service
FTU	field training unit; freight terminal unit
FUAC	functional area code

FUNCPLAN	functional plan
F/V	fishing vessel
Fv	aircraft speed correction factor
FVT	Force Validation Tool
FW	fighter wing; fixed-wing; weather correction factor
FWC	fleet weather center
FWD	forward
FWDA	friendly weapon danger area
FWF	former warring factions
FY	fiscal year
FYDP	Future Years Defense Program

G

G-1	Army or Marine Corps component manpower or personnel staff officer (Army division or higher staff, Marine Corps brigade or higher staff)
G-2	Army Deputy Chief of Staff for Intelligence; Army or Marine Corps component intelligence staff officer (Army division or higher staff, Marine Corps brigade or higher staff)
G-2X	Army counterintelligence and human intelligence staff element
G-3	Army or Marine Corps component operations staff officer (Army division or higher staff, Marine Corps brigade or higher staff); assistant chief of staff, operations
G-4	Army or Marine Corps component logistics staff officer (Army division or higher staff, Marine Corps brigade or higher staff); Assistant Chief of Staff for Logistics
G-5	assistant chief of staff, plans
G-6	Army or Marine Corps component command, control, communications, and computer systems staff officer; assistant chief of staff for communications; signal staff officer (Army)
G-7	Army component information operations staff officer; assistant chief of staff, information engagement; information operations staff officer (ARFOR)
G/A	ground to air
GA	geospatial analyst; Tabun, a nerve agent
GAA	general agency agreement; geospatial intelligence assessment activity
GAFS	General Accounting and Finance System
GAMSS	Global Air Mobility Support System
GAO	General Accounting Office; Government Accountability Office

GAR	gateway access request
GARS	Global Area Reference System
GAT	governmental assistance team
GATB	guidance, apportionment, and targeting board
GATES	Global Air Transportation Execution System
GB	group buffer; Sarin, a nerve agent
GBI	ground-based interceptor
GBL	government bill of lading
GBMD	global ballistic missile defense
GBR	ground-based radar
GBS	Global Broadcast Service; Global Broadcast System
GBU	guided bomb unit
GC	general counsel; Geneva Convention; Geneva Convention Relative to the Protection of Civilian Persons in Time of War
GC3A	global command, control, and communications assessment
GC4A	global command, control, communications, and computer assessment
GCA	ground controlled approach
GCC	geographic combatant commander; global contingency construction
GCCC	global contingency construction contract
GCCS	Global Command and Control System
GCCS-A	Global Command and Control System-Army
GCCS-I3	Global Command and Control System Integrated Imagery and Intelligence
GCCS-J	Global Command and Control System-Joint
GCCS-M	Global Command and Control System-Maritime
GCE	ground combat element (MAGTF)
GCI	ground control intercept
GCM	global container manager; Global Information Grid (GIG) Content Management
GCP	global campaign plan; ground commander's pointer
GCRI	general collective routing indicator (RI)
GCS	ground control station
GCSC	global contingency service contract
GCSS	Global Command Support System
GCSS-J	Global Combat Support System-Joint
GCTN	global combating terrorism network
GD	Soman, a nerve agent
GDF	gridded data field; Guidance for Development of the Force
GDIP	General Defense Intelligence Program
GDIPP	General Defense Intelligence Proposed Program
GDP	General Defense Plan (SACEUR): gross domestic product
GDSS	Global Decision Support System
GE	general engineering

GEF	Guidance for Employment of the Force
GEM	Global Information Grid (GIG) Enterprise Management
GENADMIN	general admin (message)
GENSER	general service (message)
GENTEXT	general text
GEO	geosynchronous Earth orbit
GEOCODE	geographic code
GEOFILE	geolocation code file; standard specified geographic location file
GEOINT	geospatial intelligence
GEOLOC	geographic location; geographic location code
GEOREF	geographic reference; world geographic reference system
GETS	geospatial enterprise tasking, processing, exploitation, and dissemination service
GF	a nerve agent
GFE	government-furnished equipment
GFI	government-furnished information
GFM	global force management; global freight management; government-furnished material
GFMAP	Global Force Management Allocation Plan
GFMB	Global Force Management Board
GFMIG	Global Force Management Implementation Guidance
GFMPL	Graphics Fleet Mission Program Library
GFOAR	global family of operation plans assessment report
GFS	global fleet station
GFU	group framing unit
GHz	gigahertz
GI	geomatics and imagery
GI&S	geospatial information and services
GIAC	graphic input aggregate control
GIANT	Global Positioning System Interference and Navigation Tool
GIBCO	geospatial intelligence base for contingency operations
GIC	(*gabarit international de chargement*) international loading gauge
GIE	global information environment
GIG	Global Information Grid
GIMS	Geospatial Intelligence Information Management Services
GIO	Geospatial Intelligence Organisation (New Zealand)
GIP	gridded installation photograph
GIS	geographic information system; geospatial information systems
GL	government leased
GLCM	ground-launched cruise missile
GLINT	gated laser intensifier
GLO	ground liaison officer

GLTD	ground laser target designator
GM	group modem
GMD	global missile defense; ground-based midcourse defense; group mux and/or demux
GMDSS	Global Maritime Distress and Safety System
GMF	ground mobile force
GMFP	global military force policy
GMI	general military intelligence
GMLRS	Global Positioning System Multiple Launch Rocket System
GMP	global maritime partnership
GMR	graduated mobilization response; ground mobile radar
GMRS	global mobility readiness squadron
GMS	global mobility squadron
GMTI	ground moving target indicator
GNA	Global Information Grid (GIG) Network Assurance
GNC	Global Network Operations Center
GNCC	global network operations (NETOPS) center
GND	Global Information Grid (GIG) Network Defense
GNSC	global network operations (NETOPS) support center
GNSS	global navigation satellite system
GO	government owned
GOCO	government-owned, contractor-operated
GOES	geostationary operational environmental satellite
GOGO	government-owned, government-operated
GOS	grade of service
GOSG	general officer steering group
GOTS	government off-the-shelf
GP	general purpose; group
GPC	geospatial planning cell; government purchase card
GPD	gallons per day
GPE	geospatial intelligence preparation of the environment
GPEE	general purpose encryption equipment
GPL	Geospatial Product Library
GPM	gallons per minute; global pallet manager
GPMDM	group modem
GPMIC	Global Patient Movement Integration Center
GPMJAB	Global Patient Movement Joint Advisory Board
GPMRC	Global Patient Movement Requirements Center
GPS	Global Positioning System
GPSOC	Global Positioning System Operations Center
GPW	Geneva Convention Relative to the Treatment of Prisoners of War
GR	graduated response
GRASP	general retrieval and sort processor
GRCA	ground reference coverage area

GRG	gridded reference graphic
GRL	global reach laydown
GRREG	graves registration
GS	general service; general support; group separator
GSA	General Services Administration
GSE	ground support equipment
GSI	glide slope indicator
GSM	ground station module
GSO	general services officer
GSORTS	Global Status of Resources and Training System
GS-R	general support-reinforcing
GSR	general support-reinforcing; ground surveillance radar
GSSA	general supply support area
GSSC	global satellite communications (SATCOM) support center
GST	geospatial support team
gt	gross ton
GTAS	ground-to-air signals
GTL	gun-target line
GTM	global transportation management
GUARD	US National Guard and Air Guard
GUARDS	General Unified Ammunition Reporting Data System
G/VLLD	ground/vehicle laser locator designator
GW	guerrilla warfare
GWC	global weather central
GWEN	Ground Wave Emergency Network
GWOT	global war on terrorism
GWS	Geneva Convention for the Amelioration of the Condition of the Wounded and Sick in Armed Forces in the Field
GWS Sea	Geneva Convention for the Amelioration of the Condition of the Wounded, Sick, and Shipwrecked Members of the Armed Forces at Sea

H

H&I	harassing and interdicting
H&S	headquarters and service
HA	holding area; humanitarian assistance
HAARS	high-altitude airdrop resupply system
HAC	helicopter aircraft commander; human intelligence analysis cell
HACC	humanitarian assistance coordination center
HAHO	high-altitude high-opening parachute technique
HAP	humanitarian assistance program
HAP-EP	humanitarian assistance program-excess property

HARC	human intelligence analysis and reporting cell
HARM	high-speed antiradiation missile
HARP	high altitude release point
HAST	humanitarian assistance survey team
HATR	hazardous air traffic report
HAZ	hazardous cargo
HAZMAT	hazardous materials
HB	heavy boat
HBCT	heavy brigade combat team
HC	Directorate for Human Capital (DIA)
HCA	head of contracting activity; humanitarian and civic assistance
HCAS	hostile casualty
HCL	hydrochloride
HCO	helicopter control officer
HCP	hardcopy printer
HCS	helicopter combat support (Navy); helicopter coordination section
HCT	human intelligence (HUMINT) collection team
HD	a mustard agent; harbor defense; harmonic distortion; homeland defense
HDC	harbor defense commander; helicopter direction center
HDCU	harbor defense command unit
HDM	humanitarian demining
HDO	humanitarian demining operations
HDPLX	half duplex
HDR	humanitarian daily ration
HDTC	Humanitarian Demining Training Center
HE	heavy equipment; high explosives
HEAT	helicopter external air transport; high explosive antitank
HEC	helicopter element coordinator
HEFOE	hydraulic electrical fuel oxygen engine
HEI	high explosives incendiary
HEL-H	heavy helicopter
HEL-L	light helicopter
HEL-M	medium helicopter
HELO	helicopter
HEMP	high-altitude electromagnetic pulse
HEMTT	heavy expanded mobile tactical truck
HEO	highly elliptical orbit
HEPA	high efficiency particulate air
HERF	hazards of electromagnetic radiation to fuels
HERO	electromagnetic radiation hazards; hazards of electromagnetic radiation to ordnance
HERP	hazards of electromagnetic radiation to personnel

HET	heavy equipment transporter; human intelligence exploitation team
HEWSweb	Humanitarian Early Warning Service
HF	high frequency
HFA	human factors analysis
HFDF	high-frequency direction finding
HFP	hostile fire pay
HFRB	high frequency regional broadcast
HH	homing pattern
HHC	headquarters and headquarters company
HHD	headquarters and headquarters detachment
H-hour	seaborne assault landing hour; specific time an operation or exercise begins
HHQ	higher headquarters
HHS	Department of Health and Human Services
HIC	humanitarian information center
HICAP	high-capacity firefighting foam station
HIDACZ	high-density airspace control zone
HIDTA	high-intensity drug trafficking area
HIFR	helicopter in-flight refueling
HIMAD	high-to-medium-altitude air defense
HIMARS	High Mobility Artillery Rocket System
HIMEZ	high-altitude missile engagement zone
HIRSS	hover infrared suppressor subsystem
HIRTA	high intensity radio transmission area
HIU	humanitarian information unit (DOS)
HIV	human immunodeficiency virus
HJ	crypto key change
HLPS	heavy-lift pre-position ship
HM	hazardous material
HMA	humanitarian mine action
HMH	Marine heavy helicopter squadron
HMIRS	Hazardous Material Information Resource System
HMIS	Hazardous Material Information System
HMLA	Marine light/attack helicopter squadron
HMM	Marine medium helicopter squadron
HMMWV	high mobility multipurpose wheeled vehicle
HMOD	harbormaster operations detachment
HMW	health, morale, and welfare
HN	host nation
HNC	host-nation coordination
HNCC	host nation coordination center
HNS	host-nation support
HNSCC	host-nation support coordination cell
HNSF	host-nation security forces
HOB	height of burst

HOC	human intelligence operations cell; humanitarian operations center
HOCC	humanitarian operations coordination center
HOD	head of delegation
HOGE	hover out of ground effect
HOIS	hostile intelligence service
HOM	head of mission
HOSTAC	helicopter operations from ships other than aircraft carriers (USN publication)
HPA	high power amplifier
HPM	high-power microwave
HPMSK	high priority mission support kit
HPT	high-payoff target
HQ	HAVE QUICK; headquarters
HQCOMDT	headquarters commandant
HQDA	Headquarters, Department of the Army
HQFM-net	HAVE QUICK frequency modulation net
HQFMT-net	HAVE QUICK frequency modulation training net
HQMC	Headquarters, Marine Corps
HR	helicopter request; hostage rescue
HRB	high-risk billet
HRC	high-risk-of-capture; Human Resources Command
HRF	homeland response force
HRJTF	humanitarian relief joint task force
HRO	humanitarian relief organizations
HRP	high-risk personnel; human remains pouch
HRS	horizon reference system
HRT	hostage rescue team
HS	health services; helicopter antisubmarine (Navy); homeland security; homing single-unit
HSAC	Homeland Security Advisory Council
HSAS	Homeland Security Advisory System
HSB	high speed boat
HSC	helicopter sea combat (Navy); Homeland Security Council
HSCDM	high speed cable driver modem
HSC/PC	Homeland Security Council Principals Committee
HSC/PCC	Homeland Security Council Policy Coordination Committee
HSD	health service delivery; human intelligence support detachment
HSE	headquarters support element; human intelligence support element (DIA)
HSEP	hospital surgical expansion package (USAF)
HSI	hyperspectral imagery; Office of Homeland Security Investigations (DHS)
HSIN	Homeland Security Information Network (DHS)
HSIP	Homeland Security Infrastructure Program

HSM	humanitarian service medal
HSPD	homeland security Presidential directive
HSPR	high speed pulse restorer
HSS	health service support
HSSDB	high speed serial data buffer
HST	helicopter support team
HSV	high-speed vessel
HT	hatch team
HTERRCAS	hostile terrorist casualty
HTG	hard target graphic
HTH	high test hypochlorite
HU	hospital unit
HUD	head-up display
HUMINT	human intelligence
HUMRO	humanitarian relief operation
HUMRO OCP	humanitarian relief operation operational capability package
HUS	hardened unique storage
HVA	high-value asset
HVAA	high value airborne asset
HVAC	heating, ventilation, and air conditioning
HVCDS	high-velocity container delivery system
HVI	high-value individual
HVT	high-value target
HW	hazardous waste
HWM	high water mark
HYE	high-yield explosives
Hz	hertz

I

I	immediate; individual
I2	identity intelligence
I2WD	Intelligence and Information Warfare Division (Army)
I&A	Office of Intelligence and Analysis (DHS)
IA	imagery analyst; implementing arrangement; individual augmentee; information assurance; initial assessment
IAA	incident area assessment; incident awareness and assessment
IAC	Interagency Advisory Council
IACG	interagency coordination group
IADB	Inter-American Defense Board
IADS	integrated air defense system
IAEA	International Atomic Energy Agency (UN)
IAF	initial approach fix
IAIP	Information Analysis and Infrastructure Protection

IAM	inertially aided munition
IAMD	integrated air and missile defense
IAMSAR	International Aeronautical and Maritime Search and Rescue manual
IAP	incident action plan; international airport
IAR	interoperability assessment report
IAS	International Assistance System
IASC	Inter-Agency Standing Committee (UN); interim acting service chief
IATA	International Air Transport Association
IATACS	Improved Army Tactical Communications System
IATO	interim authority to operate
IAVM	information assurance vulnerability management
IAW	in accordance with
I/B	inboard
IBB	International Broadcasting Bureau
IBCT	infantry brigade combat team
IBES	intelligence budget estimate submission
IBET	integrated border enforcement team
IBM	International Business Machines
IBS	Integrated Booking System; integrated broadcast service; Integrated Broadcast System
IBS-I	Integrated Broadcast Service-Interactive
IBS-S	Integrated Broadcast Service-Simplex
IBU	inshore boat unit
IC	incident commander; intelligence community; intercept
IC3	integrated command, control, and communications
ICAD	individual concern and deficiency
ICAF	Interagency Conflict Assessment Framework
ICAO	International Civil Aviation Organization
ICAT	interagency conflict assessment team
ICBM	intercontinental ballistic missile
ICC	information confidence convention; information coordination center; Intelligence Coordination Center (USCG); International Criminal Court; Interstate Commerce Commission
ICD	international classifications of diseases; International Cooperation and Development Program (USDA)
ICDC	Intelligence Community Deputies Committee
ICDS	improved container delivery system
ICE	Immigration and Customs Enforcement
ICEDEFOR	Iceland Defense Forces
ICEPP	Incident Communications Emergency Policy and Procedures
IC/EXCOM	Intelligence Community Executive Committee
ICG	interagency core group

ICIS	integrated consumable item support
ICITAP	International Criminal Investigative Training Assistance Program (DOJ)
ICM	image city map; improved conventional munitions; integrated collection management
ICN	idle channel noise; interface control net
ICNIA	integrated communications, navigation, and identification avionics
ICOD	intelligence cutoff data
ICODES	integrated computerized deployment system
ICON	imagery communications and operations node; intermediate coordination node
ICP	intertheater communications security package; interface change proposal; inventory control point
ICPC	Intelligence Community Principals Committee
ICR	Intelligence Collection Requirements
ICRC	International Committee of the Red Cross
ICRI	interswitch collective routing indicator
ICS	incident command system; internal communications system; inter-Service chaplain support
ICSF	integrated command communications system framework
ICSAR	interagency committee on search and rescue
ICT	information and communications technology
ICU	intensive care unit; interface control unit
ICVA	International Council of Voluntary Agencies
ICW	in coordination with
ID	identification; identifier; initiating directive
IDAD	internal defense and development
IDB	integrated database
IDCA	International Development Cooperation Agency
IDDF	intermediate data distribution facility
IDEAS	Intelligence Data Elements Authorized Standards
IDEX	imagery data exploitation system
IDF	intermediate distribution frame
ID/IQ	indefinite delivery/indefinite quantity
IDL	integrated distribution lane
IDM	improved data modem; information dissemination management
IDNDR	International Decade for Natural Disaster Reduction (UN)
IDO	installation deployment officer
IDP	imagery derived product; imminent danger pay; internally displaced person
IDRA	infectious disease risk assessment
IDS	individual deployment site; integrated deployment system; interface design standards; intrusion detection system

IDSRS	Integrated Defense Source Registration System
IDSS	interoperability decision support system
IDT	inactive duty training
IDZ	inner defense zone
IEB	intelligence exploitation base
IED	improvised explosive device
IEDD	improvised explosive device defeat
IEEE	Institute of Electrical and Electronics Engineers
IEL	illustrative evaluation scenario
IEM	installation emergency management
IEMATS	improved emergency message automatic transmission system
IER	information exchange requirement
IES	imagery exploitation system
IESS	imagery exploitation support system
IEW	intelligence and electronic warfare
IF	intermediate frequency
IFC	intelligence fusion center
IFCS	improved fire control system
IFF	identification, friend or foe
IFFN	identification, friend, foe, or neutral
IFF/SIF	identification, friend or foe/selective identification feature
IFO	integrated financial operations
IFP	integrated force package
IFR	instrument flight rules
IFRC	International Federation of Red Cross and Red Crescent Societies
IFSAR	interferometric synthetic aperture radar
IG	inspector general
IGC	Integrated Data Environment/Global Transportation Network Convergence
IGE	independent government estimate
IGL	intelligence gain/loss
IGO	intergovernmental organization
IGSM	interim ground station module (JSTARS)
IHADSS	integrated helmet and display sight system (Army)
IHC	international humanitarian community
IHO	industrial hygiene officer
IHS	international health specialist
IIB	interagency information bureau
IICL	Institute of International Container Lessors
IICT	Interagency Intelligence Committee on Counterterrorism
IIM	intelligence information management

IIP	Bureau of International Information Programs (DOS); interagency implementation plan; international information program; interoperability improvement program
IIR	imagery interpretation report; imaging infrared; intelligence information report
IJC3S	initial joint command, control, and communications system; Integrated Joint Command, Control, and Communications System
IL	intermediate location
ILAB	Bureau of International Labor Affairs (DOL)
ILO	in lieu of; International Labor Organization (UN)
ILOC	integrated line of communications
ILS	integrated logistic support
IM	information management; intermediate module
IMA	individual mobilization augmentee
IMAAC	Interagency Modeling and Atmospheric Assessment Center
IMC	instrument meteorological conditions; International Medical Corps
IMDC	isolated, missing, detained, or captured
IMDG	international maritime dangerous goods (UN)
IMET	international military education and training
IMETS	Integrated Meteorological System
IMF	International Monetary Fund (UN)
IMI	international military information
IMINT	imagery intelligence
IMIT	international military information team
IMLTU	intermatrix line termination unit
IMM	intelligence mission management
IMMDELREQ	immediate delivery required
IMO	information management officer; International Maritime Organization
IMOSAR	International Maritime Organization (IMO) search and rescue manual
IMOSS	interim mobile oceanographic support system
IMP	implementation; information management plan; inventory management plan
IMPP	integrated mission planning process
IMPT	incident management planning team; integrated mission planning team
IMRL	individual material requirements list
IMS	information management system; interagency management system; Interagency Management System for Reconstruction and Stabilization; international military staff; international military standardization

IMSP	information management support plan
IMSU	installation medical support unit
IMTF	integrated mission task force
IMU	inertial measuring unit; intermatrix unit
IN	Air Force component intelligence officer (staff); impulse noise; instructor
INCERFA	uncertainty phase (ICAO)
INCNR	increment number
INCSEA	incidents at sea
IND	improvised nuclear device
INDRAC	Interagency Combating Weapons of Mass Destruction Database of Responsibilities, Authorities, and Capabilities
INF	infantry
INFOCON	information operations condition
ING	Inactive National Guard
INID	intercept network in dialing
INJILL	injured or ill
INL	Bureau for International Narcotics and Law Enforcement Affairs (DOS)
INLS	improved Navy lighterage system
INM	international narcotics matters
INMARSAT	international maritime satellite
INR	Bureau of Intelligence and Research (DOS)
INREQ	information request
INRP	Initial National Response Plan
INS	Immigration and Naturalization Service; inertial navigation system; insert code
INSARAG	International Search and Rescue Advisory Group
INSCOM	United States Army Intelligence and Security Command
INTAC	individual terrorism awareness course
INTACS	integrated tactical communications system
INTELSAT	International Telecommunications Satellite Organization
INTELSITSUM	intelligence situation summary
InterAction	American Council for Voluntary International Action
INTERCO	International Code of signals
INTERPOL	International Criminal Police Organization
INTERPOL-USNCB	International Criminal Police Organization, United States National Central Bureau (DOJ)
INTREP	intelligence report
INU	inertial navigation unit; integration unit
INV	invalid
INVOL	involuntary
I/O	input/output
IO	information objectives; information operations; intelligence oversight
IOB	intelligence oversight board

IOC	Industrial Operations Command; initial operational capability; intelligence operations center; investigations operations center
IOCB	information operations coordination board
IO force	information operations force
IOI	injured other than hostilities or illness
IOII	information operations intelligence integration
IOM	installation, operation, and maintenance; International Organization for Migration
IOP	interface operating procedure
IOSS	Interagency Operations Security Support Staff
IOT	information operations team
IOU	input/output unit
IOW	information operations wing
IOWG	information operations working group
IP	initial point; initial position; instructor pilot; intelligence planning; internet protocol
IPA	intelligence production agency
IPB	intelligence preparation of the battlespace
IPBD	intelligence program budget decision
IPC	initial planning conference; integration planning cell; interagency planning cell; interagency policy committee
IPDM	intelligence program decision memorandum
IPDP	inland petroleum distribution plan
IPDS	imagery processing and dissemination system; inland petroleum distribution system (Army)
IPE	individual protective equipment; industrial plant equipment
IPG	isolated personnel guidance
IPI	indigenous populations and institutions
IPL	imagery product library; integrated priority list
IPO	International Program Office
IPOE	intelligence preparation of the operational environment
IPOM	intelligence program objective memorandum
IPP	impact point prediction; industrial preparedness program
IPR	in-progress review; intelligence production requirement
IPR F	plan approval in-progress review
IPRG	intelligence program review group
IPS	illustrative planning scenario; Integrated Planning System (DHS); Interim Polar System; interoperability planning system
IPSG	intelligence program support group
IPSP	intelligence priorities for strategic planning
IPT	integrated planning team; integrated process team; Integrated Product Team; intelligence planning team
I/R	internment/resettlement

IR	incident report; information rate; information requirements; infrared; intelligence requirement
IRA	Provisional Irish Republican Army
IRAC	Interdepartment Radio Advisory Committee (DOC)
IRBM	intermediate-range ballistic missile
I/R BN	internment/resettlement battalion
IRC	information-related capability; International Red Cross; International Rescue Committee; internet relay chat
IRCCM	infrared counter countermeasures
IRCM	infrared countermeasures
IRDS	infrared detection set
IRF	Immediate Reaction Forces (NATO); incident response force
IRINT	infrared intelligence
IRISA	Intelligence Report Index Summary File
IRO	international relief organization
IR pointer	infrared pointer
IRR	Individual Ready Reserve; integrated readiness report
IRS	Internal Revenue Service
IRSCC	interagency remote sensing coordination cell
IRST	infrared search and track
IRSTS	infrared search and track sensor; Infrared Search and Track System
IRT	Initial Response Team
IRTPA	Intelligence Reform and Terrorism Prevention Act
IS	information superiority; information system; interswitch
ISA	international standardization agreement; inter-Service agreement
ISAC	information sharing and analysis center
ISAF	International Security Assistance Force
ISB	intermediate staging base
ISDB	integrated satellite communications (SATCOM) database
ISE	Information Sharing Environment ;intelligence support element
ISG	information synchronization group; isolated soldier guidance
ISI2R	identify, separate, isolate, influence, and reintegrate
ISMCS	international station meteorological climatic summary
ISMMP	integrated continental United States (CONUS) medical mobilization plan
ISN	Bureau of International Security and Nonproliferation; internment serial number
ISO	International Organization for Standardization; isolation
ISOO	Information Security Oversight Office
ISOPAK	International Organization for Standardization package
ISOPREP	isolated personnel report
ISP	internet service provider
ISR	intelligence, surveillance, and reconnaissance

ISRD	intelligence, surveillance, and reconnaissance division
ISR WG	Intelligence, Surveillance, and Reconnaissance Wing
ISS	in-system select
ISSA	inter-Service support agreement
ISSG	Intelligence Senior Steering Group
ISSM	information system security manager
ISSO	information systems security organization
IST	integrated system test; interswitch trunk
ISU	internal airlift or helicopter slingable container unit
I/T	interpreter and translator
IT	information system technician; information technology
ITA	international telegraphic alphabet; International Trade Administration (DOC)
ITAC	intelligence and threat analysis center (Army)
ITALD	improved tactical air-launched decoy
ITAR	international traffic in arms regulation (coassembly)
ITF	intelligence task force (DIA)
ITG	infrared target graphic
ITL	intelligence task list
ITO	installation transportation officer
ITRO	inter-Service training organization
ITU	International Telecommunications Union
ITV	in-transit visibility
ITW/AA	integrated tactical warning and attack assessment
IUWG	inshore undersea warfare group
IV	intravenous
IVR	initial voice report
IVSN	Initial Voice Switched Network
IW	irregular warfare
IWC	information operations warfare commander
IW-D	defensive information warfare
IWG	intelligence working group; interagency working group
IWSC	Information Warfare Support Center
IWW	inland waterway
IWWS	inland waterway system

J

J-1	manpower and personnel directorate of a joint staff; manpower and personnel staff section
J-2	intelligence directorate of a joint staff; intelligence staff section
J-2A	deputy directorate for administration of a joint staff
J2-CI	Joint Counterintelligence Office
J-2J	deputy directorate for support of a joint staff
J-2M	deputy directorate for crisis management of a joint staff

J-2O	deputy directorate for crisis operations of a joint staff
J-2P	deputy directorate for assessment, doctrine, requirements, and capabilities of a joint staff
J-2T	Deputy Directorate for Targeting, Joint Staff Intelligence Directorate
J-2T-1	joint staff target operations division
J-2T-2	Target Plans Division
J-2X	joint force counterintelligence and human intelligence staff element
J-3	operations directorate of a joint staff; operations staff section
J-4	logistics directorate of a joint staff; logistics staff section
J-5	plans directorate of a joint staff; plans staff section
J-6	communications system directorate of a joint staff; command, control, communications, and computer systems staff section
J-7	engineering staff section; Joint Staff Directorate for Joint Force Development; operational plans and interoperability directorate of a joint staff
J-7/JED	exercises and training directorate of a joint staff
J-8	Joint Staff Directorate for Force Structure, Resource, and Assessment; force structure, resource, and assessment directorate of a joint staff
J-9	civil-military operations directorate of a joint staff; civil-military operations staff section
J-31	Joint Force Coordinator (Joint Staff)
J-35	future operations
J-39 DDGO	Joint Staff, Deputy Director for Global Operations
JA	judge advocate
J-A	judge advocate directorate of a joint staff
JAAR	joint after-action report
JAARS	Joint After-Action Reporting System
JAAT	joint air attack team
JA/ATT	joint airborne and air transportability training
JAC	joint analysis center
JACC	joint airspace control center
JCCA	joint combat capability assessment
JACCC	joint airlift coordination and control cell
JACC/CP	joint airborne communications center/command post
JACCE	joint air component coordination element
JACS	joint automated communication-electronics operating instructions system
JADO	joint air defense operations
JADOC	Joint Air Defense Operations Center (NORAD)
JADOCS	Joint Automated Deep Operations Coordination System
JAFWIN	JWICS Air Force weather information network
JAG	judge advocate general

JAGMAN	Manual of the Judge Advocate General (US Navy)
JAI	joint administrative instruction; joint airdrop inspection
JAIC	joint air intelligence center
JAIEG	joint atomic information exchange group
JAMMS	Joint Asset Movement Management System
JAMPS	Joint Interoperability of Tactical Command and Control Systems (JINTACCS) automated message preparation system
JANAP	Joint Army, Navy, Air Force publication
JAO	joint air operations
JAOC	joint air operations center
JAOP	joint air operations plan
JAPO	joint area petroleum office
JAR	joint activity report
JARB	joint acquisition review board
JARCC	joint air reconnaissance control center
JARN	joint air request net
JARS	joint automated readiness system
JASC	joint action steering committee
JASSM	joint air-to-surface standoff missile
JAT	joint acceptance test; joint assessment team
JATACS	joint advanced tactical cryptological support
JAT Guide	Joint Antiterrorism Program Manager's Guide
JAWS	Joint Munitions Effectiveness Manual (JMEM)/air-to-surface weaponeering system
JBP	Joint Blood Program
JBPO	joint blood program office
JC2WC	joint command and control warfare center
JCA	jamming control authority; joint capability area
JCASREP	joint casualty report
JCAT	joint crisis action team
JCC	joint collaboration cell; joint command center; joint contracting center; joint course catalog; joint cyberspace center
JCCB	Joint Configuration Control Board
JCCC	joint combat camera center; joint communications control center
JCCP	joint casualty collection point
JCCSE	Joint Continental United States Communications Support Environment
JCE	Joint Intelligence Virtual Architecture (JIVA) Collaborative Environment
JCEOI	joint communications-electronics operating instructions
JCET	joint combined exchange training; joint combined exercise for training
JCEWR	joint coordination of electronic warfare reprogramming

JCEWS	joint force commander's electronic warfare staff
JCGRO	joint central graves registration office
JCIDO	Joint Combat Identification Office
JCIOC	joint counterintelligence operations center
JCISA	Joint Command Information Systems Activity
JCISB	Joint Counterintelligence Support Branch
JCIU	joint counterintelligence unit
JCLL	joint center for lessons learned
JCMA	joint communications security monitoring activity
JCMB	joint collection management board
JCMC	joint crisis management capability
JCMEB	joint civil-military engineer board
JCMEC	Joint Captured Materiel Exploitation Center (DIA)
JCMO	joint communications security management office
JCMOTF	joint civil-military operations task force
JCMPO	Joint Cruise Missile Project Office
JCMT	joint collection management tools
JCN	joint communications network
JCRM	Joint Capabilities Requirements Manager
JCS	Joint Chiefs of Staff
JCSAN	Joint Chiefs of Staff Alerting Network
JCSAR	joint combat search and rescue
JCSB	joint contracting support board
JCSC	joint communications satellite center
JCSE	joint communications support element; Joint Communications Support Element (USTRANSCOM)
JCSM	Joint Chiefs of Staff memorandum
JCSP	joint contracting support plan
JCSS	joint communications support squadron
JCTN	joint composite track network
JDA	joint duty assignment
JDAAP	Joint Doctrine Awareness Action Plan
JDAL	Joint Duty Assignment List
JDAM	Joint Direct Attack Munition
JDAMIS	Joint Duty Assignment Management Information System
JDAT	joint deployable analysis team
JDC	joint deployment community; Joint Doctrine Center
JDD	joint doctrine distribution
JDDC	joint doctrine development community
JDDE	joint deployment and distribution enterprise
JDDOC	joint deployment and distribution operations center
JDDT	joint doctrine development tool
JDEC	joint document exploitation center
JDEIS	Joint Doctrine, Education, and Training Electronic Information System
JDIG	Joint Drug Intelligence Group

JDIGS	Joint Digital Information Gathering System
JDISS	joint deployable intelligence support system
JDN	joint data network
JDNC	joint data network operations cell
JDNO	joint data network operations officer
JDOG	joint detention operations group
JDOMS	Joint Director of Military Support
JDPC	Joint Doctrine Planning Conference
JDPI	joint desired point of impact
JDPO	joint deployment process owner
JDSS	Joint Decision Support System
JDSSC	Joint Data Systems Support Center
JDTC	Joint Deployment Training Center
JE	joint experimentation
JEAP	Joint Electronic Intelligence (ELINT) Analysis Program
JECC	joint enabling capabilities command; Joint Enabling Capabilities Command (USTRANSCOM)
JECE	Joint Elimination Coordination Element
JECG	joint exercise control group
JECPO	Joint Electronic Commerce Program Office
JEDD	Joint Education and Doctrine Division
JEEP	joint emergency evacuation plan
JEFF	Joint Expeditionary Forensic Facility (Army)
JEL	Joint Electronic Library
JEM	joint exercise manual
JEMB	joint environmental management board
JEMP	joint exercise management package
JEMSMO	joint electromagnetic spectrum management operations
JEMSO	joint electromagnetic spectrum operations
JENM	joint enterprise network manager
JEP	Joint Exercise Program
JEPES	joint engineer planning and execution system
JET	joint expeditionary team; Joint Operation Planning and Execution System (JOPES) editing tool
JEWC	Joint Electronic Warfare Center
JEWCS	Joint Electronic Warfare Core Staff (NATO)
JEZ	joint engagement zone
JFA	joint field activity
JFACC	joint force air component commander
JFAST	Joint Flow and Analysis System for Transportation
JFC	joint force commander
JFCC	joint functional component command
JFCC-GS	Joint Functional Component Command for Global Strike
JFCC-IMD	Joint Functional Component Command for Integrated Missile Defense

JFCC-ISR	Joint Functional Component Command for Intelligence, Surveillance, and Reconnaissance (USSTRATCOM)
JFCC NW	Joint Functional Component Command for Network Warfare
JFCC-Space	Joint Functional Component Command for Space (USSTRATCOM)
JFCH	joint force chaplain
JFE	joint fires element
JFHQ	joint force headquarters
JFHQ–NCR	Joint Force Headquarters-National Capital Region
JFHQ–State	joint force headquarters-state
JFIIT	Joint Fires Integration and Interoperability Team
JFIP	Japanese facilities improvement project
JFLCC	joint force land component commander
JFMC	joint fleet mail center
JFMCC	joint force maritime component commander
JFMO	joint frequency management office
JFO	joint field office; joint fires observer
JFP	joint force package (packaging); joint force provider; Joint Frequency Panel (MCEB)
JFRB	Joint Foreign Release Board
JFRG	joint force requirements generator
JFRG II	joint force requirements generator II
JFS	joint force surgeon
JFSOC	joint force special operations component
JFSOCC	joint force special operations component commander
JFTR	joint Federal travel regulations
JFUB	joint facilities utilization board
JGWE	joint global warning enterprise
JHMCS	joint helmet-mounted cueing system
JHSV	joint high-speed vessel
JI	joint inspection
JIA	joint individual augmentation
JIACG	joint interagency coordination group
JIADS	joint integrated air defense system
JIATF	joint interagency task force
JIATF-E	joint interagency task force - East
JIATF-S	Joint Interagency Task Force-South
JIATF-W	Joint Interagency Task Force-West
JIC	joint information center
JICC	joint information coordination center; joint interface control cell
JICO	joint interface control officer
JICPAC	Joint Intelligence Center, Pacific
JICTRANS	Joint Intelligence Center for Transportation

JIDC	joint intelligence and debriefing center; joint interrogation and debriefing center
JIEDDO	Joint Improvised Explosive Device Defeat Organization
JIEE	Joint Information Exchange Environment
JIEO	joint interoperability engineering organization
JIEP	joint intelligence estimate for planning
JIES	joint interoperability evaluation system
JIG	joint interrogation group
JILE	joint intelligence liaison element
JIMB	joint information management board
JIMP	joint implementation master plan
JIMPP	joint industrial mobilization planning process
JIMS	joint information management system
JINTACCS	Joint Interoperability of Tactical Command and Control Systems
JIO	joint interrogation operations
JIOC	joint information operations center; joint intelligence operations center
JIOCPAC	Joint Intelligence Operations Center, Pacific
JIOC-SOUTH	Joint Intelligence Operations Center, South
JIOC-TRANS	Joint Intelligence Operations Center - Transportation
JIOP	joint interface operational procedures
JIOP-MTF	joint interface operating procedures-message text formats
JIOWC	Joint Information Operations Warfare Center
JIPC	joint imagery production complex
JIPCL	joint integrated prioritized collection list
JIPOE	joint intelligence preparation of the operational environment
JIPTL	joint integrated prioritized target list
JIS	joint information system
JISE	joint intelligence support element
JITC	joint interoperability test command
JITF-CT	Joint Intelligence Task Force for Combating Terrorism
JIVA	Joint Intelligence Virtual Architecture
JIVU	Joint Intelligence Virtual University
JKDDC	Joint Knowledge Development and Distribution Capability
JKnIFE	Joint Improvised Explosive Device Defeat Organization Knowledge and Information Fusion Exchange
JLCC	joint lighterage control center; joint logistics coordination center
JLE	joint logistics environment
JLEnt	joint logistics enterprise
JLLIS	Joint Lessons Learned Information System
JLLP	Joint Lessons Learned Program
JLNCHREP	joint launch report
JLOA	joint logistics over-the-shore operation area
JLOC	joint logistics operations center

JLOTS	joint logistics over-the-shore
JLRC	joint logistics readiness center
JLSB	joint line of communications security board
JLSE	joint legal support element
JM&S	joint modeling and simulation
JMAARS	joint model after-action review system
JMAC	Joint Mortuary Affairs Center (Army)
JMAG	Joint METOC Advisory Group
JMAO	joint mortuary affairs office; joint mortuary affairs officer
JMAR	joint medical asset repository
JMAS	joint manpower automation system
JMAT	joint medical analysis tool; joint mobility assistance team
JMB	joint meteorology and oceanography board
JMC	joint military command; joint movement center
JMCC	joint meteorological and oceanographic coordination cell
JMCG	joint movement control group
JMCIS	Joint Maritime Command Information System
JMCO	joint meteorological and oceanographic coordination organization
JMCOMS	joint maritime communications system
JMD	joint manning document
JMeDSAF	joint medical semi-automated forces
JMEEL	joint mission-essential equipment list
JMEM	Joint Munitions Effectiveness Manual
JMET	joint mission-essential task
JMETL	joint mission-essential task list
JMIC	Joint Military Intelligence College; joint modular intermodal container
JMICS	Joint Worldwide Intelligence Communications System mobile integrated communications system
JMIE	joint maritime information element
JMIP	joint military intelligence program
JMISC	Joint Military Information Support Command
JMISTF	joint military information support task force
JMITC	Joint Military Intelligence Training Center
JMLO	joint medical logistics officer
JMMC	Joint Material Management Center
JMMT	joint military mail terminal
JMNA	joint military net assessment
JMO	joint maritime operations; joint meteorological and oceanographic officer; joint munitions office
JMO(AIR)	joint maritime operations (air)
JMOC	joint medical operations center
JMP	joint manpower program
JMPA	joint military postal activity; joint military satellite communications (MILSATCOM) panel administrator

JMPAB	Joint Materiel Priorities and Allocation Board
JMPT	Joint Medical Planning Tool
JMRC	joint mobile relay center
JMRO	Joint Medical Regulating Office
JMRR	Joint Monthly Readiness Review
JMSEP	joint modeling and simulation executive panel
JMSWG	Joint Multi-Tactical Digital Information Link (Multi-TADIL) Standards Working Group
JMT	joint military training
JMTCA	joint munitions transportation coordinating activity
JMTCSS	Joint Maritime Tactical Communications Switching System
JMTG	joint military information support operations task group
JMUA	Joint Meritorious Unit Award
JMV	joint METOC viewer
JMWG	joint medical working group
JNCC	joint network operations control center
JNMS	joint network management system
JNOCC	Joint Operation Planning and Execution System (JOPES) Network Operation Control Center
JNPE	joint nuclear planning element
JNSC	Joint Navigation Warfare Center Navigation Warfare Support Cell
JNWC	Joint Navigation Warfare Center
JOA	joint operations area
JOAF	joint operations area forecast
JOC	joint operations center; joint oversight committee
JOCC	joint operations command center
JOERAD	joint spectrum center ordnance electromagnetic environmental effects risk assessment database
JOG	joint operations graphic
JOGS	joint operation graphics system
JOPES	Joint Operation Planning and Execution System
JOPESIR	Joint Operation Planning and Execution System Incident Reporting System
JOPESREP	Joint Operation Planning and Execution System Reporting System
JOPG	joint operations planning group
JOPP	joint operation planning process
JOPPA	joint operation planning process for air
JOR	joint operational requirement
JORD	joint operational requirements document
JOSE	joint operations security support element
JOSG	joint operational steering group
JOT&E	joint operational test and evaluation
JOTC	joint operations tasking center

JOTS	Joint Operational Tactical System
JP	joint publication
JP4	jet propulsion fuel, type 4
JP5	jet propulsion fuel, type 5
JP8	jet propulsion fuel, type 8
JPAC	joint planning augmentation cell; Joint POW/MIA Accounting Command
JPADS	joint precision airdrop system
JPAG	Joint Planning Advisory Group
JPARR	joint personnel accountability reconciliation and reporting
JPASE	joint public affairs support element; Joint Public Affairs Support Element (USTRANSCOM)
JPATS	joint primary aircraft training system
JPAV	joint personnel asset visibility
JPC	joint planning cell; joint postal cell
JPD	joint planning document
JPEC	joint planning and execution community
JPED	joint personal effects depot
JPEG	Joint Photographic Experts Group
JPERSTAT	joint personnel status and casualty report
JPG	joint planning group
JPME	joint professional military education
JPMRC	joint patient movement requirements center
JPMT	joint patient movement team
JPN	joint planning network
JPO	joint petroleum office; Joint Program Office
JPOC	joint personnel operations center; joint planning orientation course
JPOI	joint program of instruction
JPOM	joint preparation and onward movement
JPO-STC	Joint Program Office for Special Technology Countermeasures
JPRA	Joint Personnel Recovery Agency
JPRC	joint personnel reception center ; joint personnel recovery center
JPRSP	joint personnel recovery support product
JPS	joint processing system
JPSE	Joint Planning Support Element (USTRANSCOM)
JPTTA	joint personnel training and tracking activity
JQR	joint qualification requirements
JQRR	joint quarterly readiness review
JRADS	Joint Resource Assessment Data System
JRB	Joint Requirements Oversight Council (JROC) Review Board
JRC	joint reconnaissance center
JRCC	joint reception coordination center

JRERP	Joint Radiological Emergency Response Plan
JRFL	joint restricted frequency list
JRG	joint review group
JRIC	joint reserve intelligence center
JRIP	Joint Reserve Intelligence Program
JRMB	Joint Requirements and Management Board
JROC	Joint Requirements Oversight Council
JRRB	joint requirements review board
JRS	joint reporting structure
JRSC	jam-resistant secure communications; joint rescue sub-center
JRSOI	joint reception, staging, onward movement, and integration
JRTC	joint readiness training center
JRX	joint readiness exercise
JS	Joint Staff; the Joint Staff
JSA	joint security area
JSAC	joint strike analysis cell; joint strike analysis center
JSAM	joint security assistance memorandum; Joint Service Achievement Medal; joint standoff surface attack missile
JSAN	Joint Staff automation for the nineties
JSAP	Joint Staff action package
JSAS	joint strike analysis system
JSC	joint security coordinator; Joint Spectrum Center
JSCAT	joint staff crisis action team
JSCC	joint security coordination center; joint Services coordination committee
JSCM	Joint Service Commendation Medal
JSCP	Joint Strategic Capabilities Plan
JSDS	Joint Staff doctrine sponsor
JSEC	joint strategic exploitation center
JSETS	Joint Search and Rescue Satellite-Aided Tracking Electronic Tracking System
JSF	joint support force
JSHO	joint shipboard helicopter operations
JSIDS	joint Services imagery digitizing system
JSIR	joint spectrum interference resolution
JSISC	Joint Staff Information Service Center
JSIT	Joint Operation Planning and Execution System (JOPES) information trace
JSIVA	Joint Staff Integrated Vulnerability Assessment
JSM	Joint Staff Manual
JSME	joint spectrum management element
JSMS	joint spectrum management system
JSO	joint security operations
JSOA	joint special operations area

JSOAC	joint special operations air component; joint special operations aviation component
JSOACC	joint special operations air component commander
JSOC	joint special operations command
JSOFI	Joint Special Operations Forces Institute
JSOTF	joint special operations task force
JSOU	Joint Special Operations University
JSOW	joint stand-off weapon
JSPA	joint satellite communications panel administrator
JSPD	joint strategic planning document
JSPDSA	joint strategic planning document supporting analyses
JSPOC	Joint Space Operations Center
JSPS	Joint Strategic Planning System
JSR	joint strategy review
JSRC	joint subregional command (NATO)
JSS	joint surveillance system
JSSA	joint Services survival, evasion, resistance, and escape (SERE) agency
JSSIS	joint staff support information system
JSST	joint space support team
JSTAR	joint system threat assessment report
JSTARS	Joint Surveillance Target Attack Radar System
JSTE	joint system training exercise
JSTO	joint space tasking order
JT&E	joint test and evaluation
JTA	joint technical architecture
JTAC	joint terminal attack controller; Joint Terrorism Analysis Center
JTACE	joint technical advisory chemical, biological, radiological, and nuclear element
JTADS	Joint Tactical Air Defense System (Army); Joint Tactical Display System
JTAGS	joint tactical ground station (Army); joint tactical ground station (Army and Navy); joint tactical ground system
JTAIC	Joint Technical Analysis and Integration Cell (Army)
JTAO	joint tactical air operations
JTAR	joint tactical air strike request
JTASC	joint training analysis and simulation center
JTASG	Joint Targeting Automation Steering Group
JTAV-IT	joint total asset visibility-in theater
JTB	Joint Transportation Board
JTC	joint technical committee; Joint Training Confederation
JTCB	joint targeting coordination board

JTCC	joint transportation coordination cell; joint transportation coordination center; joint transportation corporate information management center
JTCG/ME	Joint Technical Coordinating Group for Munitions Effectiveness
JTCOIC	Joint Training Counter-Improvised Explosive Device Operations Integration Center
JTD	joint table of distribution; joint theater distribution
JTDC	joint track data coordinator
JTF	joint task force
JTF-6	joint task force-6
JTF-AK	Joint Task Force-Alaska
JTF-B	joint task force-Bravo
JTFCEM	joint task force contingency engineering management
JTF-CM	joint task force - consequence management
JTF-CS	Joint Task Force-Civil Support
JTF-E	joint task force - elimination
JTF-GNO	Joint Task Force-Global Network Operations
JTF-GTMO	Joint Task Force-Guantanamo
JTF-HD	Joint Task Force-Homeland Defense
JTF HQ	joint task force headquarters
JTF-MAO	joint task force - mortuary affairs office
JTF-N	Joint Task Force-North
JTFP	Joint Tactical Fusion Program
JTF-PO	joint task force-port opening
JTFS	joint task force surgeon
JTF-State	joint task force-state
JTIC	joint transportation intelligence center
JTIDS	Joint Tactical Information Distribution System
JTL	joint target list
JTLM	joint theater logistics management
JTLS	joint theater-level simulation
JTM	joint training manual
JTMD	joint table of mobilization and distribution; Joint Terminology Master Database
JTMP	joint training master plan
JTMS	joint theater movement staff; joint training master schedule
JTP	joint test publication; joint training plan
JTR	joint travel regulations
JTRB	joint telecommunication resources board
JTS	Joint Targeting School; Joint Training System
JTSSCCB	Joint Tactical Switched Systems Configuration Control Board
JTSST	joint training system support team
JTT	joint targeting toolbox; joint training team
JTTF	joint terrorism task force

JTWG	joint targeting working group
JUH-MTF	Joint User Handbook-Message Text Formats
JUIC	joint unit identification code
JUO	joint urban operation
JUSMAG	Joint United States Military Advisory Group
JUWTF	joint unconventional warfare task force
JV	Joint Vision
JV 2020	Joint Vision 2020
JVB	Joint Visitors Bureau
JVIDS	Joint Visual Integrated Display System
JVSEAS	Joint Virtual Security Environment Assessment System
JWAC	Joint Warfare Analysis Center
JWARS	Joint Warfare Analysis and Requirements System
JWC	Joint Warfare Center
JWFC	Joint Warfighting Center
JWG	joint working group
JWICS	Joint Worldwide Intelligence Communications System
JWID	joint warrior interoperability demonstration

K

k	thousand
Ka	Kurtz-above band
KAL	key assets list
KAPP	Key Assets Protection Program
kb	kilobit
kbps	kilobits per second
KC-135	Stratotanker
KDE	key doctrine element
KEK	key encryption key
KG	key generator
kg	kilogram
kHz	kilohertz
K-Kill	catastrophic kill
KLE	key leader engagement
KLIP	key doctrine element-linked information package
km	kilometer
KMC	knowledge management center
KNP	Korean National Police
KP	key pulse
kph	kilometers per hour
KPP	key performance parameter
KQ ID	tactical location identifier
kt	kiloton(s); knot (nautical miles per hour)
Ku	Kurtz-under band
kVA	kilo Volt-Amps

KVG	key variable generator
kW	kilowatt
KWOC	keyword-out-of-context

L

L	length
l	search subarea length
LA	lead agent; legal adviser; line amplifier; loop key generator (LKG) adapter
LAADS	low altitude air defense system
LAAM	light anti-aircraft missile
LABS	laser airborne bathymetry system
LACV	lighter, air cushioned vehicle
LAD	latest arrival date; launch area denied
LAMPS	Light Airborne Multipurpose System (helicopter)
LAN	local area network
LANDCENT	Allied Land Forces Central Europe (NATO)
LANDSAT	land satellite
LANDSOUTH	Allied Land Forces Southern Europe (NATO)
LANTIRN	low-altitude navigation and targeting infrared for night
LAO	limited attack option
LARC	lighter, amphibious resupply, cargo
LARC-V	lighter, amphibious resupply, cargo, 5 ton
LARS	lightweight airborne recovery system
LASH	lighter aboard ship
LAT	latitude
LAV	light armored vehicle
lb	pound
lbs.	pounds
LBR	Laser Beam Rider
LC	lake current; legal counsel
LCAC	landing craft, air cushion
LCADS	low-cost aerial delivery system
LCAP	low combat air patrol
LCB	line of constant bearing
LCC	amphibious command ship; land component commander; launch control center; lighterage control center; link communications circuit; logistics component command
LCCS	landing craft control ship
LCE	logistics capability estimator; logistics combat element (MAGTF); logistics combat element (Marine)
LCES	line conditioning equipment scanner
LCM	landing craft, mechanized; letter-class mail; life-cycle management
LCMC	life cycle management command

LCO	landing craft air cushion control officer; lighterage control officer
LCP	lighterage control point
LCPL	landing craft, personnel, light
LCS	landing craft air cushion control ship
LCSR	life cycle systems readiness
LCU	landing craft, utility; launch correlation unit
LCVP	landing craft, vehicle, personnel
LD	line of departure
LDA	limited depository account
LDF	lightweight digital facsimile
LDI	line driver interface
LDO	laser designator operator
LDR	leader; low data rate
LE	law enforcement; low-order explosives
LEA	law enforcement agency
LEAP	Light ExoAtmospheric Projectile
LEASAT	leased satellite
LEAU	Law Enforcement Assistance Unit (FAA)
LECIC	Law Enforcement and Counterintelligence Center (DOD)
LED	law enforcement desk; light emitting diode
LEDET	law enforcement detachment (USCG)
LEGAT	legal attaché
LEIP	Law Enforcement Intelligence Program (USCG)
LEMP	low-altitude electromagnetic pulse
LEO	law enforcement operations; low Earth orbit
LEP	laser eye protection; law enforcement professional; linear error probable
LERSM	lower echelon reporting and surveillance module
LERTCON	alert condition
LES	law enforcement sensitive; leave and earnings statement; Lincoln Laboratories Experimental Satellite
LESO	Law Enforcement Support Office
LET	light equipment transport
L-EWE	land-electronic warfare element
LF	landing force; low frequency
LFA	lead federal agency
LFORM	landing force operational reserve material
LFSP	landing force support party
LfV	*Landesamt für Verfassungsschutz* (regional authority for constitutional protection)
LG	deputy chief of staff for logistics
LGB	laser-guided bomb
LGM	laser-guided missile; loop group multiplexer
LGM-30	Minuteman
LGW	laser-guided weapon

LHA	amphibious assault ship (general purpose)
LHD	amphibious assault ship (multipurpose)
L-hour	specific hour on C-day at which a deployment operation commences or is to commence
LHT	line-haul tractor
LIDAR	light detection and ranging
LIF	light interference filter
LIMDIS	limited distribution
LIMFAC	limiting factor
LIPS	Logistics Information Processing System
LIS	logistics information system
LIWA	land information warfare activity
LKG	loop key generator
LKP	last known position
LL	lessons learned
LLLGB	low-level laser-guided bomb
LLLTV	low-light level television
LLSO	low-level source operation
LLTR	low-level transit route
LM	loop modem
LMARS	Logistics Metrics Analysis Reporting System
LMAV	laser MAVERICK
LMF	language media format
LMSR	large, medium-speed roll-on/roll-off
LN	lead nation
LNA	low voice amplifier
LNI	Library of National Intelligence
LNO	liaison officer
LO	low observable
LOA	Lead Operational Authority; letter of assist; letter of authorization; letter of offer and acceptance; lodgment operational area; logistics over-the-shore operation area
LOAC	law of armed conflict
LOAL	lock-on after launch
LOBL	lock-on before launch
LOC	line of communications; logistics operations center
LOC ACC	location accuracy
LOCAP	low combat air patrol
LOCE	Linked Operational Intelligence Centers Europe; Linked Operations-Intelligence Centers Europe
LOD	line of departure
LOE	letter of evaluation; line of effort
LOG	logistics
LOGAIR	logistics aircraft
LOGAIS	logistics automated information system

LOGCAP	logistics civil augmentation program (Army)
LOGCAT	logistics capability assessment tool
LOGDET	logistics detail
LOGEX	logistics exercise
LOGFAC	Logistics Feasibility Assessment Capability
LOGFOR	logistics force packaging system
LOGMARS	logistics applications of automated marking and reading symbols
LOGMOD	logistics module
LOGPLAN	logistics planning system
LOGSAFE	logistic sustainment analysis and feasibility estimator
LOI	letter of instruction; loss of input
LO/LO	lift-on/lift-off
LOMEZ	low-altitude missile engagement zone
LONG	longitude
LOO	line of operation
LOP	line of position
LORAN	long-range aid to navigation
LO/RO	lift-on/roll-off
LOROP	long range oblique photography
LOS	line of sight
LOTS	logistics over-the-shore
LOX	liquid oxygen
LP	listening post
LPD	amphibious transport dock; low probability of detection
LPH	amphibious assault ship, landing platform helicopter
LPI	low probability of intercept
LPSB	logistics procurement support board
LPU	line printer unit
LPV	laser-protective visor
LRC	logistics readiness center
LRD	laser range finder-detector
LRF	laser range finder
LRF/D	laser range finder/detector
LRG	long-range aircraft
LRM	low rate multiplexer
LRN	Laboratory Response Network (DHHS)
LRO	lighterage repair officer
LRP	load and roll pallet
LRRP	long range reconnaissance patrol
LRS	launch and recovery site
LRST	long-range surveillance team
LRSU	long-range surveillance unit
LSA	logistic support analysis; logistics supportability analysis
LSB	landing support battalion; lower sideband
LSCDM	low speed cable driver modem

LSD	dock landing ship; least significant digit
LSE	landing signalman enlisted; logistic support element
LSO	landing safety officer; landing signals officer
LSPR	low speed pulse restorer
LSS	laser spot search; local sensor subsystem
LST	laser spot tracker; tank landing ship
LSU	logistics civil augmentation program support unit
LSV	logistics support vessel
LT	large tug; local terminal; long ton
L/T	long ton
LTD	laser target designator
LTD/R	laser target designator/ranger
LTF	logistics task force
LTG	local timing generator
LTIOV	latest time information is of value
LTL	laser-to-target line
LTON	long ton
LTS	low-altitude navigation and targeting infrared for night (LANTIRN) targeting system
LTT	loss to theater
LTU	line termination unit
LUT	local user terminal
LVS	Logistics Vehicle System (USMC)
LW	leeway
LWR	Lutheran World Relief
LZ	landing zone
LZCO	landing zone control officer

M

M	million
M&E	monitoring and evaluation
M&S	modeling and simulation
M88A1	recovery vehicle
MA	master; medical attendant; military action; mortuary affairs
mA	milliampere(s)
MAAG	military assistance advisory group
MAAP	master air attack plan
MAC	mobility assault company; Mortuary Affairs Center
MACB	multinational acquisition and contracting board
MACCS	Marine air command and control system
MACG	Marine air control group
MACOM	major command (Army)
MACP	mortuary affairs collection point
MACRMS	mortuary affairs contaminated remains mitigation site

MACS	Marine air control squadron
MACSAT	multiple access commercial satellite
MAD	*Militärischer Abschirmdienst* (military protection service); military air distress
MADCP	mortuary affairs decontamination collection point
MADS	military information support operations automated data system
MAEB	mean area of effectiveness for blast
MAEF	mean area of effectiveness for fragments
MAF	mobility air forces
MAFC	Marine air-ground task force (MAGTF) all-source fusion center
MAG	Marine aircraft group; maritime assessment group; military assignment group
MAGTF	Marine air-ground task force
MAGTF ACE	Marine air-ground task force aviation combat element
MAJCOM	major command (USAF)
MANFOR	manpower force packaging system
MANPADS	man-portable air defense system
MANPER	manpower and personnel module
MAOC-N	Maritime Analysis and Operations Center-Narcotics
MAP	Military Assistance Program; missed approach point; missed approach procedure
MAR	METOC assistance request
MARAD	Maritime Administration
MARAD RRF	Maritime Administration Ready Reserve Force
MARCORMATCOM	Marine Corps Materiel Command
MARCORSYSCOM	Marine Corps Systems Command
MARDIV	Marine division
MARFOR	Marine Corps forces
MARFOREUR	Marine Corps Forces, Europe
MARFORLANT	Marine Corps Forces, Atlantic
MARFORNORTH	Marine Corps Forces, North
MARFORPAC	Marine Corps Forces, Pacific
MARFORSOUTH	Marine Corps Forces, South
MARFORSTRAT	United States Marine Corps Forces, United States Strategic Command
MARINCEN	Maritime Intelligence Center
MARLE	Marine liaison element
MARLO	Marine liaison officer
MARO	mass atrocity response operations
MAROP	marine operators
MARPOL	International Convention for the Prevention of Pollution from Ships
MARS	Military Auxiliary Radio System
MARSA	military assumes responsibility for separation of aircraft

MARSOC	Marine Corps special operations command
MARSOF	Marine Corps special operations forces
MART	mobile Automatic Digital Network (AUTODIN) remote terminal
MARTS	Mortuary Affairs Reporting and Tracking System
MAS	maritime air support; military information support operations automated system
MASCAL	mass casualty
MASF	mobile aeromedical staging facility
MASH	mobile Army surgical hospital
MASINT	measurement and signature intelligence
MASLO	measurement and signature intelligence liaison officer
MAST	military assistance to safety and traffic; mobile ashore support terminal
MAT	medical analysis tool
MATCALS	Marine air traffic control and landing system
MATCS	Marine air traffic control squadron
M/ATMP	Missiles/Air Target Materials Program
MAW	Marine aircraft wing
MAX	maximum
MAXORD	maximum ordinate
MB	medium boat; megabyte; military information support operations battalion
MBBLs	thousands of barrels
MBCDM	medical biological chemical defense materiel
MBI	major budget issue
Mbps	megabytes per second
Mbs	megabits per second
MC	Military Committee (NATO); military community; mission-critical
MC-130	Combat Talon (I and II)
MCA	mail control activity; maximum calling area; military civic action; mission concept approval; movement control agency
MCAG	maritime civil affairs group
MCAP	maximum calling area precedence
MCAS	Marine Corps air station
MCAST	maritime civil affairs and security training
MCAT	maritime civil affairs team
MCB	movement control battalion
MCBAT	medical chemical biological advisory team
MCC	Marine component commander; maritime component commander; master control center; military cooperation committee; military coordinating committee; military counterintelligence collections; mission control center; mobility control center; movement control center

MCCC	mobile consolidated command center
MCCDC	Marine Corps Combat Development Command
MCCISWG	military command, control, and information systems working group
MCD	medical crew director
MCDA	military and civil defense assets (UN)
MCDP	Marine Corps doctrine publication
MCDS	modular cargo delivery system
MCEB	Military Communications-Electronics Board
MCEWG	Military Communications-Electronics Working Group
MC/FI	mass casualty/fatality incident
MCI	multinational communications integration
MCIA	Marine Corps Intelligence Activity
MCIO	military criminal investigative organization
MCIOC	Marine Corps Information Operations Center
MCIP	Marine Corps interim publication; military command inspection program; military customs inspection program
MCJSB	Military Committee Joint Standardization Board
MCM	Manual for Courts-Martial; military classification manual; mine countermeasures
MCMC	mine countermeasures commander
MCMG	Military Committee Meteorological Group (NATO)
MCMO	medical civil-military operations
MCMOPS	mine countermeasures operations
M/CM/S	mobility, countermobility, and/or survivability
MCMREP	mine countermeasure report
MCMRON	mine countermeasures squadron
MCO	major combat operation; Mapping Customer Operations (Defense Logistics Agency); Marine Corps order
MCOO	modified combined obstacle overlay
MCRP	Marine Corps reference publication
MCS	maneuver control system; Military Capabilities Study; mine countermeasures ship; modular causeway system
MCSB	Marine Cryptologic Support Battalion
MCSF	mobile cryptologic support facility
MCSFB	Marine Corps security force battalion
MCSFR	Marine Corps security force regiment
MCT	movement control team
MCTC	Midwest Counterdrug Training Center
MCTFT	Multijurisdictional Counterdrug Task Force Training
MCTOG	Marine Corps Tactics and Operations Group
MCU	maintenance communications unit
MCW	modulated carrier wave
MCWL	Marine Corps Warfighting Lab
MCWP	Marine Corps warfighting publication

MCX	Marine Corps Exchange
MDA	Magen David Adom (Israeli equivalent of the Red Cross); maritime domain awareness; Missile Defense Agency
M-DARC	military direct access radar channel
M-day	mobilization day; unnamed day on which mobilization of forces begins
MDBS	blood support medical detachment
MDCI	multidiscipline counterintelligence
MDCO	Military Department counterintelligence organization
MDDOC	Marine air-ground task force deployment and distribution operations center
MDF	Main Defense Forces (NATO); main distribution frame
MDITDS	migration defense intelligence threat data system; Modernized Defense Intelligence Threat Data System
MDMA	methylenedioxymethamphetamine
MDR	medium data rate
MDRO	mission disaster response officer
MDS	Message Dissemination Subsystem; mission design series
MDSS II	Marine air-ground task force Deployment Support System II
MDSU	mobile diving and salvage unit
MDU	military information support operations distribution unit
MDW	Military District of Washington
MDZ	maritime defense zone
MEA	munitions effect assessment; munitions effectiveness assessment
MEAS	military information support operations effects analysis subsystem
MEB	Marine expeditionary brigade
MEBU	mission essential backup
MEC	medium endurance cutter
ME/C	medical examiner and/or coroner
MED	manipulative electronic deception
MEDAL	Mine Warfare Environmental Decision Aids Library
MEDCAP	medical civic action program
MEDCC	medical coordination cell
MEDCOM	medical command; US Army Medical Command
MEDEVAC	medical evacuation
MEDINT	medical intelligence
MEDLOG	medical logistics (USAF AIS)
MEDLOGCO	medical logistics company
MEDLOG JR	medical logistics, junior (USAF)
MEDLOG support	medical logistics support
MEDMOB	Medical Mobilization Planning and Execution System
MEDNEO	medical noncombatant evacuation operation

MEDREG	medical regulating
MEDREGREP	medical regulating report
MEDRETE	medical readiness training exercise
MEDS	meteorological data system
MEDSOM	medical supply, optical, and maintenance unit
MEDSTAT	medical status
MEF	Marine expeditionary force
MEFPAKA	manpower and equipment force packaging
MEL	maintenance expenditure limit; minimum equipment list
MEO	medium Earth orbit; military equal opportunity
MEP	mobile electric power
MEPCOM	military entrance processing command
MEPES	Medical Planning and Execution System
MEPRS	Military Entrance Processing and Reporting System
MERCO	merchant ship reporting and control
MERSHIPS	merchant ships
MES	medical equipment set
MESAR	minimum-essential security assistance requirements
MESF	maritime expeditionary security force
MESFC	maritime expeditionary security force commander
MESO	maritime expeditionary security operations
MET	medium equipment transporter; mobile environmental team
METAR	meteorological airfield report; meteorological aviation report
METARS	routine aviation weather report (roughly translated from French; international standard code format for hourly surface weather observations)
METCON	control of meteorological information (roughly translated from French); meteorological control (Navy)
METL	mission-essential task list
METMF	meteorological mobile facility
METMR(R)	meteorological mobile facility (replacement)
METOC	meteorological and oceanographic
METSAT	meteorological satellite
METT-T	mission, enemy, terrain and weather, troops and support available-time available
METT-TC	mission, enemy, terrain and weather, troops and support available-time available and civil considerations (Army)
METWATCH	meteorological watch
MEU	Marine expeditionary unit
MEVA	mission essential vulnerable area
MEWSG	Multi-Service Electronic Warfare Support Group (NATO)
MEZ	missile engagement zone
MF	medium frequency; mobile facility; multi-frequency
MFC	multinational force commander
MFDS	Modular Fuel Delivery System
MFE	manpower force element; mobile field exchange

MFFIMS	mass fatality field information management system
MFO	multinational force and observers
MFP	major force program
MFPC	maritime future plans center
MFPF	minefield planning folder
MFS	multifunction switch
MG	military information support operations group
MGB	medium girder bridge
MGM	master group multiplexer
MGRS	military grid reference system
MGS	mobile ground system
MGT	management
MGW	maximum gross weight
MHC	management headquarters ceiling
MHD	maritime homeland defense
MHE	materials handling equipment
MHS	maritime homeland security; Military Health System
MHU	modular heat unit
MHW	mean high water
MHz	megahertz
MI	military intelligence; movement instructions
MIA	missing in action
MIAC	maritime intelligence and analysis center
MIB	Military Intelligence Board
MIC	Multinational Interoperability Council
MICAP	mission capable/mission capability
MICON	mission concept
MICRO-MICS	micro-medical inventory control system
MICRO-SNAP	micro-shipboard non-tactical automated data processing system
MIDAS	model for intertheater deployment by air and sea
MIDB	modernized integrated database; modernized intelligence database
MIDDS-T	Meteorological and Oceanographic (METOC) Integrated Data Display System-Tactical
MIF	maritime interception force
MIJI	meaconing, interference, jamming, and intrusion
MILAIR	military airlift
MILALOC	military air line of communications
MILCON	military construction
MILDEC	military deception
MILDEP	Military Department
MILGP	military group (assigned to American Embassy in host nation)
MILOB	military observer
MILOC	military oceanography group (NATO)

MILPERS	military personnel
MILSATCOM	military satellite communications
MILSPEC	military specification
MILSTAMP	military standard transportation and movement procedures
MILSTAR	military strategic and tactical relay system
MIL-STD	military standard
MILSTRAP	military standard transaction reporting and accounting procedure
MILSTRIP	military standard requisitioning and issue procedure
MILTECH	military technician
MILU	multinational integrated logistic unit
MILVAN	military van (container)
MIM	maintenance instruction manual
MIMP	Mobilization Information Management Plan
MINEOPS	joint minelaying operations
MIO	maritime interception operations
MIOC	maritime interception operations commander
MIO-9	information operations threat analysis division (DIA)
MIP	military intelligence program
MIPE	mobile intelligence processing element
MIPOE	medical intelligence preparation of the operational environment
MIPR	military interdepartmental purchase request
MIRCS	mobile integrated remains collection system
MIS	maritime intelligence summary; military information support
MISAS	military information support automated system
MISCAP	mission capability
MISG	military information support group
MISO	military information support operations
MISREP	mission report
MIST	military information support team
MISTF	military information support task force
MITASK	mission tasking
MITO	minimum interval takeoff
MITT	mobile integrated tactical terminal
MIUW	mobile inshore undersea warfare
MIW	mine warfare
MIWC	mine warfare commander
MIWG	multinational interoperability working group
MJCS	Joint Chiefs of Staff memorandum
MJLC	multinational joint logistic center
M-Kill	mobility kill
MLA	mission load allowance
MLAYREP	mine laying report
MLE	maritime law enforcement
MLEA	Maritime Law Enforcement Academy

MLEM	Maritime Law Enforcement Manual
MLG	Marine logistics group
MLI	munitions list item
MLMC	medical logistics management center
MLO	military liaison office
MLP	message load plan
MLPP	multilevel precedence and preemption
MLPS	Medical Logistics Proponent Subcommittee
MLRS	multiple launch rocket system
MLS	microwave landing system; multilevel security
MLSA	mutual logistics support agreement
MLW	mean low water
MMA	military mission area
MMAC	military mine action center
MMC	materiel management center
MMG	DOD Master Mobilization Guide
MMI	man/machine interface
MMIS	military information support operations management information subsystem
MMLS	mobile microwave landing system
MMS	mast-mounted sight
MMT	military mail terminal
MNC	multinational corporation
MNCC	multinational coordination center
MNF	multinational force
MNFACC	multinational force air component commander
MNFC	multinational force commander
MNFLCC	multinational force land component commander
MNFMCC	multinational force maritime component commander
MNFSOCC	multinational force special operations component commander
MNJLC	multinational joint logistics component
MNL	master net list; multinational logistics
MNLC	multinational logistic center
MNP	master navigation plan
MNS	mine neutralization system (USN); mission needs statement
MNTF	multinational task force
MO	month
MOA	memorandum of agreement; military operating area
MOADS	maneuver-oriented ammunition distribution system
MOB	main operating base; main operations base; mobilization
MOBCON	mobilization control
MOBREP	military manpower mobilization and accession status report; mobilization report
MOC	maritime operations center; media operations center

MOCC	measurement and signature intelligence (MASINT) operations coordination center; mobile operations control center
MOD	Minister of Defense; ministry of defense; modification
MODEM	modulator/demodulator
MODLOC	miscellaneous operational details, local operations
MOD T-AGOS	modified tactical auxiliary general ocean surveillance
MOE	measure of effectiveness
MOEI	measure of effectiveness indicator
MOG	maximum (aircraft) on ground; movement on ground (aircraft); multinational observer group
MOGAS	motor gasoline
MOLE	multichannel operational line evaluator
MOMAT	mobility matting
MOMSS	mode and message selection system
MOP	measure of performance; memorandum of policy
MOPP	mission-oriented protective posture
MOR	memorandum of record
MOS	military occupational specialty
MOSC	meteorological and oceanographic operations support community
MOTR	maritime operational threat response
MOU	memorandum of understanding
MOUT	military operations in urban terrain; military operations on urbanized terrain
MOVREP	movement report
MOW	maintenance orderwire
MP	military police (Army and Marine); multinational publication
MPA	maritime patrol aircraft; mission and payload assessment; mission planning agent
MPAT	military patient administration team; Multinational Planning Augmentation Team
MPC	mid-planning conference; military personnel center
MPE/S	maritime pre-positioning equipment and supplies
MPF	maritime pre-positioning force
MPFUB	maritime pre-positioning force utility boat
MPG	maritime planning group; mensurated point graphic
mph	miles per hour
MPICE	measuring progress in conflict environments
MPLAN	Marine Corps Mobilization Management Plan
MPM	medical planning module
MPNTP	Master Positioning Navigation and Timing Plan
MPO	military post office
MPP	maritime procedural publication
MPR	maritime patrol and reconnaissance

MPRS	multipoint refueling system
MPS	maritime pre-positioning ship; message processor shelter; Military Planning Service (UN); Military Postal Service
MPSA	Military Postal Service Agency
MPSRON	maritime pre-positioning ships squadron
MR	milliradian; mobile reserve
MRAALS	Marine remote area approach and landing system
MRAT	medical radiobiology advisory team
MRBM	medium-range ballistic missile
MRCI	maximum rescue coverage intercept
MRE	meal, ready to eat
MRG	movement requirements generator
MRI	magnetic resonance imaging
MRMC	US Army Medical Research and Materiel Command
MRO	mass rescue operation; materiel release order; medical regulating office; medical regulating officer
MROC	multicommand required operational capability
MRR	minimum-risk route
MRRR	mobility requirement resource roster
MRS	measurement and signature intelligence (MASINT) requirements system; meteorological radar subsystem; movement report system
MRSA	Materiel Readiness Support Agency
MRT	maintenance recovery team
MRU	mountain rescue unit
MRX	mission readiness exercise
MS	message switch
ms	millisecond
MSA	Maritime Security Act
MSC	major subordinate command; maritime support center; Military Sealift Command; military staff committee; mission support confirmation
MSCA	military support to civilian authorities
MSCD	military support to civil defense
MSCO	Military Sealift Command Office
MSD	marginal support date; mobile security division
MS-DOS	Microsoft disk operating system
MSE	mission support element; mobile subscriber equipment
MSECR	HIS 6000 security module
MSEL	master scenario events list
MSF	*Medicins Sans Frontieres* ("Doctors Without Borders"); mission support force; mobile security force; multiplex signal format
MSG	Marine security guard; message
MSGID	message identification

MSHARPP	mission, symbolism, history, accessibility, recognizability, population, and proximity
MSI	modified surface index; multispectral imagery
MSIC	Missile and Space Intelligence Center
MSIS	Marine safety information system
MSK	mission support kit
MSL	master station log; military shipping label
MSNAP	merchant ship naval augmentation program
MSO	map support office; marine safety office(r); maritime security operations; military satellite communications (MILSATCOM) systems organization; military source operation; military strategic objective; military support operations; mobilization staff officer
MSOAG	Marine special operations advisor group
MSOC	Marine special operations company
MSP	maritime security program; mission support plan; mobile sensor platform
MSPES	mobilization stationing, planning, and execution system
MSPS	military information support operations studies program subsystem; mobilization stationing and planning system
MSR	main supply route; maritime support request; mission support request
MSRON	maritime expeditionary security squadron
MSRP	mission strategic resource plan
MSRR	modeling and simulation resource repository
MSRT	Maritime Security Response Team (USCG)
MSRV	message switch rekeying variable
MSS	medical surveillance system; meteorological satellite subsystem
MSSG	Marine expeditionary unit (MEU) service support group
MST	Marine expeditionary force (MEF) weather support team; meteorological and oceanographic support team; mission support team
MT	military technician; ministry team
MTA	military training agreement
MTAC	Multiple Threat Alert Center (DON)
MTBF	mean time between failures
MT Bn	motor transport battalion
MTCR	missile technology control regime
MT/D	measurement tons per day
mtDNA	mitochondrial deoxyribonucleic acid
MTF	medical treatment facility; message text format; military information support operations task force
MTG	master timing generator; military information support operations task group
MTI	moving target indicator

MTIC	Military Targeting Intelligence Committee
MTL	mission tasking letter
MTMS	maritime tactical message system
MTN	multi-tactical data link network
MTO	message to observer; mission type order
MTOE	modified table of organization and equipment
MTON	measurement ton
MTP	Marine tactical publication; maritime task plan; mission tasking packet
MTS	Movement Tracking System
MTS/SOF-IRIS	multifunction system
MTT	magnetic tape transport; mobile training team
MTTP	multi-Service tactics, techniques, and procedures
MTW	major theater war
MTX	message text format
MU	marry up
MUL	master urgency list (DOD)
MULE	modular universal laser equipment
MUOS	Mobile Users Object System
MUREP	munitions report
MUSARC	major United States Army reserve commands
MUSE	mobile utilities support equipment
MUST	medical unit, self-contained, transportable
MUX	multiplex
MV	merchant vessel; motor vessel
mV	millivolt
MWBP	missile warning bypass
MWC	Missile Warning Center (NORAD)
MWD	military working dog
MWDT	military working dog team
MWF	medical working file
MWG	mobilization working group
MWOD	multiple word-of-day
MWR	missile warning receiver; morale, welfare, and recreation
MWSG	Marine wing support group
MWSS	Marine wing support squadron
MWT	modular warping tug

N

N	number of required track spacings; number of search and rescue units (SRUs)
N-1	Navy component manpower or personnel staff officer
N-2	Director of Naval Intelligence; Navy component intelligence staff officer
N-3	Navy component operations staff officer

N-4	Navy component logistics staff officer
N-5	Navy component plans staff officer
N-6	Navy component communications staff officer
NA	nation assistance
NA5CRO	non-Article 5 crisis response operation (NATO)
NAAG	North Atlantic Treaty Organization (NATO) Army Armaments Group
NAC	North American Aerospace Defense Command (NORAD) Air Center; North Atlantic Council (NATO)
NACE	National Military Command System (NMCS) Automated Control Executive
NACISA	North Atlantic Treaty Organization (NATO) Communications and Information Systems Agency
NACISC	North Atlantic Treaty Organization (NATO) Communications and Information Systems Committee
NACSEM	National Communications Security/Emanations Security (COMSEC/EMSEC) Information Memorandum
NACSI	national communications security (COMSEC) instruction
NACSIM	national communications security (COMSEC) information memorandum
NAD 83	North American Datum 1983
NADEFCOL	North Atlantic Treaty Organization (NATO) Defense College
NADEP	naval aircraft depot
NAE	Navy acquisition executive
NAEC-ENG	Naval Air Engineering Center - Engineering
NAF	naval air facility; nonappropriated funds; numbered air force
NAFAG	North Atlantic Treaty Organization (NATO) Air Force Armaments Group
NAI	named area of interest
NAIC	National Air Intelligence Center
NAK	negative acknowledgement
NALC	Navy ammunition logistics code
NALE	naval and amphibious liaison element
NALSS	naval advanced logistic support site
NAMP	North Atlantic Treaty Organization (NATO) Annual Manpower Plan
NAMS	National Air Mobility System
NAMTO	Navy material transportation office
NAOC	national airborne operations center (E-4B aircraft)
NAPCAP	North Atlantic Treaty Organization (NATO) Allied Pre-Committed Civil Aircraft Program
NAPMA	North Atlantic Treaty Organization (NATO) Airborne Early Warning and Control Program Management Agency

NAPMIS	Navy Preventive Medicine Information System
NAR	nonconventional assisted recovery; notice of ammunition reclassification
NARAC	national atmospheric release advisory capability; National Atmospheric Release Advisory Center (DOE)
NARC	non-automatic relay center
NARP	Nuclear Weapon Accident Response Procedures
NAS	naval air station
NASA	National Aeronautics and Space Administration
NASAR	National Association for Search and Rescue
NAS computer	national airspace system computer
NASIC	National Air and Space Intelligence Center
NAT	nonair-transportable (cargo)
NATO	North Atlantic Treaty Organization
NATOPS	Naval Air Training and Operating Procedures Standardization
NAU	Narcotics Assistance Unit
NAVAID	navigation aid
NAVAIDS	navigational aids
NAVAIR	naval air; Naval Air Systems Command
NAVAIRSYSCOM	Naval Air Systems Command (Also called NAVAIR)
NAVATAC	Navy Antiterrorism Analysis Center; Navy Antiterrorist Alert Center
NAVCHAPDET	naval cargo handling and port group detachment
NAVCHAPGRU	Navy cargo handling and port group
NAVCOMSTA	naval communications station
NAVCYBERFOR	Navy Cyber Forces
NAVELSG	Navy expeditionary logistic support group
NAVEODTECHDIV	Naval Explosives Ordnance Disposal Technology Division
NAVEURMETOCCEN	Naval Europe Meteorology and Oceanography Center
NAVFAC	Naval Facilities Engineering Command
NAVFACENGCOM	Navy Facilities Engineering Command
NAVFAC-X	Naval Facilities Engineering Command-expeditionary
NAVFAX	Navy facsimile
NAVFOR	Navy forces
NAVICECEN	Naval Ice Center
NAVLANTMETOCCEN	Naval Atlantic Meteorology and Oceanography Center
NAVMAG	naval magazine
NAVMED	Navy Medical; Navy medicine
NAVMEDCOMINST	Navy medical command instruction
NAVMEDLOGCOM	Navy Medical Logistics Command
NAVMEDP	Navy medical pamphlet
NAVMETOCCOM	Naval Meteorology and Oceanography Command
NAVMTO	naval military transportation office; Navy Material Transportation Office

NAVOCEANO	Naval Oceanographic Office
NAVORD	naval ordnance
NAVORDSTA	naval ordnance station
NAVPACMETOCCEN	Naval Pacific Meteorology and Oceanography Center
NAVSAFECEN	naval safety center
NAVSAT	navigation satellite
NAVSEA	Naval Sea Systems Command
NAVSEAINST	Naval Sea Systems Command instruction
NAVSEALOGCEN	naval sea logistics center
NAVSEASYSCOM	Naval Sea Systems Command
NAVSO	United States Navy Forces, Southern Command
NAVSOC	Naval Satellite Operations Center; naval special operations command; naval special operations component; Navy special operations component
NAVSOF	Navy special operations forces
NAVSPACECOM	Naval Space Command
NAVSPECWARCOM	Naval Special Warfare Command
NAVSPOC	Naval Space Operations Center
NAVSUP	Naval Supply Systems Command
NAVSUPINST	Navy Support Instruction
NAVSUPSYSCOM	Naval Supply Systems Command
NAVWAR	navigation warfare
NAWCAD	Naval Air Warfare Center, Aircraft Division
NB	narrowband
NBC	nuclear, biological, and chemical
NBCCS	nuclear, biological, and chemical (NBC) contamination survivability
NBDP	narrow band direct printing
NBG	naval beach group
NBI	nonbattle injury
NBS	National Bureau of Standards
NBST	narrowband secure terminal
NBVC	Naval Base Ventura County
NC3A	nuclear command, control, and communications (C3) assessment
NCAA	North Atlantic Treaty Organization (NATO) Civil Airlift Agency
NCAGS	naval cooperation and guidance for shipping
NCAPS	naval coordination and protection of shipping
NCB	national central bureau; naval construction brigade; noncompliant boarding
NCC	National Coordinating Center; naval component commander; Navy component command; Navy component commander; network control center; North American Aerospace Defense Command (NORAD) Command Center

NCCS	Nuclear Command and Control System
NCD	net control device
NCDC	National Climatic Data Center
NCES	Net-Centric Enterprise Services
NCESGR	National Committee of Employer Support for the Guard and Reserve
NCF	naval construction force
NCFSU	naval construction force support unit
NCHB	Navy cargo-handling battalion
NCIC	National Crime Information Center
NCIJTF-AG	National Cyber Investigative Joint Task Force-Analytical Group (DOD)
NCIS	Naval Criminal Investigative Service
NCISRA	Naval Criminal Investigative Service resident agent
NCISRO	Naval Criminal Investigative Service regional office
NCISRU	Naval Criminal Investigative Service resident unit
NCIX	National Counterintelligence Executive
NCMI	National Center for Medical Intelligence
NCMP	Navy Capabilities and Mobilization Plan
NCO	noncombat operations; noncommissioned officer
NCOB	National Counterintelligence Operations Board
NCOIC	noncommissioned officer in charge
NCOS	naval control of shipping
NCP	National Oil and Hazardous Substances Pollution Contingency Plan
NCPC	National Counterproliferation Center
NCR	National Capital Region (US); national cryptologic representative; National Security Agency/Central Security Service representative; naval construction regiment
NCRCC	National Capital Region Coordination Center; United States Northern Command Rescue Coordination Center
NCRCG	National Cyber Response Coordination Group
NCRDEF	national cryptologic representative defense
NCR-IADS	National Capital Region-Integrated Air Defense System
NCS	National Clandestine Service; National Communications System; naval control of shipping
NCSC	National Computer Security Center
NCSD	National Cyber Security Division (DHS)
NCSE	national intelligence support team (NIST) communications support element
NCT	network control terminal
NCTAMS	naval computer and telecommunications area master station
NCTC	National Counterterrorism Center; North East Counterdrug Training Center

NCTS	naval computer and telecommunications station
NCWS	naval coastal warfare squadron
NDAA	national defense authorization act
NDAF	Navy, Defense Logistics Agency, Air Force
N-day	day an active duty unit is notified for deployment or redeployment
NDB	nondirectional beacon
NDCS	national drug control strategy
NDDOC	North American Aerospace Defense Command and United States Northern Command Deployment and Distribution Operations Cell
NDHQ	National Defence Headquarters, Canada
NDIC	National Defense Intelligence College; National Drug Intelligence Center
NDL	national desired ground zero list
NDMC	North Atlantic Treaty Organization (NATO) Defense Manpower Committee
NDMS	National Disaster Medical System (DHHS)
NDOC	National Defense Operations Center
NDP	national disclosure policy
NDPB	National Drug Policy Board
NDPC	National Disclosure Policy Committee
NDRC	National Detainee Reporting Center
NDRF	National Defense Reserve Fleet
NDS	national defense strategy
NDSF	National Defense Sealift Fund
NDT	nuclear disablement team
NDU	National Defense University
NEA	Northeast Asia
NEAT	naval embarked advisory team
NEC	National Economic Council
NECC	Navy Expeditionary Combat Command
NEIC	Navy Expeditionary Intelligence Command
NELR	Navy expeditionary logistics regiment
NEMT	National Emergency Management Team
NEO	noncombatant evacuation operation
NEOCC	noncombatant evacuation operation coordination center
NEP	National Exercise Program
NEPA	National Environmental Policy Act
NEREP	Nuclear Execution and Reporting Plan
NES	National Exploitation System
NESDIS	National Environmental Satellite, Data and Information Service (DOC)
NEST	nuclear emergency support team (DOE)
NETOPS	network operations
NETS	Nationwide Emergency Telecommunications System

NETT	new equipment training team
NETWARCOM	Naval Network Warfare Command
NEW	net explosive weight
NEWAC	North Atlantic Treaty Organization (NATO) Electronic Warfare Advisory Committee
NEWCS	NATO electronic warfare core staff
NEXCOM	Navy Exchange Command
NFA	no-fire area
NFC	numbered fleet commander
NFD	nodal fault diagnostics
NFELC	Naval Facilities Expeditionary Logistics Center
NFESC	Naval Facilities Engineering Service Center
NFI	national foreign intelligence
NFIB	National Foreign Intelligence Board
NFIP	National Flood Insurance Program (FEMA); National Foreign Intelligence Program
NFLIR	navigation forward-looking infrared
NFLS	naval forward logistic site
NFN	national file number
NFO	naval flight officer
NG	National Guard
NGA	National Geospatial-Intelligence Agency
NGB	National Guard Bureau
NGB-OC	National Guard Bureau-Office of the Chaplain
NGCC	National Guard coordination center
NGCDP	National Guard Counterdrug Program
NGCSP	National Guard Counterdrug Support Program
NGF	naval gun fire
NGFS	naval gunfire support
NGIC	National Ground Intelligence Center
NG JFHQ-State	National Guard joint force headquarters-state
NGLO	naval gunfire liaison officer
NGO	nongovernmental organization
NGP	National Geospatial-Intelligence Agency Program
NGRF	National Guard reaction force
NI	national identification (number); noted item
NIBRS	National Incident-Based Reporting System
NIC	National Intelligence Council; naval intelligence center
NICC	National Intelligence Coordination Center
NICCL	National Incident Communications Conference Line
NICCP	National Interdiction Command and Control Plan
NICI	National Interagency Counternarcotics Institute
NID	naval intelligence database
NIDMS	National Military Command System (NMCS) Information for Decision Makers System

NIDS	National Military Command Center (NMCC) information display system
NIE	national intelligence estimate
NIEX	no-notice interoperability exercise
NIEXPG	No-Notice Interoperability Exercise Planning Group
NIIB	National Geospatial Intelligence Agency imagery intelligence brief
NIL	National Information Library
NIMCAMP	National Information Management and Communications Master Plan
NIMS	National Incident Management System
NIOC	Navy Information Operations Command
NIP	National Intelligence Program
NIPF	National Intelligence Priority Framework
NIPRNET	Nonsecure Internet Protocol Router Network
NIPS	Naval Intelligence Processing System
NIRT	Nuclear Incident Response Team
NISH	noncombatant evacuation operation (NEO) intelligence support handbook
NISP	national intelligence support plan; Nuclear Weapons Intelligence Support Plan
NIST	National Institute of Standards and Technology
NIT	nuclear incident team
NITES	Navy Integrated Tactical Environmental System
NITF	national imagery transmission format
NIU	North Atlantic Treaty Organization (NATO) interface unit
NIWA	naval information warfare activity
NJOIC	National Joint Operations and Intelligence Center
NJTTF	National Joint Terrorism Task Force
NL	Navy lighterage
NLO	naval liaison officer
.NL.	not less than
NLT	not later than
NLW	nonlethal weapon
NM	network management
nm	nautical mile
NMAWC	Naval Mine and Anti-Submarine Warfare Command
NMB	North Atlantic Treaty Organization (NATO) military body
NMC	Navy Munitions Command
NMCB	naval mobile construction battalion
NMCC	National Military Command Center
NMCM	not mission capable, maintenance
NMCS	National Military Command System; not mission capable, supply
NMCSO	Navy and Marine Corps spectrum office

NMD	national missile defense
NMEC	National Media Exploitation Center
NMFS	National Marine Fisheries Services
NMIC	National Maritime Intelligence Center
NMIO	National Maritime Intelligence-Integration Office
NMIST	National Military Intelligence Support Team (DIA)
NMO	National Measurement and Signature Intelligence Office
NMOC	network management operations center
NMOSW	Naval METOC Operational Support Web
NMP	national media pool
NMPS	Navy mobilization processing site
NMR	news media representative
NMRC	Naval Medical Research Center
NMS	national military strategy
NMSA	North Atlantic Treaty Organization (NATO) Mutual Support Act
NMSC	Navy and Marine Corps Spectrum Center
NMS-CO	National Military Strategy for Cyberspace Operations
NMS-CWMD	National Military Strategy to Combat Weapons of Mass Destruction
NMSP-WOT	National Military Strategic Plan for the War on Terrorism
NNAG	North Atlantic Treaty Organization (NATO) Naval Armaments Group
NNSA	National Nuclear Security Administration (DOE)
NNWC	Naval Network Warfare Command
NOAA	National Oceanic and Atmospheric Administration
NOACT	Navy overseas air cargo terminal
NOC	National Operations Center (DHS); network operations center
NOCONTRACT	not releasable to contractors or consultants
NODDS	Naval Oceanographic Data Distribution System
NOE	nap-of-the-earth
NOEA	nuclear operations emergency action
NOFORN	not releasable to foreign nationals
NOG	Nuclear Operations Group
NOGAPS	Navy Operational Global Atmospheric Prediction System
NOHD	nominal ocular hazard distance
NOIC	Naval Operational Intelligence Center
NOK	next of kin
NOLSC	Naval Operational Logistics Support Center
NOMS	Nuclear Operations Monitoring System
NOMWC	Navy Oceanographic Mine Warfare Center
NOP	nuclear operations
NOPLAN	no operation plan available or prepared
NORAD	North American Aerospace Defense Command
NORM	normal; not operationally ready, maintenance

NORS	not operationally ready, supply
NOSC	network operations and security center
NOSSA	Navy Ordnance Safety and Security Activity
NOTMAR	notice to mariners
NP	nonproliferation
NPC	Nonproliferation Center
NPES	Nuclear Planning and Execution System
NPG	nonunit personnel generator
NPOESS	National Polar-orbiting Operational Environmental Satellite System
NPPD	National Protection and Programs Directorate (DHS)
NPS	National Park Service; nonprior service; Nuclear Planning System
NPT	national pipe thread; Treaty on the Nonproliferation of Nuclear Weapons
NPWIC	National Prisoner of War Information Center
NQ	nonquota
NR	North Atlantic Treaty Organization (NATO) restricted; number
NRAT	nuclear/radiological advisory team
NRC	National Response Center (USCG); non-unit-related cargo
NRCHB	Naval Reserve cargo-handling battalion
NRCHF	Naval Reserve cargo handling force
NRCHTB	Naval Reserve cargo handling training battalion
NRF	National Response Framework
NRFI	not ready for issue
NRG	notional requirements generator
NRL	nuclear weapons (NUWEP) reconnaissance list
NRO	National Reconnaissance Office
NROC	Northern Regional Operations Center (CARIBROC-CBRN)
NRP	National Response Plan; non-unit-related personnel
NRPC	Naval Reserve Personnel Center
NRT	near real time
NRTD	near-real-time dissemination
NRZ	non-return-to-zero
NS	nuclear survivability
NSA	national security act; National Security Agency; national security area; national shipping authority; North Atlantic Treaty Organization (NATO) Standardization Agency
NSA/CSS	National Security Agency/Central Security Service
NSARC	National Search and Rescue Committee
NSAT	United States Northern Command situational awareness team
NSAWC	Naval Strike and Air Warfare Center
NSC	National Security Council
NSC/DC	National Security Council/Deputies Committee
NSCID	National Security Council intelligence directive

NSC/IPC	National Security Council/interagency policy committee
NSC/IWG	National Security Council/Interagency Working Group
NSC/PC	National Security Council/Principals Committee
NSC/PCC	National Security Council Policy Coordinating Committee
NSCS	National Security Council System
NSCTI	Naval Special Clearance Team One
NS-CWMD	National Strategy to Combat Weapons of Mass Destruction
NSD	National Security Directive; National Security Division (FBI)
NSDA	non-self deployment aircraft
NSDD	national security decision directive
NSDM	national security decision memorandum
NSDS-E	Navy Satellite Display System-Enhanced
NSE	Navy support element
NSEP	national security emergency preparedness
NSF	National Science Foundation; national security forces; National Strike Force (USCG)
NSFS	naval surface fire support
NSG	National System for Geospatial Intelligence; north-seeking gyro
NSGI	National System for Geospatial Intelligence
NSHS	National Strategy for Homeland Security
NSL	no-strike list
NSM	national search and rescue (SAR) manual
NSMS	National Strategy for Maritime Security
NSN	national stock number
NSO	non-Single Integrated Operational Plan (SIOP) option
NSOC	National Security Operations Center; National Signals Intelligence (SIGINT) Operations Center; Navy Satellite Operations Center
NSOOC	North Atlantic Treaty Organization (NATO) Staff Officer Orientation Course
NSP	national search and rescue plan
N-Sp/CC	North American Aerospace Defense Command (NORAD)-US Space Command/Command Center
NSPD	national security Presidential directive
NSPI	National Strategy for Pandemic Influenza
NSRL	national signals intelligence (SIGINT) requirements list
NSS	National Search and Rescue Supplement; national security strategy; national security system; non-self-sustaining
NSSA	National Security Space Architect
NSSE	national special security event
NSST	naval space support team
NST	National Geospatial-Intelligence Agency support team
NSTAC	National Security Telecommunications Advisory Committee

NSTISSC	National Security Telecommunications and Information Systems Security Committee
NSTL	national strategic targets list
NSTS	National Secure Telephone System
NSW	naval special warfare
NSWCDD	Naval Surface Warfare Center Dahlgren Division
NSWCOM	Naval Special Warfare Command
NSWG	naval special warfare group
NSWTE	naval special warfare task element
NSWTF	naval special warfare task force
NSWTG	naval special warfare task group
NSWTU	naval special warfare task unit
NSWU	naval special warfare unit
NT	nodal terminal
NTACS	Navy tactical air control system
NTAP	National Track Analysis Program
NTB	national target base
NTBC	National Military Joint Intelligence Center Targeting and Battle Damage Assessment Cell
NTC	National Training Center
NTCS-A	Navy Tactical Command System-Afloat
NTDS	naval tactical data system
NTF	nuclear task force
N-TFS	New Tactical Forecast System
NTIA	National Telecommunications and Information Administration
NTIC	Navy Tactical Intelligence Center
NTISS	National Telecommunications and Information Security System
NTISSI	National Telecommunications and Information Security System (NTISS) Instruction
NTISSP	National Telecommunications and Information Security System (NTISS) Policy
NTM	national or multinational technical means of verification; notice to mariners
NTMPDE	National Telecommunications Master Plan for Drug Enforcement
NTMS	national telecommunications management structure
NTPS	near-term pre-positioned ships
NTRP	Navy tactical reference publication
NTS	night targeting system; noncombatant evacuation operations tracking system
NTSB	National Transportation Safety Board
NTSS	National Time-Sensitive System
NTTP	Navy tactics, techniques, and procedures
NTU	new threat upgrade

NUC	non-unit-related cargo; nuclear
NUDET	nuclear detonation
NUFEA	Navy-unique fleet essential aircraft
NUP	non-unit-related personnel
NURP	non-unit-related personnel
NUWEP	policy guidance for the employment of nuclear weapons
NVD	night vision device
NVDT	National Geospatial-Intelligence Agency voluntary deployment team
NVG	night vision goggle(s)
NVS	night vision system
NW	network warfare; not waiverable
NWARS	National Wargaming System
NWB	normal wideband
NWBLTU	normal wideband line termination unit
NWDC	Navy Warfare Development Command
NWFP	Northwest Frontier Province (Pakistan)
NWP	Navy warfare publication; numerical weather prediction
NWREP	nuclear weapons report
NWS	National Weather Service
NWT	normal wideband terminal

O

1MC	general announcing system
1NCD	1st Naval Construction Division
O	contour pattern
O&I	operations and intelligence
O&M	operation and maintenance
OA	objective area; operating assembly; operational area; Operations Aerology shipboard METOC division
OADR	originating agency's determination required
OAE	operational area evaluation
OAF	Operation ALLIED FORCE
OAFME	Office of the Armed Forces Medical Examiner
OAG	operations advisory group
OAI	oceanographic area of interest
OAJCG	Operation Alliance joint control group
OAM	Office of Air and Marine (DHS)
OAP	offset aimpoint
OAR	Chairman of the Joint Chiefs of Staff operation plans assessment report
OAS	offensive air support; Organization of American States
OASD	Office of the Assistant Secretary of Defense
OASD(NII/CIO)	Office of the Assistant Secretary of Defense (Networks and Information Integration/Chief Information Officer)

OASD(PA)	Office of the Assistant Secretary of Defense (Public Affairs)
OASD(RA)	Office of the Assistant Secretary of Defense (Reserve Affairs)
OAU	Organization of African Unity
O/B	outboard
OB	operating base; order of battle
OBA	oxygen breathing apparatus
OBFS	offshore bulk fuel system
OBST	obstacle
OBSTINT	obstacle intelligence
OC	oleoresin capsicum ; operations center
OCA	offensive counterair; operational control authority
OCBD	Office of Capacity Building and Development (USDA)
OCC	Operations Computer Center (USCG)
OCD	orderwire clock distributor
OCDETF	Organized Crime and Drug Enforcement Task Force
OCE	officer conducting the exercise
OCEANCON	control of oceanographic information
OCHA	Office for the Coordination of Humanitarian Affairs
OCJCS	Office of the Chairman of the Joint Chiefs of Staff
OCJCS-PA	Office of the Chairman of the Joint Chiefs of Staff-Public Affairs
OCMI	officer in charge, Marine inspection
OCO	offensive cyberspace operations; offload control officer
OCONUS	outside the continental United States
OCOP	outline contingency operation plan
OCP	operational capability package; operational configuration processing
OCR	Office of Collateral Responsibility
OCS	operational contract support
OCU	orderwire control unit (Types I, II, and III)
OCU-1	orderwire control unit-1
OD	operational detachment; other detainee
ODA	operational detachment-Alpha
ODATE	organization date
O-Day	off-load day
ODB	operational detachment-Bravo
ODC	Office of Defense Cooperation
ODCSLOG	Office of the Deputy Chief of Staff for Logistics (Army)
ODCSOPS	Office of the Deputy Chief of Staff for Operations and Plans (Army)
ODCSPER	Office of the Deputy Chief of Staff for Personnel (Army)
ODIN	Operational Digital Network
ODJS	Office of the Director, Joint Staff
ODNI	Office of the Director of National Intelligence
ODR	Office of Defense representative
ODZ	outer defense zone

OE	operational environment
OE&AS	organization for embarkation and assignment to shipping
OEBGD	Overseas Environmental Baseline Guidance Document
OECD	Organisation for Economic Co-operation and Development
OEF	Operation ENDURING FREEDOM
OEG	operational experts group; operational exposure guidance operations security executive group
OEH	occupational and environmental health
OEM	original equipment manufacturer
OER	officer evaluation report; operational electronic intelligence (ELINT) requirements
OES	office of emergency services
OET	Office of Emergency Transportation (DOT)
OF	officer (NATO)
OFAC	Office of Foreign Assets Control (TREAS)
OFCO	offensive counterintelligence operation
OFDA	Office of United States Foreign Disaster Assistance (USAID)
OFHIS	operational fleet hospital information system
OFOESA	Office of Field Operational and External Support Activities
OGS	overseas ground station
OH	overhead
OHDACA	Overseas Humanitarian, Disaster, and Civic Aid (DSCA)
OHDM	Office of Humanitarian Assistance, Disaster Relief, and Mine Action
OI	Office of Intelligence (USCS); operating instruction; operational interest
OIA	Office of International Affairs (TREAS)
OI&A	Office of Intelligence and Analysis (DHS)
OIC	officer in charge
OICC	officer in charge of construction; operational intelligence coordination center
OID	operation order (OPORD) identification
OIF	Operation IRAQI FREEDOM
OIIL	Office of Intelligence and Investigative Liaison (CBP)
OIR	operational intelligence requirements; other intelligence requirements
OJT	on-the-job training
OL	operating location
OLD	on-line tests and diagnostics
OLS	operational linescan system; optical landing system
OM	contour multiunit
OMA	Office of Military Affairs (CIA and USAID)
OMB	Office of Management and Budget; operations management branch
OMC	Office of Military Cooperation; optical memory card

OMF	officer master file
OMS	Office of Mission Support
OMSPH	Office of Medicine, Science, and Public Health (DHHS)
OMT	operations management team; orthogonal mode transducer
OMT/OMTP	operational maintenance test(ing)/test plan
ONDCP	Office of National Drug Control Policy
ONE	Operation NOBLE EAGLE
ONI	Office of Naval Intelligence
OOB	order of battle
OOD	officer of the deck
OODA	observe, orient, decide, act
OOS	out of service
OP	operational publication (USN); ordnance publication
OPARS	Optimum Path Aircraft Routing System
OPBAT	Operation Bahamas, Turks, and Caicos
OPC	Ocean Prediction Center (DOC)
OPCEN	operations center (USCG)
OPCOM	operational command (NATO)
OPCON	operational control
OPDAT	Office of Overseas Prosecutorial Development, Assistance, and Training (DOJ)
OPDEC	operational deception
OPDS	offshore petroleum discharge system (Navy)
OPDS-Future	offshore petroleum discharge system-future (Navy)
OPDS-L	offshore petroleum discharge system-legacy (Navy)
OPE	operational preparation of the environment
OPELINT	operational electronic intelligence
OPEO	Office of Preparedness and Emergency Operations (DHHS)
OPFOR	opposing force; opposition force
OPG	operations planning group
OPGEN	operation general matter
OPIR	overhead persistent infrared
OPLAN	operation plan
OPLAW	operational law
OPM	Office of Personnel Management; operations per minute
OPMG	Office of the Provost Marshal General
OPNAVINST	Chief of Naval Operations instruction
OPORD	operation order
OPP	orderwire patch panel
OPR	office of primary responsibility
OPREP	operational report
OPROJ	operational project
OPS	operational project stock; operations; operations center
OPSCOM	Operations Committee
OPSDEPS	Service Operations Deputies
OPSEC	operations security

OPSTK	operational stock
OPSUM	operation summary
OPT	operational planning team
OPTAR	operating target
OPTASK	operation task
OPTASKLINK	operations task link
OPTEMPO	operating tempo
OPTINT	optical intelligence
OPZONE	operation zone
OR	operational readiness; other rank(s) (NATO)
ORBAT	order of battle
ORCON	originator controlled
ORDREF	order reference
ORDTYP	order type
ORG	origin (GEOLOC)
ORIG	origin
ORM	operational risk management
ORP	ocean reception point
ORS	operationally responsive space
ORSA	operations research and systems analysis
OS	operating system
OSA	operational support airlift
OSAT	out-of-service analog test
OSC	offensive space control; on-scene commander; on-site commander; Open Source Center (CIA); operational support command; operations support center
OSCE	Organization for Security and Cooperation in Europe
OSD	Office of the Secretary of Defense
OSD/DMDPO	Office of the Secretary of Defense, Defense Military Deception Program Office
OSE	on scene endurance; operations support element
OSEI	operational significant event imagery
OSG	operational support group
OSI	open system interconnection; operational subsystem interface
OSIA	on-site inspection activity
OSINT	open-source intelligence
OSIS	open-source information system
OSM	Office of Spectrum Management (NTIA)
OSO	operational support office
OSOCC	on-site operations coordination center
OSP	operations support package
OSPG	overseas security policy group
OSRI	originating station routing indicator
OSV	ocean station vessel
OT	operational test

OT&E	operational test and evaluation
OTA	Office of Technical Assistance (TREAS)
OTC	officer in tactical command; over the counter
OTG	operational target graphic
OTH	other; over the horizon
OTH-B	over-the-horizon backscatter (radar)
OTHT	over-the-horizon targeting
OTI	Office of Transition Initiatives (USAID)
OTS	Officer Training School; one-time source
OUB	offshore petroleum discharge system utility boat
OUSD	Office of the Under Secretary of Defense
OUSD(AT&L)	Office of the Under Secretary of Defense (Acquisition, Technology, and Logistics)
OUSD(C/CFO)	Office of the Under Secretary of Defense (Comptroller/Chief Financial Officer)
OUSD(I)	Office of the Under Secretary of Defense (Intelligence)
OUSD(P)	Office of the Under Secretary of Defense for Policy
OUT	outsize cargo
OVE	on-vehicle equipment
OVER	oversize cargo
OVM	Operation VIGILANT MARINER
OW	orderwire
OWS	operational weather squadron

P

P	parallel pattern; priority; publication
PA	parent relay; physician assistant; primary agency; probability of arrival; public affairs
PAA	position area of artillery
PABX	private automatic branch exchange (telephone)
PACAF	Pacific Air Forces
PAD	patient administration director; positional adjustment; precision aircraft direction
PADD	person authorized to direct disposition of human remains
PADRU	Pan American Disaster Response Unit
PADS	position azimuth determining system
PAG	public affairs guidance
PAL	personnel allowance list; program assembler language
PALCON	pallet container
PALS	precision approach landing system
PAM	preventive and aerospace medicine; pulse amplitude modulation
PaM	passage material
PANS	procedures for air navigation services
PAO	public affairs office; public affairs officer

PAR	performance assessment report; population at risk; precision approach radar
PARC	principal assistant for contracting
PARKHILL	high frequency cryptological device
PARPRO	peacetime application of reconnaissance programs
PARS	Personnel and Accountability System
PAS	personnel accounting symbol
PAT	public affairs team
PAV	policy assessment visit
PAWS	phased array warning system
PAX	passengers; public affairs plans
PB	particle beam; patrol boat; peace building; President's budget
PB4T	planning board for training
PBA	performance-based agreement; production base analysis
PBCR	portable bar code recorder
PBD	program budget decision
PBIED	person-borne improvised explosive device
PBOS	Planning Board for Ocean Shipping
PC	patrol craft; personal computer; pilot in command; preliminary coordination; Principals Committee
Pc	cumulative probability of detection
P,C,&H	packing, crating, and handling
PC&S	post, camp, and station
PCA	Posse Comitatus Act
PCC	policy coordination committee; primary control center
PCF	personnel control facility
PCL	positive control launch
PC-LITE	processor, laptop imagery transmission equipment
PCM	pulse code modulation
PCO	primary control officer; procuring contracting officer
PCRTS	primary casualty receiving and treatment ship
PCS	permanent change of station; personal communications system; primary control ship; processing subsystem; processor controlled strapping
PCT	personnel control team
PCTC	pure car and truck carrier
PCZ	physical control zone
PD	position description; Presidential directive; priority designator; probability of damage; probability of detection; procedures description; program definition; program directive; program director; public diplomacy
Pd	drift compensated parallelogram pattern
PDA	preliminary damage assessment
PDAI	primary development/test aircraft inventory

PDC	Pacific Disaster Center
PDD	Presidential decision directive
PDDA	power driven decontamination apparatus
PDDG	program directive development group
PDG	positional data graphic
PDM	program decision memorandum
PDOP	position dilution of precision
PDS	position determining system; primary distribution site; protected distribution system
PDSC	public diplomacy and strategic communication
PDSS	predeployment site survey
PDUSD(P&R)	Principal Deputy Under Secretary of Defense (Personnel and Readiness)
PE	peace enforcement; peacetime establishment; personal effects; preparation of the environment; program element
PEAD	Presidential emergency action document
PEC	program element code
PECK	patient evacuation contingency kit
PECP	precision engagement collaboration process
PED	processing, exploitation, and dissemination
PEDB	planning and execution database
PEGEO	personnel geographic location
PEI	principal end item
PEIO	personnel effects inventory officer
PEM	program element monitor
PEO	peace enforcement operations; program executive officer
PEP	personnel exchange program
PER	personnel
PERE	person eligible to receive effects
PERID	period
PERMREP	permanent representative (NATO)
PERSCO	personnel support for contingency operations
PERSCOM	personnel command (Army)
PERSINS	personnel information system
PES	preparedness evaluation system
PFA	primary federal agency
PFD	personal flotation device
PFDB	planning factors database
PFIAB	President's Foreign Intelligence Advisory Board
PFID	positive friendly identification
PFO	principal federal official
PfP	Partnership for Peace (NATO)
PGI	procedures, guidance, and information
PGM	precision-guided munition
pH	potential of hydrogen

PHEO	public health emergency officer
PHIBCB	amphibious construction battalion
PHIBGRU	amphibious group
PHIBOP	amphibious operation
PHIBRON	amphibious squadron
PHIT	port handling/in-land transportation
PHO	posthostilities operations
PHS	Public Health Service
PI	pandemic influenza; point of impact; probability of incapacitation; procedural item; purposeful interference
PIAB	President's Intelligence Advisory Board
PIC	parent indicator code; payment in cash; person identification code; pilot in command; press information center (NATO)
PID	plan identification number; positive identification
PIDD	planned inactivation or discontinued date
PIF	problem identification flag
PII	pre-incident indicators
PIM	pretrained individual manpower
PIN	personnel increment number
PINS	precise integrated navigation system
PIO	press information officer; public information officer
PIPS	plans integration partitioning system
PIR	priority intelligence requirement
PIREP	pilot report
PIRT	purposeful interference response team
PIW	person in water
PJ	pararescue jumper
PK	peacekeeping; probability of kill
PKG-POL	packaged petroleum, oils, and lubricants
PKI	public key infrastructure
PKO	peacekeeping operations
PKP	purple k powder
PKSOI	Peacekeeping and Stability Operations Institute
PL	phase line; public law
PLA	plain language address
PLAD	plain language address directory
PLANORD	planning order
PLAT	pilot's landing aid television
PLB	personal locator beacon
PLC	power line conditioner
PLGR	precise lightweight global positioning system (GPS) receiver
PLL	phase locked loop
PLL/ASL	prescribed load list/authorized stock level
PLRS	position location reporting system

PLS	palletized load system; personal locator system; personal locator system; pillars of logistic support; precision location system
PLT	platoon; program library tape
PM	Bureau of Political-Military Affairs (DOS); parallel track multiunit; passage material; patient movement; peacemaking; political-military affairs; preventive medicine; program management; program manager; provost marshal
PMA	political/military assessment
PMAA	Production Management Alternative Architecture
PMAI	primary mission aircraft inventory
P/M/C	passengers/mail/cargo
PMC	parallel multiunit circle; private military company
PMCF	post maintenance check flight
PMCT	port movement control team
PMD	program management directive
PME	professional military education
PMEL	precision measurement equipment laboratory
PMESII	political, military, economic, social, information, and infrastructure
PMGM	program manager's guidance memorandum
PMI	patient movement item; prevention of mutual interference
PMITS	Patient Movement Item Tracking System
PMN	parallel track multiunit non-return
PMO	production management office(r); program management office
PMOS	primary military occupational specialty
PMR	parallel track multiunit return; patient movement request; patient movement requirement
PMRC	patient movement requirements center
PMS	portable meteorological subsystem
PN	partner nation; pseudonoise
PNA	postal net alert
PNID	precedence network in dialing
PNT	positioning, navigation, and timing
PNVS	pilot night vision system
P/O	part of
PO	peace operations; petty officer
POA	plan of action
POAI	primary other aircraft inventory
POB	persons on board
POC	point of contact
POCD	port operations cargo detachment
POD	plan of the day; port of debarkation; probability of detection

POE	port of embarkation; port of entry
POES	polar operational environment satellite
POF	priority of fires
POG	port operations group
POI	program of instruction
POL	petroleum, oils, and lubricants
POLAD	policy advisor; political advisor
POLCAP	bulk petroleum capabilities report
POLMIL	political-military
POM	program objective memorandum
POMCUS	pre-positioning of materiel configured to unit sets
POMSO	Plans, Operations, and Military Support Office(r) (NG)
POP	performance oriented packaging
POPS	port operational performance simulator
POR	proposed operational requirement
PORTS	portable remote telecommunications system
PORTSIM	port simulation model
POS	peacetime operating stocks; point of sale; probability of success
POSF	port of support file
POSSUB	possible submarine
POSTMOB	post mobilization
POTUS	President of the United States
POV	privately owned vehicle
POW	prisoner of war
P/P	patch panel
p-p	peak-to-peak
PPA	personnel information system (PERSINS) personnel activity
PPAG	proposed public affairs guidance
PPBE	Planning, Programming, Budgeting, and Execution
PPD	Presidential policy directive; program planning document
PPDB	point positioning database
PPE	personal protective equipment
PPF	personnel processing file
Pplan	programming plan
PPLI	precise participant location and identification
ppm	parts per million
PPP	power projection platform; primary patch panel; priority placement program
PPR	prior permission required
PPS	precise positioning service
PPTO	petroleum pipeline and terminal operating
PR	personnel recovery; Phoenix Raven; primary zone; production requirement; program review
PRA	patient reception area; primary review authority

PRANG	Puerto Rican Air National Guard
PRBS	pseudorandom binary sequence
PRC	populace and resources control; Presidential Reserve Call-up
PRCC	personnel recovery coordination cell; personnel recovery coordination center
PRCS	personnel recovery coordination section
PRD	personnel readiness division; Presidential review directive
PRDO	personnel recovery duty officer
PREMOB	pre-mobilization
PREPO	pre-positioned force, equipment, or supplies; pre-positioning
PREREP	pre-arrival report
PRF	personnel resources file; pulse repetition frequency
PRG	personnel recovery guidance; program review group
PRI	movement priority for forces having the same latest arrival date (LAD); priority; progressive routing indicator
PRIFLY	primary flight control
Prime BEEF	Prime Base Engineer Emergency Force
PRISM	Planning Tool for Resource, Integration, Synchronization, and Management
PRM	Bureau of Population, Refugees, and Migration (DOS); Presidential review memorandum
PRMFL	perm file
PRMS	personnel recovery mission software
PRN	pseudorandom noise
PRO	personnel recovery officer
PROBSUB	probable submarine
PROC	processor; Puerto Rican Operations Center
PROFIS	professional officer filler information system
PROM	programmable read-only memory
PROPIN	caution - proprietary information involved
PROVORG	providing organization
proword	procedure word
PRP	personnel reliability program; Personnel Retrieval and Processing
PRRIS	Puerto Rican radar integration system
PRSL	primary zone/switch location
PRT	patient reception team; provincial reconstruction team
PRTF	personnel recovery task force
PRU	pararescue unit; primary reporting unit
PS	parallel track single-unit; port security; processing subsystem
PSA	port support activity; principal staff assistant
PSB	poststrike base

PSC	port security company; principal subordinate command; private security contractor
PSD	planning systems division; port security detachment
PSE	peculiar support equipment
PS/HD	port security/harbor defense
PSHDGRU	port security and harbor defense group
PSI	personnel security investigation; Proliferation Security Initiative
psi	pounds per square inch
PSK	phase-shift keying
PSL	parallel track single-unit long-range aid to navigation (LORAN)
PSMS	Personnel Status Monitoring System
PSN	packet switching node; public switch network
PSO	peace support operations (NATO); post security officer
PSP	perforated steel planking; portable sensor platform; power support platform
PSS	parallel single-unit spiral; personnel services support
P-STATIC	precipitation static
PSTN	public switched telephone network
PSU	port security unit
PSV	pseudosynthetic video
PTA	position, time, altitude
PTAI	primary training aircraft inventory
PTC	peace through confrontation; primary traffic channel
PTDO	prepare to deploy order
PTM	personnel transport module
PTT	postal telephone and telegraph; public telephone and telegraph; push-to-talk
PTTI	precise time and time interval
pub	publication
PUK	packup kit
PUL	parent unit level
PV	prime vendor
PVNTMED	preventive medicine
PVT	positioning, velocity, and timing
PW	prisoner of war
pW	picowatt
PWB	printed wiring board (assembly)
PWD	programmed warhead detonation
PWF	personnel working file
PWG	protection working group
PWIS	Prisoner of War Information System
PWR	pre-positioned wartime reserves
PWRMR	pre-positioned war materiel requirement
PWRMS	pre-positioned war reserve materiel stock

PWRR	petroleum war reserve requirements
PWRS	petroleum war reserve stocks; pre-positioned war reserve stock
PWS	performance work statement
PZ	pickup zone

Q

QA	quality assurance
QAM	quadrature amplitude modulation
QAT	quality assurance team
QC	quality control
QD	quality distance
QDR	quadrennial defense review; quality deficiency report
QEEM	quick erect expandable mast
QHDA	qualified hazardous duty area
QIP	quick impact project
QM	quartermaster
QPSK	quadrature phase shift keying
QRA	quick reaction antenna
QRCT	quick reaction communications terminal
QRE	quick reaction element
QRF	quick reaction force; quick response force
QRG	quick response graphic
QRP	quick response posture
QRS	quick reaction strike
QRSA	quick reaction satellite antenna
QRT	quick reaction team
QS	quality surveillance
QSR	quality surveillance representative
QSTAG	quadripartite standardization agreement
QTY	quantity
QUADCON	quadruple container

R

R	routine
R&D	research and development
R&R	rest and recuperation
R&S	reconnaissance and surveillance; reconstruction and stabilization
R2P2	rapid response planning process
RA	response action; risk analysis; risk assessment
RAA	redeployment assembly area
RABFAC	radar beacon forward air controller
RAC	responsible analytic center

RAC-OT	readiness assessment system - output tool
RAD	routine aerial distribution
RADAY	radio day
RADBN	radio battalion
RADC	regional air defense commander
RADCON	radiological control team
RADF	radarfind
RADHAZ	electromagnetic radiation hazards
RADS	rapid area distribution support (USAF)
RAE	right of assistance entry
RAF	Royal Air Force (UK)
R-AFF	regimental affiliation
RAM	raised angle marker; random access memory; random antiterrorism measure
RAMCC	regional air movement control center
RAOB	rawindsonde observation
RAOC	rear area operations center; regional air operations center
RAP	Radiological Assistance Program (DOE); rear area protection; Remedial Action Projects Program (JCS)
RAS	replenishment at sea
RAS-OT	readiness assessment system - output tool
RAST	recovery assistance, securing, and traversing systems
RASU	random access storage unit
RATE	refine, adapt, terminate, execute
RATT	radio teletype
RB	short-range coastal or river boat
RBA	reimbursable budget authority
RBC	red blood cell
RBE	remain-behind equipment
RBECS	Revised Battlefield Electronic Communications, Electronics, Intelligence, and Operations (CEIO) System
RBI	RED/BLACK isolator
RB std	rubidium standard
RC	receive clock; regional coordinator; Reserve Component; resident coordinator (UN); river current
RCA	riot control agent
RCAT	regional counterdrug analysis team
RCC	regional contracting center; relocation coordination center; rescue coordination center
RCCPDS	Reserve Component common personnel data system
RCD	regional collection detachment
RCEM	regional contingency engineering management
RCHB	reserve cargo handling battalion
RCIED	radio-controlled improvised explosive device
RCM	Rules for Courts-Martial
RCMP	Royal Canadian Mounted Police

RC NORTH	Regional Command North (NATO)
RCO	regional contracting office
RCP	resynchronization control panel
RCS	radar cross section
RC SOUTH	Regional Command South (NATO)
RCSP	remote call service position
RCT	regimental combat team; rescue coordination team (Navy)
RCTA	Regional Counterdrug Training Academy
RCU	rate changes unit; remote control unit
RCVR	receiver
RD	receive data; ringdown
RDA	research, development, and acquisition
R-day	redeployment day
RDCFP	Regional Defense Counterterrorism Fellowship Program
RDCTFP	Regional Defense Combating Terrorism Fellowship Program
RDD	radiological dispersal device; required delivery date
RDECOM	US Army Research, Development, and Engineering Command
RDF	radio direction finder; rapid deployment force
RDO	request for deployment order
RDT&E	research, development, test and evaluation
REACT	rapid execution and combat targeting
REAC/TS	radiation emergency assistance center/training site (DOE)
READY	resource augmentation duty program
RECA	Residual Capability Assessment
RECAS	residual capability assessment system
RECAT	residual capability assessment team
RECCE	reconnaissance
RECMOB	reconstitution-mobilization
RECON	reconnaissance
RED	radiological exposure device
RED HORSE	Rapid Engineer Deployable Heavy Operational Repair Squadron Engineer
REF	Rapid Equipping Force (Army); reference(s)
REGT	regiment
REL	relative
RELCAN	releasable to Canada
REMT	regional emergency management team
REMUS	remote environmental monitoring unit system
REPOL	bulk petroleum contingency report; petroleum damage and deficiency report; reporting emergency petroleum, oils, and lubricants
REPSHIP	report of shipment
REPUNIT	reporting unit
REQCONF	request confirmation

REQSTATASK	air mission request status tasking
RES	radiation exposure status
RESA	research, evaluation, and system analysis
RESCAP	rescue combat air patrol
RESCORT	rescue escort
RESPROD	responsible production
RET	retired
RF	radio frequency; reserve force; response force
RFA	radio frequency authorization; request for assistance; restrictive fire area
RFC	request for capabilities; response force commander revision final coordination
RF CM	radio frequency countermeasures
RFD	revision first draft
RF/EMPINT	radio frequency/electromagnetic pulse intelligence
RFF	request for feedback; request for forces
RFI	radio frequency interference; ready for issue; request for information
RFID	radio frequency identification
RFL	restrictive fire line
RFP	request for proposal
RFS	request for service; request for support
RFW	request for waiver
RG	reconstitution group
RGR	Rangers
RGS	remote geospatial intelligence services
RH	reentry home
Rh	Rhesus
RHIB	rigid hull inflatable boat
RI	Refugees International; routing indicator
RIB	rubberized inflatable boat
RIC	routing indicator code
RICO	regional interface control officer
RIG	recognition identification group
RIK	replacement in kind
RIMS	registrant information management system
RIP	register of intelligence publications
RIS	reconnaissance information system
RISOP	red integrated strategic offensive plan
RISTA	reconnaissance, intelligence, surveillance, and target acquisition
RIT	remote imagery transceiver
RIVRON	riverine squadron
RJTD	reconstitution joint table of distribution
RLD	ready-to-load date
RLE	rail liaison element

RLG	regional liaison group; ring laser gyro
RLGM	remote loop group multiplexer
RLGM/CD	remote loop group multiplexer/cable driver
RLP	remote line printer
RM	ramp module; recovery mechanism; resource management; risk management
RMC	remote multiplexer combiner; rescue mission commander; Resource Management Committee (CSIF); returned to military control
RMKS	remarks
RMO	regional Marine officer
RMP	religious ministry professional
RMS	requirements management system; root-mean-square
RMU	receiver matrix unit
RNAV	area navigation
RNP	remote network processor
R/O	receive only
Ro	search radius rounded to next highest whole number
ROA	restricted operations area
ROC	regional operations center; rehearsal of concept; required operational capability
ROCU	remote orderwire control unit
ROE	rules of engagement
ROEX	rules of engagement exercise
ROG	railhead operations group
ROICC	resident officer in charge of construction
ROK	Republic of Korea
ROM	read-only memory; restriction of movement; rough order of magnitude
ROMO	range of military operations
RON	remain overnight
RO/RO	roll-on/roll-off
ROS	reduced operating status
ROTC	Reserve Officer Training Corps
ROTHR	relocatable over-the-horizon backscatter radar (USN)
ROWPU	reverse osmosis water purification unit
ROZ	restricted operations zone
RP	reconstitution priority; release point (road); religious program specialist; retained personnel
RPG	rocket propelled grenade
RPM	revolutions per minute
RPO	rendezvous and proximity operations
RPPO	Requirements, Plans, and Policy Office
RPT	report
RPTOR	reporting organization
RPV	remotely piloted vehicle

RQMT	requirement
RQT	rapid query tool
RR	reattack recommendation
RRC	regional reporting center
RRCC	regional response coordination center
RRDF	roll-on/roll-off discharge facility
RRF	rapid reaction force; rapid response force; Ready Reserve Fleet; Ready Reserve Force
RRPP	rapid response planning process
RS	rate synthesizer; religious support; requirement submission
RSA	retrograde storage area
RSC	red station clock; regional service center; rescue sub-center
RSD	reporting of supply discrepancy
RSE	retrograde support element
RSG	reference signal generator
RSI	rationalization, standardization, and interoperability
RSL	received signal level
RSN	role specialist nation
RSO	reception, staging, and onward movement; regional security officer
RSOC	regional signals intelligence (SIGINT) operations center
RSOI	reception, staging, onward movement, and integration
RSP	recognized surface picture; Red Switch Project (DOD); religious support policy
RSPA	Research and Special Programs Administration
RSS	radio subsystem; really simple syndication; remote sensors subsystem; root-sum-squared
RSSC	regional satellite communications support center; regional satellite support cell; regional signals intelligence (SIGINT) support center (NSA); regional space support center
RSSC-LO	regional satellite communications support center liaison officer; regional space support center liaison officer
RST	religious support team
RSTA	reconnaissance, surveillance, and target acquisition
RSTV	real-time synthetic video
RSU	rapid support unit; rear support unit; remote switching unit
R/T	receiver/transmitter
RT	recovery team; remote terminal; rough terrain
RTA	residual threat assessment
RTB	return to base
RTCC	rough terrain container crane
RTCH	rough terrain container handler
RTD	returned to duty
RTF	regional task force; return to force
RTFL	rough terrain forklift

RTG	radar target graphic
RTL	restricted target list
RTLP	receiver test level point
RTM	real-time mode
RTOC	rear tactical operations center
RTS	remote transfer switch
RTTY	radio teletype
RU	release unit; rescue unit
RUF	rules for the use of force
RUIC	Reserve unit identification number
RUSCOM	rapid ultrahigh frequency (UHF) satellite communications
RV	long-range seagoing rescue vessel; rekeying variable; rendezvous
RVR	runway visibility recorder
RVT	remote video terminal
RW	rotary-wing
RWCM	regional wartime construction manager
RWR	radar warning receiver
RWS	rawinsonde subsystem
RX	receive; receiver
RZ	recovery zone; return-to-zero

S

618 AOC (TACC)	618 Air Operations Center (Tanker Airlift Control Center)
S&F	store-and-forward
S&R	search and recovery
S&T	science and technology; scientific and technical
S&TI	scientific and technical intelligence
S-1	battalion or brigade manpower and personnel staff officer (Marine Corps battalion or regiment)
S-2	battalion or brigade intelligence staff officer (Army; Marine Corps battalion or regiment)
S-3	battalion or brigade operations staff officer (Army; Marine Corps battalion or regiment)
S-4	battalion or brigade logistics staff officer (Army; Marine Corps battalion or regiment)
SA	security assistance; selective availability (GPS); senior adviser; situational awareness; staging area; stand-alone switch
SAA	senior airfield authority
SAAFR	standard use Army aircraft flight route
SAAM	special assignment airlift mission
SAB	scientific advisory board (USAF)
SABER	situational awareness beacon with reply

SAC	special actions cell; special agent in charge; supporting arms coordinator
SACC	supporting arms coordination center
SACEUR	Supreme Allied Commander, Europe (NATO)
SACLANT	Supreme Allied Command, Atlantic
SACS	secure telephone unit (STU) access control system
SACT	Supreme Allied Commander Transformation
SADC	sector air defense commander
SADL	situation awareness data link
SADO	senior air defense officer
SAF	Secretary of the Air Force
SAFE	secure analyst file environment; selected area for evasion; sexual assault forensic examination
SAFE-CP	selected area for evasion-contact point
SAFER	evasion and recovery selected area for evasion (SAFE) area activation request
SAFWIN	secure Air Force weather information network
SAG	surface action group
SAI	sea-to-air interface; single agency item
SAL	small arms locker
SAL-GP	semiactive laser-guided projectile (USN)
SALM	single-anchor leg mooring
SALT	supporting arms liaison team
SALTS	streamlined automated logistics transfer system; streamlined automated logistics transmission system
SALUTE	size, activity, location, unit, time, and equipment
SAM	special airlift mission; surface-to-air missile
SAMM	security assistance management manual
SAMS	School of Advanced Military Studies
SAO	security assistance office; security assistance officer; selected attack option
SAOC	sector air operations center
SAP	special access program
SAPI	special access program for intelligence
SAPO	subarea petroleum office
SAPR	sexual assault prevention and response
SAR	satellite access request; search and rescue; site access request; special access requirement; suspicious activity report; synthetic aperture radar
SARC	sexual assault response coordinator; surveillance and reconnaissance center
SARDOT	search and rescue point
SARIR	search and rescue incident report
SARMIS	search and rescue management information system
SARNEG	search and rescue numerical encryption group
SAROPS	Search and Rescue Optimal Planning System

SARREQ	search and rescue request
SARSAT	search and rescue satellite-aided tracking
SARSIT	search and rescue situation summary report
SARTEL	search and rescue (SAR) telephone (private hotline)
SARTF	search and rescue task force
SAS	sealed authenticator system; special ammunition storage
SASP	special ammunition supply point
SASS	supporting arms special staff
SASSY	supported activities supply systems
SAT	satellite
SATCOM	satellite communications
SAW	surface acoustic wave
SB	standby base
SBCT	Stryker brigade combat team
SBL	space-based laser
SBPO	Service blood program officer
SBR	special boat squadron
SBRPT	subordinate reporting organization
SBS	senior battle staff; support battle staff
SBSS	science-based stockpile stewardship
SBT	special boat team
SBSO	sustainment brigade special operations
SBU	sensitive but unclassified; special boat unit
SC	sea current; search and rescue coordinator; security cooperation; station clock; strategic communication
SCA	sociocultural analysis; space coordinating authority; support to civil administration
SCAR	strike coordination and reconnaissance
SCAS	stability control augment system
SCATANA	security control of air traffic and navigation aids
SC ATLANTIC	Strategic Command, Atlantic (NATO)
SCATMINE	scatterable mine
SCATMINEWARN	scatterable minefield warning
SCC	security classification code; service cryptologic Component; shipping coordination center; Standards Coordinating Committee
SCC-WMD	United States Strategic Command Center for Combating Weapons of Mass Destruction
SCDL	surveillance control data link
SCE	Service cryptologic element
SC EUROPE	Strategic Command, Europe (NATO)
SCF(UK)	Save the Children Fund (United Kingdom)
SCF(US)	Save the Children Federation (United States)
SCG	Security Cooperation Guidance; switching controller group
SCHBT	shape-clear-hold-build-transition

SCI	security and counterintelligence interviews; sensitive compartmented information
SCIF	sensitive compartmented information facility
SCL	standard conventional load
SCM	security countermeasure; Service container manager
SCMP	strategic command, control, and communications (C3) master plan
SCNE	self-contained navigation equipment
SCO	secondary control officer; security cooperation organization; senior contracting official; state coordinating officer
SCOC	systems control and operations concept
SCONUM	ship control number
SCP	secure conferencing project; security cooperation plan; service control point; system change proposal
SCPT	strategic connectivity performance test
SCRB	software configuration review board
S/CRS	Office of the Coordinator for Reconstruction and Stabilization (DOS)
SCT	shipping coordination team; single channel transponder
S/CT	Office of the Coordinator for Counterterrorism (DOS)
SCTIS	single channel transponder injection system
SCTS	single channel transponder system
SCT-UR	single channel transponder ultrahigh frequency (UHF) receiver
SCUD	surface-to-surface missile system
SD	strategy division
SDA	Seventh-Day Adventist (ADRA)
S-day	day the President authorizes selective reserve call-up
SDB	Satellite Communications Database
SDDC	Surface Deployment and Distribution Command
SDDCTEA	Surface Deployment and Distribution Command Transportation Engineering Agency
SDF	self defense force
SDIO	Strategic Defense Initiative Organization
SDLS	satellite data link standards
SDMX	space division matrix
SDN	system development notification
SDNRIU	secure digital net radio interface unit
SDO	senior defense official; ship's debarkation officer
SDO/DATT	senior defense official/defense attaché
SDP	strategic distribution platform
SDR	system design review
SDSG	space division switching group
SDSM	space division switching matrix
SDV	SEAL team delivery vehicle; submerged delivery vehicle

SDZ	self-defense zone
SE	site exploitation; spherical error
SEA	Southeast Asia
SEABEE	sea barge
Seabee	Navy construction engineer
SEAD	suppression of enemy air defenses
SEC	submarine element coordinator
SECAF	Secretary of the Air Force
SECARMY	Secretary of the Army
SecDef	Secretary of Defense
SECDHS	Secretary of the Department of Homeland Security
SECHS	Secretary of Homeland Security
SECNAV	Secretary of the Navy
SECNAVINST	Secretary of the Navy instruction
SECOMP	secure en route communications package
SECORD	secure cord switchboard
SECRA	secondary radar data only
SECSTATE	Secretary of State
SECTRANS	Secretary of Transportation
SED	signals external data
SEDAS	spurious emission detection acquisition system
SEF	sealift enhancement feature
SEI	specific emitter identification
SEL	senior enlisted leader
SEL REL	selective release
SELRES	Selected Reserve
SEMA	special electronic mission aircraft
SEMS	standard embarkation management system
SEO/SEP	special enforcement operation/special enforcement program
SEP	signal entrance panel; spherical error probable
SEPLO	state emergency preparedness liaison officer
SERE	survival, evasion, resistance, and escape
SERER	survival, evasion, resistance, escape, recovery
SES	senior executive service
SETA	system engineering and technical assistance
SEW	shared early warning
S/EWCC	signals intelligence/electronic warfare coordination center
SEWG	Special Events Working Group
SEWOC	signals intelligence/electronic warfare operations centre (NATO)
SEWS	satellite early warning system
SF	security force; security forces (Air Force or Navy); single frequency; special forces; standard form
SFA	security force assistance
SFAF	standard frequency action format

SFAT	spectrum flyaway team
SFC	single-fuel concept
SFCP	shore fire control party
SFG	security forces group; special forces group
SFI	spectral composition
SFLEO	senior federal law enforcement official
SFMS	special forces medical sergeant
SFOD-A/B/C	special forces operational detachment-A/B/C
SFOR	Stabilization Force
SFS	security forces squadron
SG	steering group; strike group; supergroup; surgeon general
SGEMP	system-generated electromagnetic pulse
SGS	strategic guidance statement
SGSA	squadron group systems advisor
SGXM	Headquarters, Air Mobility Command/Surgeon
SHAPE	Supreme Headquarters Allied Powers, Europe
SHD	special handling designator
SHF	super-high frequency
SHORAD	short-range air defense
SHORADEZ	short-range air defense engagement zone
SI	special intelligence; United States Strategic Command strategic instruction
SIA	station of initial assignment
SIAGL	survey instrument azimuth gyroscope lightweight
SIC	subject identification code; supporting intelligence center
SICO	sector interface control officer
SID	standard instrument departure
SIDAC	single integrated damage analysis capability
SIDL	standard intelligence documents list
SIDO	senior intelligence duty officer
SIDS	secondary imagery dissemination system
SIF	selective identification feature; strategic internment facility
SIG	signal
SIGINT	signals intelligence
SIGSEC	signal security
SII	statement of intelligence interest
SIM	system impact message
SIMLM	single integrated medical logistics management; single integrated medical logistics manager
SINCGARS	single-channel ground and airborne radio system
SINS	ship's inertial navigation system
SIO	senior intelligence officer
SIOC	Strategic Information and Operations Center (FBI)
SIOP	Single Integrated Operational Plan
SIOP-ESI	Single Integrated Operational Plan-Extremely Sensitive Information

SIPRNET	SECRET Internet Protocol Router Network
SIR	serious incident report; specific information requirement; Strategic Military Intelligence Review
SIRADS	stored imagery repository and dissemination system
SIRMO	senior information resources management official
SIS	special information systems
SITREP	situation report
SIV	special interest vessel
SJA	staff judge advocate
SJFHQ	standing joint force headquarters
SJFHQ(CE)	standing joint force headquarters (core element)
SJFHQ-N	Standing Joint Force Headquarters - North
SJS	Secretary, Joint Staff
SKE	station-keeping equipment
SL	sea level; switch locator
SLA	service level agreement; special leave accrual
SLAM	stand-off land attack missile
SLBM	submarine-launched ballistic missile
SLC	satellite laser communications; single line concept
SLCM	sea-launched cruise missile
SLCP	ship lighterage control point; ship's loading characteristics pamphlet
SLD	system link designator
SLEP	service life extension program
SLGR	small, lightweight ground receiver (GPS)
SLIT	serial-lot item tracking
SLO	space liaison officer
SLOC	sea line of communications
SLP	seaward launch point
SLRP	survey, liaison, and reconnaissance party
SLWT	side loadable warping tug
SM	Secretary, Joint Staff, memorandum; Service manager; spectrum management; staff memorandum; system manager
SMA	special military information support operations assessment
SMART	special medical augmentation response team
SMART-AIT	special medical augmentation response - aeromedical isolation team
SMB	spectrum management branch
SMC	midpoint compromise track spacing; search and rescue mission coordinator; system master catalog
SMCA	single manager for conventional ammunition
SMCC	strategic mobile command center
SMCM	surface mine countermeasures
SMCOO	spectrum management concept of operations
SMCR	Selected Marine Corps Reserve

SMD	strategic missile defense
SMDC	Space & Missile Defense Command (Army)
SMDC/ARSTRAT	United States Army Space and Missile Defense Command/United States Army Forces Strategic Command
SME	subject matter expert
SMEB	significant military exercise brief
SMEO	small end office
SMFT	semi-trailer mounted fabric tank
SMI	security management infrastructure
SMIO	search and rescue (SAR) mission information officer
SMO	senior meteorological and oceanographic officer; strategic mobility office(r); support to military operations
SMP	sub-motor pool
SMPT	School of Military Packaging Technology
SMRC	Specialized Medical Response Capabilities
SMRI	service message routing indicator
SMS	single mobility system; special military information support operations study
SMTP	simple message transfer protocol
SMU	special mission unit; supported activities supply system (SASSY) management unit
S/N	signal to noise
SN	serial number
SNCO	staff noncommissioned officer
SNF	strategic nuclear forces
SNIE	special national intelligence estimates
SNLC	Senior North Atlantic Treaty Organization (NATO) Logisticians Conference
SNM	system notification message
SNOI	signal not of interest
SO	safety observer; special operations
SOA	separate operating agency; special operations aviation (Army); status of action; sustained operations ashore
SOAF	status of action file
SOAGS	special operations air-ground system
SOC	security operations center; special operations commander; special operations component
SOCA	special operations communications assembly
SOCC	Sector Operations Control Center (NORAD)
SOCCE	special operations command and control element
SOCCENT	Special Operations Component, United States Central Command
SOCCET	special operations critical care evacuation team

SOCCT	special operations combat control team
SOCEUR	Special Operations Component, United States European Command
SOCEX	special operations capable exercise
SOCJFCOM	Special Operations Command, Joint Forces Command
SOCM	special operations combat medic
SOCOORD	special operations coordination element
SOCP	special operations communication package
SOCPAC	Special Operations Command, Pacific
SOCRATES	Special Operations Command, Research, Analysis, and Threat Evaluation System
SOCSOUTH	Special Operations Component, United States Southern Command
SOD	special operations division; strategy and options decision (Planning, Programming, and Budgeting System)
SODARS	special operations debrief and retrieval system
SOE	special operations executive
SOF	special operations forces; supervisor of flying
SOFA	status-of-forces agreement
SOFAR	sound fixing and ranging
SOFLAM	special operations laser marker
SOFME	special operations forces medical element
SOFSA	special operations forces support activity
SOG	special operations group
SOI	signal of interest; signal operating instructions; space object identification
SOIC	senior officer of the intelligence community
SOJTF	special operations joint task force
SOLAS	safety of life at sea
SOLE	special operations liaison element
SOLIS	signals intelligence (SIGINT) On-line Information System
SOLL	special operations low-level
SOM	satellite communications (SATCOM) operational manager; start of message; system operational manager
SOMA	status of mission agreement
SOMARDS	Standard Operation and Maintenance Army Research and Development System
SOMARDS NT	Standard Operation and Maintenance Army Research and Development System Non-Technical
SOMPF	special operations mission planning folder
SONMET	special operations naval mobile environment team
SoO	ship of opportunity
SOOP	Center for Operations, Plans, and Policy
SOP	standard operating procedure; standing operating procedure
SO-peculiar	special operations-peculiar

SOR	statement of requirement
SORTIEALOT	sortie allotment message
SORTS	Status of Resources and Training System
SOS	special operations squadron
SOSB	special operations support battalion
SOSC	special operations support command (theater army)
SOSCOM	special operations support command
SOSE	special operations staff element
SOSG	station operations support group
SOSR	suppress, obscure, secure, and reduce
SOST	special operations support team
SOTA	signals intelligence operational tasking authority
SOTF	special operations task force
SOTSE	special operations theater support element
SOUTHAF	Southern Command Air Forces
SOUTHROC	Southern Region Operational Center (USSOUTHCOM)
SOW	special operations wing; standoff weapon; statement of work
SOWT	special operations weather team
SP	security police
SPACEAF	Space Air Forces
SPACECON	control of space information
SPCC	ships parts control center (USN)
SPEAR	strike protection evaluation and antiair research
SPEC	specified
SPECAT	special category
SPECWAR	special warfare
SPG	Strategic Planning Guidance
SPI	special investigative (USAF)
SPINS	special instructions
SPINTCOMM	special intelligence communications handling system
SPIREP	spot intelligence report
SPLX	simplex
SPM	service postal manager; single point mooring; single port manager
SPO	system program office
SPOC	search and rescue (SAR) points of contact; space command operations center
SPOD	seaport of debarkation
SPOE	seaport of embarkation
SPOT	Synchronized Predeployment and Operational Tracker
SPOTREP	spot report
SPP	Security and Prosperity Partnership of North America; shared production program; State Partnership Program (NG)
SPR	software problem report
SPRINT	special psychiatric rapid intervention team

SPS	standard positioning service
SPSC	system planning and system control
SPTCONF	support confirmation
SPTD CMD	supported command
SPTG CMD	supporting command
SPTREQ	support request
sqft	square feet
SR	special reconnaissance
SRA	specialized-repair activity
SRAM	short-range air-to-surface attack missile; system replacement and modernization
SRB	software release bulletin; system review board (JOPES)
SRBM	short-range ballistic missile
SRC	security risk category; service reception center; Single Integrated Operational Plan (SIOP) response cell; standard requirements code; survival recovery center
SRCC	service reserve coordination center
SRF	secure Reserve force
SRG	Seabee readiness group; short-range aircraft
SRI	surveillance, reconnaissance, and intelligence (Marine Corps)
SRIG	surveillance, reconnaissance, and intelligence group (USMC)
S/RM	sustainment, restoration, and modernization
SROC	Senior Readiness Oversight Council; Southern Region Operational Center, United States Southern Command
SROE	standing rules of engagement
SRP	sealift reserve program; seaward recovery point; Single Integrated Operational Plan (SIOP) reconnaissance plan
SRP/PDS	stabilization reference package/position determining system
SRR	search and rescue region
SRS	search and rescue sector
SRSG	special representative of the Secretary-General
SRT	scheduled return time; special reaction team; standard remote terminal; strategic relocatable target
SRTD	signals research and target development
S/RTF	search and recovery task force
SRU	search and rescue unit
SR-UAV	short-range unmanned aerial vehicle
SRUF	standing rules for the use of force
SRWBR	short range wide band radio
S/S	steamship
SS	submarine

SSA	software support activity; space situational awareness; special support activity (NSA); strapdown sensor assembly; supply support activity; supply support area
SSB	single side band; support services branch; surveillance support branch
SSB-SC	single sideband-suppressed carrier
SSC	small scale contingency; special security center surveillance support center
SSCO	shipper's service control office
SSCRA	Soldiers and Sailors Civil Relief Act
SSD	strategic studies detachment
SSE	satellite communications (SATCOM) systems expert; space support element
SSF	software support facility
SSI	standing signal instruction
SSM	surface-to-surface missile
SSMI	special sensor microwave imager
SSMS	single shelter message switch
SSN	attack submarine, nuclear; Social Security number; space surveillance network
SS (number)	sea state (number)
SSO	special security office(r); spot security office
SSP	signals intelligence (SIGINT) support plan
SSPM	single-service postal manager
SSPO	strategic systems program office
SSR	security sector reform
SSS	Selective Service System; shelter subsystem
SSSC	surface, subsurface search surveillance coordination
SST	special support team (National Security Agency)
SSTR	stability, security, transition, and reconstruction
SSWG	space support working group
ST	short ton; small tug; strike team
S/T	short ton
ST&E	security test and evaluation
STA	system tape A
STAB	space tactical awareness brief
STA clk	station clock
STAMMIS	standard Army multi-command management information system
STAMP	standard air munitions package (USAF)
STANAG	standardization agreement (NATO)
STANAVFORLANT	Standing Naval Forces, Atlantic (NATO)
STAR	scheduled theater airlift route; sensitive target approval and review; standard attribute reference; standard terminal arrival route; surface-to-air recovery; system threat assessment report

STARC	state area coordinators
STARS	Standard Accounting and Reporting System
START	Strategic Arms Reduction Treaty
STARTEX	start of exercise
STB	super tropical bleach
STC	secondary traffic channel
STD	sexually transmitted disease
STDM	synchronous time division multiplexer
STE	secure telephone equipment
S-Team	staff augmentation team
STEL STU III	Standford telecommunications (secure telephone)
STEP	software test and evaluation program; standardized tactical entry point; standard tool for employment planning
STG	seasonal target graphic
STICS	scalable transportable intelligence communications system
STO	special technical operations
STOC	special technical operations coordinator
STOD	special technical operations division
STOL	short takeoff and landing
STOMPS	stand-alone tactical operational message processing system
STON	short ton
STP	security technical procedure
STR	strength
STRAPP	standard tanks, racks and pylons packages (USAF)
STRATOPS	strategic operations division
STREAM	standard tensioned replenishment alongside method
STS	special tactics squadron
STT	small tactical terminal; special tactics team
STU	secure telephone unit
STU-III	secure telephone unit III
STW	strike warfare
STWC	strike warfare commander
STX	start of text
SU	search unit
SUBJ	subject
sub-JIB	subordinate-joint information bureau
SUBOPAUTH	submarine operating authority
sub-PIC	subordinate-press information center
SUBROC	submarine rocket
SUC	surf current
SUIC	service unit identification code
SUMMITS	scenario unrestricted mobility model of intratheater simulation
SUPE	supervisory commands program
SUPP	supplement
SUPPO	supply officer

SURG	surgeon
SUROBS	surf observation
SURPIC	surface picture
SUST BDE	sustainment brigade
SUW	surface warfare
SUWC	surface warfare commander
S/V	sailboat
SVC	Service; stored value card
SVIP	secure voice improvement program
SVLTU	service line termination unit
SVR	surface vessel radar
SVS	secure voice system
Sw	switch
SWA	Southwest Asia
SWAT	special weapons and tactics
SWBD	switchboard
SWC	strike warfare commander; swell/wave current
SWI	special weather intelligence
SWO	staff weather officer
SWORD	submarine warfare operations research division
SWPC	Space Weather Prediction Center
SWSOCC	Southwest Sector Operation Control Center North American Aerospace Defense Command (NORAD)
SWXS	Space Weather Squadron
SXXI	SPECTRUM XXI
SXXI-O	SPECTRUM XXI-Online
SYDP	six year defense plan
SYG	Secretary-General (UN)
SYNC	synchronization
SYS	system
SYSCOM	systems command
SYSCON	systems control
SZ	surf zone

T

2-D	two-dimensional
2E	Role 2 enhanced
2LM	Role 2 light maneuver
3-D	three-dimensional
T	search time available; short ton; trackline pattern
T&DE	test and diagnostic equipment
T&E	test and evaluation
T2	technology transfer
TA	target acquisition; target audience; technical arrangement; theater Army; threat assessment

TAA	tactical assembly area; target audience analysis
TAACOM	theater Army area command
TAADS	The Army Authorization Document System
TAAMDCOORD	theater Army air and missile defense coordinator
TAB	tactical air base
TAC	tactical advanced computer; terminal access controller; terminal attack control; terminal attack controller
TAC(A)	tactical air coordinator (airborne)
TACAIR	tactical air
TACAMO	take charge and move out (E-6A/B aircraft)
TACAN	tactical air navigation
TACC	tactical air command center (USMC); tactical air control center (USN); tanker airlift control center
TAC-D	tactical deception
TACDAR	tactical detection and reporting
TACINTEL	tactical intelligence
TACLAN	tactical local area network
TACLOG	tactical-logistical
TACM	tactical air command manual
TACO	theater allied contracting office
TACON	tactical control
TACOPDAT	tactical operational data
TA/CP	technology assessment/control plan
TACP	tactical air control party
TACRON	tactical air control squadron
T-ACS	auxiliary crane ship
TACS	tactical air control system; theater air control system
TACSAT	tactical satellite
TACSIM	tactical simulation
TACSTANS	tactical standards
TACT	tactical aviation control team
TACTRAGRULANT	Tactical Training Group, Atlantic
TAD	tactical air direction; temporary additional duty (non-unit-related personnel); theater air defense; time available for delivery
TADC	tactical air direction center
TADCS	tactical airborne digital camera system
TADS	Tactical Air Defense System; target acquisition system and designation sight
TAES	theater aeromedical evacuation system
TAF	tactical air force
TAFDS	tactical airfield fuel dispensing system
TAFIM	technical architecture framework for information management
TAFS	tactical aerodrome forecasts
TAFT	technical assistance field team

TAG	technical assessment group; technical assistance group; the adjutant general; Tomahawk land-attack missile aimpoint graphic
T-AGOS	tactical auxiliary general ocean surveillance
TAGS	theater air-ground system
T-AH	hospital ship
TAI	target area of interest; total active inventory
TAIS	transportation automated information systems
TAK	cargo ship
T-AKR	fast logistics ship
TALD	tactical air-launched decoy
TALON	Threat and Local Observation Notice
TAMCA	theater Army movement control agency
TAMCO	theater Army movement control center
TAMD	theater air and missile defense
TAMMC	theater army material management command
TAMMIS	theater Army medical management information system
TAMS	transportation analysis, modeling, and simulation
tanalt	tangent altitude
TAO	tactical action officer
TAOC	tactical air operations center (USMC)
TAP	troopship
TAR	tactical air request; Training and Administration of the Reserve
TARBS	transportable amplitude modulation and frequency modulation radio broadcast system
TARBUL	target bulletin
TARE	tactical record evaluation
TAREX	target exploitation; target plans and operations
TARS	tethered aerostat radar system
TARWI	target weather and intelligence
TAS	tactical atmospheric summary; true air speed
T-ASA	Television Audio Support Agency
TASCID	tactical Automatic Digital Network (AUTODIN) satellite compensation interface device
TASCO	tactical automatic switch control officer
TASIP	tailored analytic intelligence support to individual electronic warfare and command and control warfare projects
TASKORD	tasking order
TASMO	tactical air support for maritime operations
TASOSC	theater Army special operations support command
TASS	tactical automated security system; tactical automated switch system
TASWC	theater antisubmarine warfare commander
TAT	tactical analysis team; technical assistance team

TATC	tactical air traffic control
T-AVB	aviation logistics support ship
TAW	tactical airlift wing
TBD	to be determined
TBM	tactical ballistic missile; theater ballistic missile
TBMCS	theater battle management core system
TBMD	theater ballistic missile defense
TBP	to be published
TBSL	to be supplied later
TBTC	transportable blood transshipment center
TC	tidal current; transmit clock and/or telemetry combiner; training circular; Transportation Corps (Army)
TCA	terminal control area; time of closest approach; traditional combatant commander activity
TC-ACCIS	Transportation Coordinator's Automated Command and Control Information System
TC-AIMS	Transportation Coordinator's Automated Information for Movement System
TC-AIMS II	Transportation Coordinator's Automated Information for Movement System II
TCAM	theater Army medical management information system (TAMMIS) customer assistance module
TCC	transmission control code; transportation component command
TCCF	tactical communications control facility
TCEM	theater contingency engineering management
TCF	tactical combat force; technical control facility
TCM	theater construction manager; theater container manager
TCMD	transportation control and movement document
TCN	third country national; transportation control number; troop contributing nations
TCO	transnational criminal organization
TCP	theater campaign plan
TCPED	tasking, collection, processing, exploitation, and dissemination
TCS	theater communications system
TCSEC	trusted computer system evaluation criteria
TCSP	theater consolidation and shipping point
TD	temporary duty; theater distribution; tie down; timing distributor; total drift; transmit data
TDA	Table of Distribution and Allowance
TDAD	Table of Distribution and Allowance (TDA) designation
T-day	effective day coincident with Presidential declaration of a National Emergency and authorization of partial mobilization
TDBM	technical database management

TDBSS	Theater Defense Blood Standard System
TDC	target development cell
TDD	target desired ground zero (DGZ) designator; time-definite delivery
TDF	tactical digital facsimile
TDIC	time division interface controller
TDIG	time division interface group
TDIM	time division interface module
TDL	tactical data link
TDM	time division multiplexed
TDMA	time division multiple access
TDMC	theater distribution management cell
TDMF	time division matrix function
TDMM	time division memory module
TDMX	time division matrix
TDN	tactical data network; target development nomination
TDP	theater distribution plan
TDR	transportation discrepancy report
TDRC	theater detainee reporting center
TDSG	time division switching group
TDSGM	time division switching group modified
TDT	theater display terminal
TDY	temporary duty
TE	transaction editor
TEA	Transportation Engineering Agency
TEC	theater engineer command
tech	technical
TECHCON	technical control
TECHDOC	technical documentation
TECHELINT	technical electronic intelligence
TECHEVAL	technical evaluation
TECHINT	technical intelligence
TECHOPDAT	technical operational data
TECS II	Treasury Enforcement Communications System
TED	trunk encryption device
TEDAC	Terrorist Explosive Device Analytical Center (FBI)
TEK	TeleEngineering Kit
TEL	transporter-erector-launcher
TELEX	teletype
TELINT	telemetry intelligence
TELNET	telecommunication network
TEMPER	tent extendible modular personnel
TENCAP	tactical exploitation of national capabilities program
TEO	team embarkation officer
TEOB	tactical electronic order of battle
TEP	test and evaluation plan; theater engagement plan

TERCOM	terrain contour matching
TERF	terrain flight
TERPES	tactical electronic reconnaissance processing and evaluation system
TERPROM	terrain profile matching
TERS	tactical event reporting system
TES	theater event system
TESS	Tactical Environmental Support System
TET	targeting effects team
TETK	TeleEngineering Toolkit
TEU	technical escort unit; twenty-foot equivalent unit
TEWLS	Theater Enterprise Wide Logistics System
TF	task force
TFA	toxic free area
TFADS	Table Formatted Aeronautic Data Set
TFC	threat finance cell
TFCICA	task force counterintelligence coordinating authority
TFE	tactical field exchange; threat finance exploitation; transportation feasibility estimator
TFF	total force fitness
TFLIR	targeting forward-looking infrared
TFMS-M	Transportation Financial Management System-Military
TFR	temporary flight restriction
TFS	tactical fighter squadron; Tactical Forecast System
TG	task group
TGC	trunk group cluster
TGEN	table generate
TGM	terminally guided munitions; trunk group multiplexer
TGMOW	transmission group module and/or orderwire
TGO	terminal guidance operations
TGT	target
TGTINFOREP	target information report
TGU	trunk compatibility unit
THAAD	Terminal High Altitude Area Defense
THT	tactical human intelligence team
THX	theater express
TI	threat identification; training instructor
TIA	theater intelligence assessment
TIAP	theater intelligence architecture program
TIB	theater intelligence brigade; toxic industrial biological
TIBS	tactical information broadcast service
TIC	target information center; toxic industrial chemical
TIDE	Terrorist Identities Datamart Environment
TIDP	technical interface design plan
TIDS	tactical imagery dissemination system
TIF	theater internment facility

TIFF	tagged image file format
TII	total inactive inventory
TIM	theater information management; toxic industrial material
TIO	target intelligence officer
TIP	target intelligence package; trafficking in persons
TIPG	telephone interface planning guide
TIPI	tactical information processing interpretation
TIPS	tactical optical surveillance system (TOSS) imagery processing system
TIR	toxic industrial radiological
TIROS	television infrared observation satellite
TIS	technical interface specification; thermal imaging system
TISG	technical interoperability standards group
TISS	thermal imaging sensor system
TJAG	the judge advocate general
T-JMC	theater-joint movement center
T-JTB	theater-joint transportation board
TJTN	theater joint tactical network
TL	team leader
TLA	theater logistics analysis
TLAM	Tomahawk land-attack missile
TLAMM	theater lead agent for medical materiel
TLAM/N	Tomahawk land attack missile/nuclear
TLC	traffic load control
TLE	target location error
TLM	target list management; topographic line map
TLO	theater logistics overview
TLP	transmission level point
TLR	trailer
TLX	teletype
TM	tactical missile; target materials; team member; technical manual; TROPO modem
TMAO	theater mortuary affairs office; theater mortuary affairs officer
TMB	tactical military information support operations battalion
TMD	tactical munitions dispenser; theater missile defense
TMEP	theater mortuary evacuation point
TMG	timing
TMIP	theater medical information program
TMIS	theater medical information system
TML	terminal
TMLMC	theater medical logistic management center
TMMMC	theater medical materiel management center
TMN	trackline multiunit non-return
TMO	traffic management office; transportation management office

TMP	target materials program; telecommunications management program; theater manpower forces
TMR	trackline multiunit return
T/M/S	type, model, and/or series
TMT	time-phased force and deployment data management tool
TNAPS	tactical network analysis and planning system
TNAPS+	tactical network analysis and planning system plus
TNC	theater network operations center
TNCC	theater network operations control center
TNCO	transnational criminal organization
T-net	training net
TNF	theater nuclear force
TNL	target nomination list
T/O	table of organization
TO	technical order; theater of operations
TO&E	table of organization and equipment
TOA	table of allowance
TOC	tactical operations center; tanker airlift control center (TALCE) operations center
TOCU	tropospheric scatter (TROPO) orderwire control unit
TOD	tactical ocean data; time of day
TOE	table of organization and equipment
TOF	time of flight
TOFC	trailer on flatcar
TOH	top of hill
TOI	track of interest
TOPINT	technical operational intelligence
TOR	term of reference; time of receipt
TOS	time on station
TOSS	tactical optical surveillance system
TOT	time on target
TOW	tube launched, optically tracked, wire guided
TP	technical publication; transportation priority; turn point
TPC	tactical pilotage chart
TPC/PC	tactical pilotage chart and/or pilotage chart
TPE	theater provided equipment
TPED	tasking, processing, exploitation, and dissemination; theater personal effects depot
TPERS	type personnel element
TPFDD	time-phased force and deployment data
TPFDL	time-phased force and deployment list
TPL	technical publications list; telephone private line
TPME	task, purpose, method, and effects
TPMRC	theater patient movement requirements center
TPO	task performance observation
TPRC	theater planning response cell

TPT	tactical petroleum terminal
TPTRL	time-phased transportation requirements list
TPU	tank pump unit
TQ	tactical questioning
TRA	technical review authority
TRAC2ES	United States Transportation Command Regulating and Command and Control Evacuation System
TRACON	terminal radar approach control facility
TRADOC	United States Army Training and Doctrine Command
TRAM	target recognition attack multisensor
Trans BDE	transportation brigade
Trans Det RPO	transportation detachment rapid port opening
TRANSEC	transmission security
TRAP	tactical recovery of aircraft and personnel (Marine Corps); tactical related applications; tanks, racks, adapters, and pylons; terrorism research and analysis program
TRC	tactical radio communication; threat reduction cooperation; transmission release code
TRCC	tactical record communications center
TRE	tactical receive equipment
TREAS	Department of the Treasury
TREE	transient radiation effects on electronics
TRIADS	Tri-Wall Aerial Distribution System
TRICON	triple container
TRI-TAC	Tri-Service Tactical Communications Program
TRK	truck; trunk
TRNG	training
TRO	training and readiness oversight
TROPO	troposphere; tropospheric scatter
TRP	target reference point
TRS	tactical reconnaissance squadron
TS	terminal service; time-sensitive; top secret
TSA	target system analysis; theater storage area; Transportation Security Administration (DHS); travel security advisory
TSB	technical support branch; trunk signaling buffer
TSBn	transportation support battalion (USMC)
TSC	theater security cooperation; theater support command; theater sustainment command (Army)
TSCIF	tactical sensitive compartmented information facility
TSCM	technical surveillance countermeasures
TSCO	target selection confusion of the operator; top secret control officer
TSCP	theater security cooperation plan
TSCR	time sensitive collection requirement
TSE	tactical support element

TSEC	transmission security
TSG	targeting support group; test signal generator
TSGCE	tri-Service group on communications and electronics
TSGCEE	tri-Service group on communications and electronic equipment (NATO)
TSM	trunk signaling message
TSN	trackline single-unit non-return; track supervision net
TSO	technical standard order; telecommunications service order
TSOC	theater special operations command
TSP	telecommunications service priority
TSR	telecommunications service request; theater source registry; theater support representative; trackline single-unit return
TSS	tactical shelter system; target sensing system; timesharing system; time signal set; traffic service station
TSSP	tactical satellite signal processor
TSSR	tropospheric scatter (TROPO)-satellite support radio
TST	tactical support team; terminal support team; theater support team; time-sensitive target
TSWA	temporary secure working area
TT	terminal transfer
TT&C	telemetry, tracking, and commanding
TTAN	transportation tracking account number
TTB	transportation terminal battalion
TTD	tactical terrain data; technical task directive
TTFACOR	targets, threats, friendlies, artillery, clearance, ordnance, restrictions
TTG	thermally tempered glass
TTL	transistor-transistor logic
TTM	threat training manual; training target material
TTN	transportation tracking number
TTP	tactics, techniques, and procedures; trailer transfer point
TTR	tactical training range
TTT	time to target
TTU	transportation terminal unit
TTY	teletype
TUBA	transition unit box assembly
TUCHA	type unit characteristics file
TUCHAREP	type unit characteristics report
TUDET	type unit equipment detail file
TV	television
TVA	Tennessee Valley Authority
TW&A	threat warning and assessment
TWC	Office for Counterterrorism Analysis (DIA); total water current
TWCF	Transportation Working Capital Fund
TWCM	theater wartime construction manager

TWD	transnational warfare counterdrug analysis
TWDS	tactical water distribution system
TWI	Office for Information Warfare Support (DIA)
TWPL	teletypewriter private line
TWX	teletypewriter exchange
TX	transmitter; transmit
TYCOM	type commander

U

U	wind speed
UA	unmanned aircraft
UAOBS	upper air observation
UAR	unconventional assisted recovery
UARCC	unconventional assisted recovery coordination cell
UAS	unmanned aircraft system
UAV	unmanned aerial vehicle
U/C	unit cost; upconverter
UCFF	Unit Type Code Consumption Factors File
UCMJ	Uniform Code of Military Justice
UCP	Unified Command Plan
UCT	underwater construction team
UDAC	unauthorized disclosure analysis center
UDC	unit descriptor code
UDESC	unit description
UDL	unit deployment list; unit designation list
UDP	unit deployment program
UDT	underwater demolition team
UE	unit equipment
UEWR	upgraded early warning radar
UFAC	Underground Facilities Analysis Center
UFC	Unified Facilities Criteria
UFO	ultrahigh frequency follow-on
UFR	unfunded requirement
UGA	ungoverned area
UGIRH	Urban Generic Information Requirements Handbook
UGM-84A	Harpoon
UGM-96A	Trident I
UGO	unified geospatial-intelligence operations
UHF	ultrahigh frequency
UHV	Upper Huallaga Valley
UIC	unit identification code
UICIO	unit identification code information officer
UIRV	unique interswitch rekeying variable
UIS	unit identification system
UJTL	Universal Joint Task List

UK	United Kingdom
UK(I)	United Kingdom and Ireland
ULC	unit level code
ULF	ultra low frequency
ULLS	unit level logistics system
ULN	unit line number
ULSD	ultra-low sulfur diesel
UMCC	unit movement control center
UMCM	underwater mine countermeasures
UMD	unit manning document; unit movement data
UMIB	urgent marine information broadcast
UMMIPS	uniform material movement and issue priority system
UMO	unit movement officer
UMPR	unit manpower personnel record
UMT	unit ministry team
UN	United Nations
UNAMIR	United Nations Assistance Mission in Rwanda
UNC	United Nations Command
UNCLOS	United Nations Convention on the Law of the Sea
UN CMCoord	United Nations humanitarian civil-military coordination
UNCTAD	United Nations Conference on Trade and Development
UND	urgency of need designator
UNDAC	United Nations disaster assessment and coordination
UNDFS	United Nations Department of Field Support
UNDHA	United Nations Department of Humanitarian Affairs
UN-DMT	United Nations disaster management team
UNDP	United Nations development programme
UNDPKO	United Nations Department for Peacekeeping Operations
UNEF	United Nations emergency force
UNEP	United Nations environment program
UNESCO	United Nations Educational, Scientific, and Cultural Organization
UNHCHR	United Nations High Commissioner for Human Rights
UNHCR	United Nations Office of the High Commissioner for Refugees
UNHQ	United Nations Headquarters
UNICEF	United Nations Children's Fund
UNIFIL	United Nations Interim Force in Lebanon
UNIL	unclassified national information library
UNITAF	unified task force
UNITAR	United Nations Institute for Training and Research
UNITREP	unit status and identity report
UNJLC	United Nations Joint Logistic Centre
UNLOC	United Nations logistic course
UNMEM	United Nations military expert on mission
UNMIH	United Nations Mission in Haiti

UNMILPOC	United Nations military police course
UNMOC	United Nations military observers course
UNMOVCC	United Nations movement control course
UNO	unit number
UNOCHA	United Nations Office for the Coordination of Humanitarian Affairs
UNODC	United Nations Office on Drugs and Crime
UNODIR	unless otherwise directed
UNOSOM	United Nations Operations in Somalia
UNPA	United Nations Participation Act
UNPROFOR	United Nations protection force
UNREP	underway replenishment
UNREP CONSOL	underway replenishment consolidation
UNRWA	United Nations Relief and Works Agency for Palestine Refugees in the Near East
UNSC	United Nations Security Council
UNSCR	United Nations Security Council resolution
UNSG	United Nations Secretary-General
UNSOC	United Nations staff officers course
UNTAC	United Nations Transition Authority in Cambodia
UNTSO	United Nations Truce and Supervision Organization
UNV	United Nations volunteer
UOF	use of force
UP&TT	unit personnel and tonnage table
URDB	user requirements database
URL	uniform resource locater
USA	United States Army
USAB	United States Army barracks
USACCSA	United States Army Command and Control Support Agency
USACE	United States Army Corps of Engineers
USACFSC	United States Army Community and Family Support Center
USACHPPM	United States Army Center for Health Promotion and Preventive Medicine
USACIDC	United States Army Criminal Investigation Command
USAEDS	United States Atomic Energy Detection System
USAF	United States Air Force
USAFE	United States Air Forces in Europe
USAFEP	United States Air Force, Europe pamphlet
USAFLANT	United States Air Force, Atlantic Command
USAFR	United States Air Force Reserve
USAFRICOM	United States Africa Command
USAFSOC	United States Air Force, Special Operations Command
USAFSOF	United States Air Force, Special Operations Forces
USAFSOS	United States Air Force Special Operations School

USAID	United States Agency for International Development
USAITAC	United States Army Intelligence Threat Analysis Center
USAJFKSWC	United States Army John F. Kennedy Special Warfare Center
USAMC	United States Army Materiel Command
USAMMA	United States Army Medical Materiel Agency
USAMPS	United States Army Military Police School
USAMRICD	United States Army Medical Research Institute of Chemical Defense
USAMRIID	United States Army Medical Research Institute of Infectious Diseases
USAMRMC	United States Army Medical Research and Materiel Command
USANCA	United States Army Nuclear and Combating Weapons of Mass Destruction Agency
USAO	United States Attorney Office
USAR	United States Army Reserve
USARCENT	United States Army, Central Command
USARDECOM	United States Army Research, Development, and Engineering Command
USAREUR	United States Army, European Command
USARIEM	United States Army Research Institute of Environmental Medicine
USARJ	United States Army, Japan
USARNORTH	United States Army, North
USARPAC	United States Army, Pacific Command
USARSO	United States Army, Southern Command
USASMDC/ARSTRAT	United States Army Space and Missile Defense Command/Army Forces Strategic Command
USASOC	United States Army Special Operations Command
USB	upper side band
USBP	United States Border Patrol
USC	United States Code; universal service contract
USCENTAF	United States Central Command Air Forces
USCENTCOM	United States Central Command
USCG	United States Coast Guard
USCGR	United States Coast Guard Reserve
USCIS	United States Citizenship and Immigration Services
USCS	United States Cryptologic System; United States Customs Service
USCYBERCOM	United States Cyber Command
USDA	United States Department of Agriculture
USD(A&T)	Under Secretary of Defense for Acquisition and Technology
USDAO	United States defense attaché office

USD(AT&L)	Under Secretary of Defense for Acquisition, Technology, and Logistics
USD(C/CFO)	Under Secretary of Defense (Comptroller/Chief Financial Officer)
USDELMC	United States Delegation to the NATO Military Committee
USD(I)	Under Secretary of Defense for Intelligence
USD(P)	Under Secretary of Defense for Policy
USD(P&R)	Under Secretary of Defense for Personnel and Readiness
USD(R&E)	Under Secretary of Defense for Research and Engineering
USELEMCMOC	United States Element Cheyenne Mountain Operations Center
USELEMNORAD	United States Element, North American Aerospace Defense Command
USERID	user identification
USERRA	Uniformed Services Employment and Reemployment Rights Act
USEUCOM	United States European Command
USFF	United States Fleet Forces Command
USFJ	United States Forces, Japan
USFK	United States Forces, Korea
USFORAZORES	United States Forces, Azores
USFS	United States Forest Service
USFWS	United States Fish and Wildlife Service
USG	United States Government
USGS	United States Geological Survey
USIA	United States Information Agency
USIC	United States interdiction coordinator
USIS	United States Information Service
USJFCOM	United States Joint Forces Command
USLANTFLT	United States Atlantic Fleet
USLO	United States liaison office; United States liaison officer
USMARFORCENT	United States Marine Component, Central Command
USMARFORLANT	United States Marine Component, Atlantic Command
USMARFORNORTH	United States Marine Corps Forces North
USMARFORPAC	United States Marine Component, Pacific Command
USMARFORSOUTH	United States Marine Component, Southern Command
USMC	United States Marine Corps
USMCEB	United States Military Communications-Electronics Board
USMCR	United States Marine Corps Reserve
USMER	United States merchant ship vessel locator reporting system
USMILGP	United States military group
USMILREP	United States military representative
USML	United States Munitions List
USMOG-W	United States Military Observer Group - Washington

USMS	United States Marshals Service
USMTF	United States message text format
USMTM	United States military training mission
USN	United States Navy
USNAVCENT	United States Naval Forces, Central Command
USNAVEUR	United States Naval Forces, Europe
USNAVSO	US Naval Forces Southern Command
USNCB	United States National Central Bureau (INTERPOL)
USNMR	United States National Military representative
USNMTG	United States North Atlantic Treaty Organization (NATO) Military Terminology Group
USNO	United States Naval Observatory
USNORTHCOM	United States Northern Command
USNR	United States Navy Reserve
USNS	United States Naval Ship
USPACAF	United States Air Forces, Pacific Command
USPACFLT	United States Pacific Fleet
USPACOM	United States Pacific Command
USPFO(P&C)	United States Property and Fiscal Office (Purchasing and Contracting)
USPHS	United States Public Health Service
USPS	United States Postal Service
USREPMC	United States representative to the military committee (NATO)
USSOCOM	United States Special Operations Command
USSOUTHAF	United States Air Force, Southern Command
USSOUTHCOM	United States Southern Command
USSS	United States Secret Service (TREAS); United States Signals Intelligence (SIGINT) System
USSTRATCOM	United States Strategic Command
USTRANSCOM	United States Transportation Command
USUN	United States Mission to the United Nations
USW	undersea warfare
USW/USWC	undersea warfare and/or undersea warfare commander
USYG	under secretary general
UT1	unit trainer; Universal Time
UTC	Coordinated Universal Time; unit type code
UTM	universal transverse mercator
UTO	unit table of organization
UTR	underwater tracking range
UUV	unmanned underwater vehicle; unmanned underwater vessel
UVEPROM	ultraviolet erasable programmable read-only memory
UW	unconventional warfare
UWOA	unconventional warfare operating area
UXO	unexploded explosive ordnance; unexploded ordnance

V

V	search and rescue unit ground speed; sector pattern; volt
v	velocity of target drift
VA	Veterans Administration; victim advocate; vulnerability assessment
V&A	valuation and availability
VAAP	vulnerability assessment and assistance program
VAC	volts, alternating current
VARVAL	vessel arrival data, list of vessels available to marine safety offices and captains of the port
VAT B	(weather) visibility (in miles), amount (of clouds, in eighths), (height of cloud) top (in thousands of feet), (height of cloud) base (in thousands of feet)
VBIED	vehicle-borne improvised explosive device
VBS	visit, board, search
VBSS	visit, board, search, and seizure
VCC	voice communications circuit
VCG	virtual coordination group
VCJCS	Vice Chairman of the Joint Chiefs of Staff
VCNOG	Vice Chairman, Nuclear Operations Group
VCO	voltage controlled oscillator
VCOPG	Vice Chairman, Operations Planners Group
VCR	violent crime report
VCXO	voltage controlled crystal oscillator; voltage controlled oscillator
VDC	volts, direct current
VDJS	Vice Director, Joint Staff
VDL	video downlink
VDR	voice digitization rate
VDS	video subsystem
VDSD	visual distress signaling device
VDU	visual display unit
VDUC	visual display unit controller
VE	vertical error
VEE	Venezuelan equine encephalitis
VEH	vehicle; vehicular cargo
VEO	violent extremist organization
VERTREP	vertical replenishment
VF	voice frequency
VFR	visual flight rules
VFS	validating flight surgeon
VFTG	voice frequency telegraph
VHF	very high frequency
VI	visual information
VICE	advice

VID	visual identification; visual identification information display
VINSON	encrypted ultrahigh frequency communications system
VIP	very important person; visual information processor
VIRS	verbally initiated release system
VIS	visual imaging system
VISA	Voluntary Intermodal Sealift Agreement
VISOBS	visual observer
VIXS	video information exchange system
VLA	vertical line array; visual landing aid
VLF	very low frequency
VLR	very-long-range aircraft
VLZ	vertical landing zone
VMap	vector map
VMAQ	Marine tactical electronic warfare squadron
VMC	visual meteorological conditions
VMF	variable message format
VMGR	Marine aerial refueler and transport squadron
VMI	vendor managed inventory
VNTK	target vulnerability indicator designating degree of hardness; susceptibility of blast; and K-factor
VO	validation office
VOCODER	voice encoder
VOCU	voice orderwire control unit
VOD	vertical onboard delivery
VOL	volunteer
vol	volume
VOLS	vertical optical landing system
VOR	very high frequency omnidirectional range station
VORTAC	very high frequency omnidirectional range station and/or tactical air navigation
VOX	voice actuation (keying)
VP	video processor
VPB	version planning board
VPD	version planning document
VPV	virtual prime vendor
VS	sector single-unit
VS&PT	vehicle summary and priority table
VSG	virtual support group
VSP	voice selection panel
VSR	sector single-unit radar
V/STOL	vertical and/or short takeoff and landing aircraft
VSW	very shallow water
VTA	voluntary tanker agreement
VTC	video teleconferencing
VTOL	vertical takeoff and landing
VTOL-UAS	vertical takeoff and landing unmanned aircraft system

VTOL-UAV	vertical takeoff and landing unmanned aerial vehicle
VTS	vessel traffic service
VTT	video teletraining
VU	volume unit
VV&A	verification, validation, and accreditation
VV&C	verification, validation, and certification
VX	nerve agent (O-Ethyl S-Diisopropylaminomethyl Methylphosphonothiolate)

W

W	sweep width
w	search subarea width
WAAR	Wartime Aircraft Activity Report
WACBE	World Area Code Basic Encyclopedia
WADS	Western Air Defense Sector
WAGB	icebreaker (USCG)
WAI	weather area of interest
WAN	wide-area network
WANGO	World Association of Non-Governmental Organizations
WARM	wartime reserve mode
WARNORD	warning order
WARP	web-based access and retrieval portal
WAS	wide area surveillance
WASP	war air service program
WATCHCON	watch condition
WB	wideband
WBGTI	wet bulb globe temperature index
WBIED	waterborne improvised explosive device
WC	wind current
WCA	water clearance authority
WCCS	Wing Command and Control System
WCDO	War Consumables Distribution Objective
WCE	weapons of mass destruction coordination element
WCO	World Customs Organization
WCS	weapons control status
W-day	declared by the President, W-day is associated with an adversary decision to prepare for war
WDCO	well deck control officer
WDT	warning and display terminal
WEAX	weather facsimile
Web SM	Web scheduling and movement
WES	weapon engagement status
WETM	weather team
WEU	Western European Union
WEZ	weapon engagement zone

WFE	warfighting environment
WFP	World Food Programme (UN)
WG	working group
WGS	Wideband Global Satellite Communications; World Geodetic System
WGS 84	World Geodetic System 1984
WH	wounded due to hostilities
WHEC	high-endurance cutter (USCG)
WHNRS	wartime host-nation religious support
WHNS	wartime host-nation support
WHNSIMS	Wartime Host Nation Support Information Management System
WHO	World Health Organization (UN)
WIA	wounded in action
WISDIM	Warfighting and Intelligence Systems Dictionary for Information Management
WISP	Wartime Information Security Program
WIT	weapons intelligence team
WLG	Washington Liaison Group
WMD	weapons of mass destruction
WMD CM	weapons of mass destruction consequence management
WMD-CST	weapons of mass destruction-civil support team
WMD-E	weapons of mass destruction-elimination
WMEC	Coast Guard medium-endurance cutter
WMO	World Meteorological Organization
WMP	Air Force War and Mobilization Plan; War and Mobilization Plan
WOC	wing operations center (USAF)
WOD	wind-over deck; word-of-day
WORM	write once read many
WOT	war on terrorism
WP	white phosphorous; working party; Working Party (NATO)
WPA	water jet propulsion assembly
WPAL	wartime personnel allowance list
WPARR	War Plans Additive Requirements Roster
WPB	Coast Guard patrol boat
WPC	Washington Planning Center
WPM	words per minute
WPN	weapon
WPR	War Powers Resolution
WR	war reserve; weapon radius
WRA	Office of Weapons Removal and Abatement (DOS); weapons release authority
WRAIR	Walter Reed Army Institute of Research
WRC	World Radiocommunication Conference
WRL	weapons release line

WRM	war reserve materiel
WRMS	war reserve materiel stock
WRR	weapons response range (as well as wpns release rg)
WRS	war reserve stock
WRSA	war reserve stocks for allies
WRSK	war readiness spares kit; war reserve spares kit
WS	weather squadron
WSE	weapon support equipment
WSES	surface effect ship (USCG)
WSESRB	Weapon System Explosive Safety Review Board
WSM	waterspace management
WSR	weapon system reliability
WSV	weapons system video
WT	warping tug; weight
WTCA	water terminal clearance authority
WTCT	weapons of mass destruction technical collection team
WTI	weapons technical intelligence
WTLO	water terminal logistic office
Wu	uncorrected sweep width
WVRD	World Vision Relief and Development, Inc.
WWABNCP	worldwide airborne command post
WWII	World War II
WWSVCS	Worldwide Secure Voice Conferencing System
WWX	worldwide express
WX	weather

X

X	initial position error
XCDS	Extracted Container Delivery System
XCVR	transceiver
XMPP	presence protocol
XO	executive officer
XSB	barrier single unit

Y

Y	search and rescue unit (SRU) error
YR	year

Z

Z	zulu
z	effort
ZF	zone of fire
Zt	total available effort
ZULU	time zone indicator for Universal Time

Intentionally Blank

APPENDIX B
ADMINISTRATIVE INSTRUCTIONS

1. User Comments

Users are highly encouraged to submit comments on this publication to the Directorate for Joint Force Development, J-7, Joint Doctrine and Education Division, Joint Doctrine Branch, ATTN: Chairman, US NATO Military Terminology Group, 7000 Joint Staff, Pentagon, Washington, DC 20318-7000; Tel (703) 692-7255, DSN 222-7255; Fax (703) 692-5224, DSN 222-5224. All comments recommending modifications, deletions, or additions to terminology in JP 1-02 must be made in accordance with DODI 5025.12, *Standardization of Military and Associated Terminology*, and CJCSI 5705.01, *Standardization of Military and Associated Terminology*.

2. Authorship

The lead agent and the Joint Staff doctrine sponsor for this publication is the Director for Joint Force Development (J-7).

3. Supersession

a. This publication supersedes JP 1-02, *Department of Defense Dictionary of Military and Associated Terms*, 12 April 2001. The terms and definitions in JP 1-02 will be updated on a monthly basis with modifications, deletions, or additions that have been approved in accordance with DODI 5025.12, and CJCSI 5705.01.

b. Record of Updates:

31 December 2010: JP 3-02.1, JP 3-07.2, and DODD 3025.18 added.

31 January 2011: JP 3-68, JP 4-03, and SecDef Memo 12401-10 added.

15 April 2011: JP 2-01.2 added.

15 May 2011: JP 3-05 added.

15 July 2011: JP 3-08, JP 3-15, and JP 3-34 added.

15 August 2011: JP 3-0 and JP 5-0 added.

15 September 2011: JP 1-04 added.

15 October 2011: JP 3-03, JP 3-07, and JP 4-06 added.

15 November 2011: JP 1-0 added.

15 January 2012: JP 2-01, JP 3-13.2, JP 3-13.3, JP 3-15,1, JP 3-50, CJCSI 5120.02C, and CJCSM 5120.01 added.

15 February 2012: JP 3-13.1 and JP 3-13.4 added.

15 March 2012: JP 1-06 added.

15 April 2012: JP 3-01, JP 4-01.5, and JP 6-01 added.

15 July 2012: JP 3-41 and JP 4-01.2 added.

15 August 2012: JP 3-07.3, JP 3-33, and JP 4-02 added.

15 November 2012: JP 2-03 added.

15 December 2012: JP 3-04, JP 3-13, JP 3-18, JP 3-59, and JP 4-01.6 added.

15 February 2013: JP 3-12, JP 3-35, and JP 3-60 added.

15 March 2013: JP 4-08 added.

15 April 2013: JP 1 added.

15 June 2013: JP 3-14 and JP 4-01 added.

16 July 2013: JP 3-16 added.

15 August 2013: JP 3-07.4, JP 3-27, JP 3-28, and JP 3-32 added.

15 September 2013: JP 3-57 added.

15 October 2013: JP 3-11 and JP 3-17 added.

15 November 2013: JP 2-0 and JP 4-0 added.

15 December 2013: JP 1-05, JP 3-06, and JP 3-24.

4. Terms Removed or Replaced by the 15 December 2013 Amendment

The following is a listing of terminology with rationale that supports removal from JP 1-02.

Terms Removed or Replaced			
Term	**Action**	**Pub**	**Rationale**
disaffected person	Removed	JP 3-24	Not used
irregular forces	Removed	JP 3-24	Not used
key facilities list	Removed	JP 3-06	Defined adequately in common dictionary
religious support plan	Removed	JP 1-05	Not required with "religious support" defined
subversive political action	Removed	JP 3-24	Defined adequately in common dictionary
urban triad	Removed	JP 3-06	Not used

Figure B-1. Terms Removed or Replaced

5. Terms Added, Sourced, or Modified by the 15 December 2013 Amendment

Terms Added, Sourced, or Modified		
Term	**Action**	**Publication**
combatant command chaplain	Modified	JP 1-05
counterinsurgency	Modified	JP 3-24
insurgency	Modified	JP 3-24
joint civil-military operations task force	Modified	JP 3-57
joint urban operations	Modified	JP 3-06
objective area	Modified	JP 3-06
religious advisement	Modified	JP 1-05
religious support	Modified	JP 1-05
religious support team	Modified	JP 1-05

Figure B-2. Terms Added, Sourced, or Modified

6.   Terminology Commonly Used in Error

Terminology Commonly Used in Error	
Correct Terminology	**Misused Terminology**
military information support operations (MISO)	psychological operations (PSYOP)
lethal/nonlethal	kinetic/nonkinetic
CCMD (combatant command)	COCOM
cyberspace	cyber
DOD Information Network (DODIN)	Global Information Grid (GIG)
create effects	achieve effects

Figure B-3. Terminology Commonly Used in Error

7.   Distribution

Joint Staff J-7 does not print copies of JP 1-02 for distribution. Electronic versions are available on JDEIS at https://jdeis.js.mil (NIPRNET) and https://jdeis.js.smil.mil (SIPRNET) and on the JEL at http://www.dtic.mil/doctrine (NIPRNET).

www.ingramcontent.com/pod-product-compliance
Lightning Source LLC
Chambersburg PA
CBHW082350270326
41935CB00013B/1564